Plant Adaptation to Extreme Environments in Drylands

Plant Adaptation to Extreme Environments in Drylands

Editors

Xiao-Dong Yang
Nai-Cheng Wu
Xue-Wei Gong

MDPI • Basel • Beijing • Wuhan • Barcelona • Belgrade • Manchester • Tokyo • Cluj • Tianjin

Editors

Xiao-Dong Yang
Department of Geography &
Spatial Information
Ningbo University
Ningbo
China

Nai-Cheng Wu
Department of Geography &
Spatial Information
Ningbo University
Ningbo
China

Xue-Wei Gong
Institute of Applied Ecology
Chinese Academy of Sciences
Shengyang
China

Editorial Office
MDPI
St. Alban-Anlage 66
4052 Basel, Switzerland

This is a reprint of articles from the Special Issue published online in the open access journal *Forests* (ISSN 1999-4907) (available at: www.mdpi.com/journal/forests/special_issues/Drylands_ Adaptation).

For citation purposes, cite each article independently as indicated on the article page online and as indicated below:

LastName, A.A.; LastName, B.B.; LastName, C.C. Article Title. *Journal Name* **Year**, *Volume Number*, Page Range.

ISBN 978-3-0365-7163-8 (Hbk)
ISBN 978-3-0365-7162-1 (PDF)

Cover image courtesy of Xiao-Dong Yang

Contents

Preface to "Plant Adaptation to Extreme Environments in Drylands"

Arid and semi-arid lands cover more than one-third of the Earth's terrestrial area and are typically characterized by rainfall scarcity, higher temperatures and evapotranspiration, salinization, nutrient-poor soil, and a paucity of vegetation cover. Climate modelling research projects that the frequency and intensity of extreme environmental events in these regions will become increasingly higher in future climate scenarios. Determining the adaptive strategies and mechanisms of dryland plants to extreme environments such as drought, salinity and heat has become a research hotspot and is of great relevance for utilizing appropriate practices for the conservation and management of dryland vegetation.

We gathered studies on plant–soil relations, water, and carbon and nutrient physiology, as well as species diversity and distribution patterns, in order to deepen the understanding of the adaption of dryland plants to more extensive and frequent environmental stresses under projected climate change. Our Special Issue comprises thirteen original articles that provide a brief overview of the latest research progress, spanning a broad range of aspects related to the adaptation of plants to extreme environments in drylands. We are indebted to all the authors contributing their work to this reprint and to all reviewers whose comments and suggestions helped to improve the quality of the accepted papers.

Despite our best efforts to collect papers that cover as many relevant topics as possible, to provide a comprehensive introduction to the basic knowledge and scientific preamble on the adaptation of plants to environmental stress in arid areas, the papers we collected in this Special Issue represent only a small portion of scientific knowledge in this field. In order to collect and display more advancedknowledge, we have started to collect papers for the new Specific Issue "Plant Adaptation to Extreme Environments in Drylands—Series II" through *Forests*. We hope that readers will contribute high-quality papers to this Special Issue.

Xiao-Dong Yang, Nai-Cheng Wu, and Xue-Wei Gong

Editors

Article

Scale Effects on the Relationship between Plant Diversity and Ecosystem Multifunctionality in Arid Desert Areas

Jiaxin Liu [1], Dong Hu [2], Hengfang Wang [1], Lamei Jiang [1] and Guanghui Lv [1,*]

1   College of the Ecology and Environment, Xinjiang University, Urumqi 830017, China
2   College of Life Science, Northwestern University, Xi'an 710069, China
*   Correspondence: guanghui_xju@sina.com; Tel.: +0991-2111427

Abstract: Understanding the relationship between biodiversity and ecosystem multifunctionality is popular topic in ecological research. Although scale is an important factor driving changes in biodiversity and ecosystem multifunctionality, we still know little about the scale effects of the relationship between the different dimensions of biodiversity and ecosystem multifunctionality. Using plant communities in the northwest of the Qira Desert Ecosystem National Field Research Station of the Chinese Academy of Sciences in Qira County, Xinjiang, as the study object, we explored the scale effects of plant diversity and ecosystem multifunctionality at different sampling scales (5 m × 5 m, 20 m × 20 m, and 50 m × 50 m) and the relative contribution of different dimensions of diversity (species diversity, functional diversity, and phylogenetic diversity) to variation in ecosystem multifunctionality. At different scales, a significant scale effect was observed in the relationship between plant diversity and ecosystem multifunctionality. Species diversity dominated ecosystem multifunctionality at large scales (50 m × 50 m), and species diversity and ecosystem multifunctionality varied linearly between scales. Functional diversity made the greatest contribution in small scales (5 m × 5 m), and the relationship between phylogenetic diversity and ecosystem multifunctionality tended to show a single-peaked variation between scales, with a dominant effect on multifunctionality at the mesoscale (20 m × 20 m). The results of the study deepen the understanding of the scale effect of the relationship between plant diversity and ecosystem multifunctionality in arid desert areas, and help to further conserve plant diversity and maintain ecosystem multifunctionality.

Keywords: desert ecosystems; scale; plant diversity; ecosystem multifunctionality

Citation: Liu, J.; Hu, D.; Wang, H.; Jiang, L.; Lv, G. Scale Effects on the Relationship between Plant Diversity and Ecosystem Multifunctionality in Arid Desert Areas. *Forests* **2022**, *13*, 1505. https://doi.org/10.3390/f13091505

Academic Editor: Takuo Nagaike

Received: 27 August 2022
Accepted: 14 September 2022
Published: 16 September 2022

Publisher's Note: MDPI stays neutral with regard to jurisdictional claims in published maps and institutional affiliations.

1. Introduction

Global climate change and habitat fragmentation play a significant negative role in biodiversity conservation and the sustainability of ecosystem functions. [1,2]. The study of the relationship between biodiversity and ecosystem function is important to enhance biodiversity and restore ecosystem function. Biodiversity includes species diversity, functional diversity, and phylogenetic diversity [3]. Early studies on the relationship between biodiversity and ecosystem function mostly considered the relationship between species diversity and single ecosystem function [4,5]. With the progress in research, scholars have found that functional diversity and phylogenetic diversity have significant effects on ecosystem function and cannot be replaced by species diversity, and considering only single ecosystem functions may underestimate the role of biodiversity in ecosystem function [6,7]. Therefore, the study of the relationship between multidimensional biodiversity and ecosystem multifunctionality contributes to a deeper understanding of the biodiversity maintenance mechanisms.

Studies on biodiversity in China and elsewhere tended to focus on different environmental gradients, disturbance levels and successional stages [8,9], with less attention paid to the scale dependence of biodiversity. However, biodiversity depends on the number, composition and distribution of community species, and these factors are scale-dependent. [10,11]. The earliest studies on the relationship between plant diversity and ecosystem multifunctionality at different scales showed that alpha diversity was significantly positively correlated with ecosystem multifunctionality, with alpha diversity playing a dominant role [12]. Studies in subtropical regions further demonstrated that the positive correlation also showed a trend of rapid increase followed by a gentle increase [13,14], but the effect of α-diversity on single ecosystem function is not significant in forest ecosystem studies [15]. Functional and phylogenetic diversity is also being studied in greater depth by researchers. A study by Dang et al. (2018) [16] in desert ecosystems has shown that functional diversity indices vary significantly between scales and shape different community structures. Changes in sampling scale have a significant effect on the divergence and aggregation of genealogical structure and the level of genealogical diversity [17]. Studies on cave plants have also shown significant differences in genealogical diversity between large and small scales [18]. Ecosystem multifunctionality and biodiversity interactions are also scale-dependent [19]. First, different species play different roles between scales, allowing inter-scale differences in ecosystem multifunctionality [20,21]. Second, as the scale increases, community differences lead to a constant exchange of materials and energy flows between communities, which affects ecosystem multifunctionality [22]. Finally, the composition and distribution of functional traits among species vary by scale. As scale changes, functional traits segregating or overlapping in trait space as scale changes, making ecosystem multifunctionality change in response [19].

Arid desert ecosystems are sensitive areas of global change and priority areas for biodiversity conservation. As an important part of terrestrial ecosystems, arid zones have distinctive climatic environments, geographical locations, and resource distribution patterns that make them unique in terms of biodiversity and ecosystem multifunctionality [23].

Located at the southern edge of the Taklamakan Desert, Xinjiang's Qira County has a dry climate with little rain and wind, and its ecosystem type is a typical temperate desert ecosystem. Due to its geographical location and topographical constraints, the region is ecologically fragile, and desertification is severe, which has led to a reduction in biodiversity and diminished ecosystem function services [24]. Considering the scale dependence of biodiversity and ecosystem multifunctionality, this study explored the relationship between multidimensional biodiversity and ecosystem multifunctionality based on three sampling scales (5 m × 5 m, 20 m × 20 m, and 50 m × 50 m), aiming to address the following scientific questions: (1) How do plant diversity and ecosystem multifunctionality relate to scale? (2) How do the relative contributions of species diversity, functional diversity, and phylogenetic diversity to ecosystem multifunctionality vary at different scales?

2. Materials and Methods

2.1. Overview of Experimental Area

The study area is located on the northern foot of the Kunlun Mountains and on the periphery of the oasis at the southern edge of the Tarim Basin (80°37′12″ E, 37°2′0″ N). The climate in the reserve is extremely arid, and water resources are scarce, with an average annual precipitation of 35.1 mm and a potential annual evaporation of 2595.3 mm [25]. The main types of soil are gray-brown desert soil, gray desert soil, and wind-sand soil, with a high degree of soil salinity [26]. The natural vegetation is dominated by perennial desert plants, with the main species including *Populus euphratica*, *Tamarix chinensis*, *Alhagi sparsifolia*, *Salsola collina*, and *Hexinia polydichotoma*.

2.2. Research Method

2.2.1. Sample Setting

A 100 m × 100 m sample plot was set up in the northwest of the Qira Desert Ecosystem National Field Research Station of the Chinese Academy of Sciences in July 2019, where quadrats of three scales (50 m × 50 m, 20 m × 20 m, and 5 m × 5 m) were set up. Using the 5 m × 5 m quadrat as the basic unit, 100 quadrats were randomly selected at each scale

within the sample plots. Where each 20 m × 20 m quadrat contains 16 quadrats of 5 m × 5 m and each 50 m × 50 m quadrat contains 100 quadrats of 5 m × 5 m (Figure 1).
Qira.

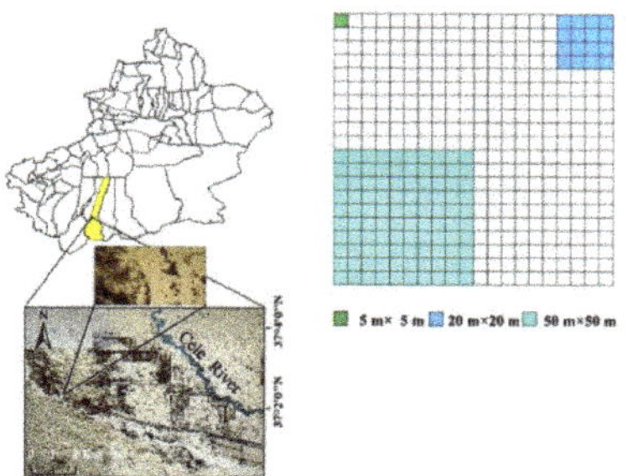

Figure 1. Location of the study area and the investigated plots.

2.2.2. Collection of Plant Samples

Plant species information, abundance, and plant height were recorded in the field at a minimum sampling scale of 5 m × 5 m. Three 1 m^2 samples were selected on the diagonal of each 5 m × 5 m square to record and calculate the herbaceous abundance within the 5 m × 5 m square. In a 5 m × 5 m sample, approximately 30 mature leaves were collected from each plant species, and three were selected to measure leaf length (LL), leaf width (LW), leaf thickness (LT), and fresh leaf weight. All leaves were taken back to the laboratory to be dried, ground, and used for the measurement of leaf dry matter content (LDMC), leaf carbon content (LC), leaf nitrogen content (LN), and leaf phosphorus content (LP) indexes.

2.2.3. Collection of Soil Samples

The soil was sampled in 5 m × 5 m units, with the diagonal method of taking a 0–20 cm surface layer of soil at the center, using an aluminum box to store the soil and calculate the soil water content, and then taking a sample in a sealing bag for the determination of other soil physical and chemical properties. The method of determination [27] is shown in Table 1.

Table 1. Soil index and determination method.

Soil Factor	Method
Soil water content	Drying and weighing method
pH	Acidimeter
Electrical conductivity	Residue method
Organic carbon	Potassium dichromate dilution heating method
Total phosphorus	Molybdenum antimony colorimetric method
Available phosphorus	Spectrophotometry
Total nitrogen	Kjeldahl method
Nitrate nitrogen	UV spectrophotometry
Ammonium nitrogen	UV spectrophotometry

2.3. Data Calculation and Analysis

2.3.1. Calculation of Plant Diversity

In this study, we selected species diversity indices, namely Shannon–Wiener diversity index, Simpson diversity index, Margalef richness index, and Pielou evenness index ([28]; functional diversity indices, including FRic richness index, FEve evenness index, FDiv divergence index, and RaoQ quadratic entropy index [29,30]; and phylogenetic diversity indices, including the mean interspecific distance index (MPD), mean nearest interspecific distance (MNTD), and Faith diversity index (PD) (for calculation methods, see Supplementary Table S1) [31,32].

2.3.2. Calculation of Ecosystem Multifunctionality

In this study, soil environmental factors (total nitrogen, total phosphorus, ammonium nitrogen, nitrate nitrogen, fast-acting phosphorus, and organic matter) were used as indicators of ecosystem multifunctionality, and the "Z-score" mean method was used to calculate ecosystem multifunctionality, represented by the formula [33,34]:

$$MF_a = \sum_i^F \frac{g\left(r_{i(f_i)}\right)}{F}.$$

In the above equation, MF_a represents ecosystem multifunctionality, f_i represents the measured value of function i, r_i is the mathematical function that converts f_i into a positive value, g represents the normalization of all measured values, and F represents the number of functions measured.

2.3.3. Data Analysis

Excel 2019 was used for the initial processing and calculation of the data. Differences in plant diversity and ecosystem multifunctionality between the three scales (5 m × 5 m, 20 m × 20 m, 50 m × 50 m) were analyzed in SPSS 26.0 using one-way analysis of variance (ANOVA). When the variance was equal, the least significant difference (LSD) method was used for the results of multiple comparisons; when the variance was not uniform, the results of multiple comparisons were tested using a non-parametric test. The Kolmogorov–Smirnov test (K–S test) were used to test the normality of ecosystem multifunctionality. Random sampling of the sample plot is done in R4.1.3. A community phylogenetic tree was created in R4.1.3 using the "V.PhyloMaker" package [35]. The species diversity index, functional diversity index, and phylogenetic diversity index were calculated using the "vegan", "FD", and "picante" packages, respectively.

The model was selected in R4.1.3 using the function "dredge" from the "MuMin" package [36], based on the corrected Akaike's information criterion (AICc; ΔAICc < 2) [37]. A selection procedure was used to select the best predictor of ecosystem multifunctionality, and when multiple models were selected, model averaging was performed based on AICc weights. The model residuals were inspected for constant variance and normality. All predictors and response variables were standardized before the model was constructed. Predictors were log-transformed as necessary before analysis to meet the assumptions. The model calculated relative explanatory rates for each diversity index and compared them with the total explanatory rates for all diversity indicators in the model, after which the explanatory rates for the indices in the model were categorized and summed by species diversity, functional diversity, and phylogenetic diversity to obtain the relative importance of different diversity dimensions (species diversity, functional diversity, and phylogenetic diversity) as drivers of ecosystem multifunctionality.

3. Results

3.1. Characteristics of Plant Diversity

Among the species diversity indices, the Shannon diversity index, Simpson diversity index, and Margalef richness index tended to increase with scale and showed significant differences between scales ($p < 0.05$), and the Pielou evenness at large scales (50 m × 50 m) was significantly smaller than at small (5 m × 5 m) and medium scales (20 m × 20 m). For the functional diversity index, the RaoQ index showed an increasing trend from small to large scales. The FRic richness index, FEve evenness index, and FDiv divergence index were significantly different between small and large scales. The MNTD index of phylogenetic diversity showed a decreasing trend with increasing scale. The PD index showed an opposite trend and was significantly different between scales, while the MPD index was not significantly different between scales ($p > 0.05$) (Figure 2).

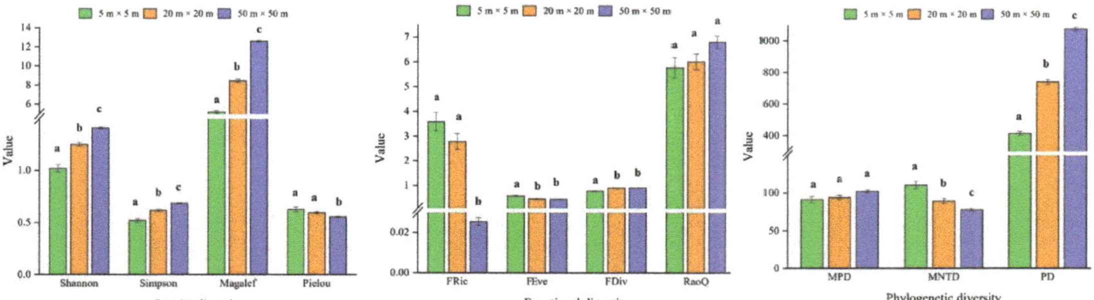

Figure 2. Characteristics and differences in plant diversity between scales (Mean ± SE). Note: error lines are standard errors; different lowercase letters on the error line for the same diversity index indicate highly significant differences between data ($p < 0.05$).

3.2. Characteristics of Ecosystem Multifunctionality

Using the K–S test and Raincloud plot (Figure 3), the ecosystem multifunctionality index calculated by the mean method was distributed normally at all three scales, with the values of the multifunctionality index varying from −1.085 to 1.129 for small-scale samples, −0.931 to 0.973 for medium-scale samples, and −0.484 to 0.718 for large-scale samples, but the ecosystem multifunctionality index did not vary significantly between scales ($p > 0.05$).

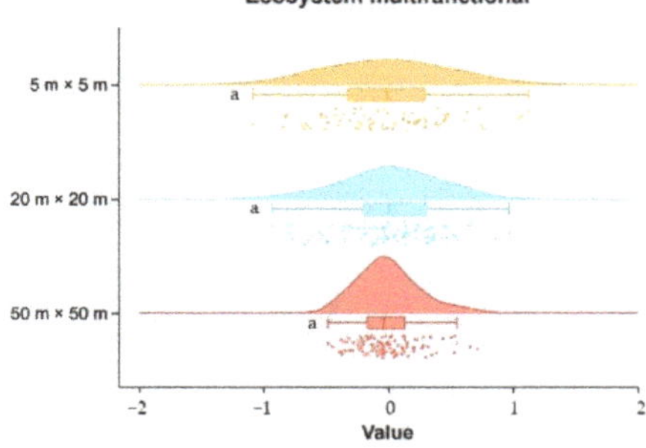

Figure 3. Characteristics and differences in ecosystem multifunctionality between scales. Note: Error lines are standard errors; same lowercase letters on the error line indicate no significant differences between data ($p > 0.05$).

3.3. Characterization of the Relationship between Diversity and Multifunctionality

The results of the variance decomposition showed that there was a scale effect on the contribution of different dimensions of diversity to ecosystem multifunctionality. The FRic, FEve, and RaoQ indexes of plant functional diversity explained 56% of the variation in multifunctionality together on small scales, and they were the main factors driving ecosystem multifunctionality. The FRic index was significantly correlated with ecosystem multifunctionality.

However, at the mesoscale, phylogenetic diversity was the main factor shaping multifunctionality, with the MPD, MNTD, and PD indices accounting for 53% of multifunctionality. The MPD index was significantly correlated with multifunctionality, while the functional diversity of plants explained less of multifunctionality, but the FRic index was still significantly correlated with multifunctionality. Species diversity had a stronger influence on multifunctionality, with the Margalef index being significantly correlated with multifunctionality.

The contribution of species diversity to multifunctionality reached a maximum of 75% on large scales, and it was the main explanatory factor for ecosystem multifunctionality. The Margalef and Pielou indexes were significantly correlated with multifunctionality. Among the functional diversity indices, the FRic, FDiv, and RaoQ indexes together contributed to 13% of the variation in multifunctionality and were all significantly correlated with multifunctionality. Among the phylogenetic diversity indices, the MNTD index was significantly correlated with multifunctionality (Figure 4).

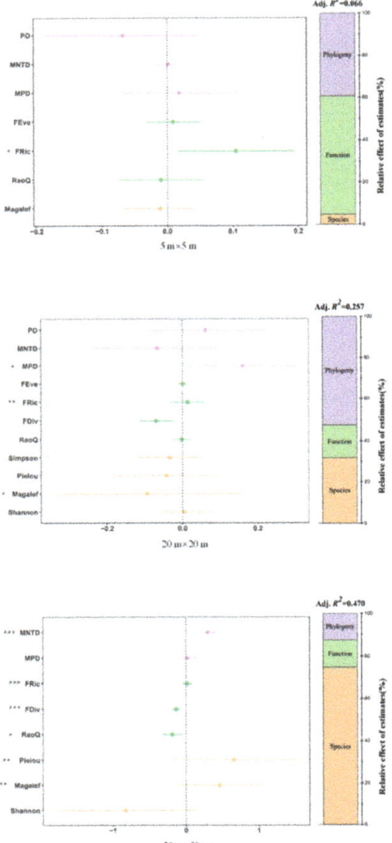

Figure 4. Relationship between plant diversity and ecosystem multifunction at different scales. Note: *, ** and *** represent $p < 0.05$, $p < 0.01$ and $p < 0.001$.

4. Discussion

4.1. Plant Diversity at Different Scales

Plants are driven by interspecific interaction and environmental influence, resulting in certain spatial distribution patterns [38]. When spatial scales change, plant community structure and diversity characteristics also change. Exploring the relationship between sampling scale and plant diversity can contribute to a more comprehensive understanding of community diversity trends and species coexistence mechanisms [39]. Our study found that species diversity was strongly scale-dependent [40]. The Margalef richness index showed an increasing trend with increasing scale. This is because the study site is highly windy and sandy, and water resources are scarce; thus, plants are clumped and aggregated to avoid wind and sand attacks and water scarcity [40], which limits the number of plant species that can be accommodated at small scales. As the sampling scale increases, the number of plant species increases, the composition of dominant species at each level becomes more diverse, and the Margalef index also increases. The Shannon diversity index and Simpson diversity index also increased with expansion in scale because, as the scale increases, habitat heterogeneity, the number of plants that can be accommodated, the number of species, and the level of species increases, and the structure of plant communities becomes more integrated and complex [41]. In addition, the Shannon diversity, Simpson diversity, and Margalef richness indexes showed a rapid increase and then a steady climb with the expansion of the scale, which is similar to the results of Deng et al. (2015) [14] in mixed coniferous forests of *Pinus radiata*. This indicates that the diversity of desert plant species increases with scale and plateaus after reaching a certain threshold [42].

Functional diversity is an extremely important part of plant diversity and plays an irreplaceable role in shaping the structure of plant communities and altering ecosystem functions [43]. Previous studies have shown that functional diversity varies with scale due to phenotypic plasticity [44–46]. Similar results were obtained in our study. The FEve evenness index and the FRic richness index both showed a decreasing trend with increasing scale and were opposite the trend in species richness, with significant differences between scales ($p < 0.05$); this indicates that species diversity and functional diversity have relatively independent trends [47]. This may be attributed to the obvious environmental filtering effect of the arid zone, where the functional composition of species is restricted to a certain range of functional traits, resulting in a more homogeneous pool of functional traits in this study area and an increase in species richness, leading to a more refined division of ecological niches rather than greater functional diversity [48], thus producing a different trend in functional diversity from that of species diversity [49,50].

The FDiv divergence index shows the degree of overlap in ecological niches between species within a community; that is, the heterogeneity of community character values [51], and a higher FDiv index indicates a high degree of ecological niche differentiation and higher resource use [52,53]. In our study, the FDiv index was significantly greater at large scales than at small scales, probably because, as scale increases, plant competition for the same or several habitat-specific resources diminishes, and ecological niches diverge further; thus, the FDiv index increases. The results of this study showed that the FRic richness index tended to decrease with increasing scale and was negatively correlated with the Shannon diversity index. This may be because functional richness is influenced not only by the functional ecological niche of the species but also by the range of functional trait values [54]. To overcome extreme drought conditions, functional traits of species in the study area are prioritized in response to selection pressure to adapt to drought [48,55], and functional traits tend to develop homogeneously, with increasing scale leading to a continuous increase in species richness followed by a deepening of functional redundancy [56,57], the FRic richness index declines, and the results of previous studies in Pinus oak forests in the Qinling Mountains are consistent with our study [58]. The negative correlation between species diversity and functional diversity because of scale expansion suggests that species diversity alone should not be considered when extrapolating functional diversity but also species differences and functional redundancy between scales [58].

In our study, the PD and MNTD indices varied significantly with scale, with the PD index increasing and the MNTD index decreasing with scale. This is probably because there is a significant correlation between the PD index and the Margalef index [59], with species richness increasing with scale, which is conducive to the maintenance of genealogical diversity at large scales. Genealogical structure is one of the most important expressions of community structure, as it reflects the process of community construction and evolution. Numerous studies have shown that the divergence or aggregation of genealogical structure is related to scale size [60–62]. The MNTD index in our study decreases with increasing scale and is significantly different between the three scales, suggesting that the community genealogical structure gradually moves from divergence to aggregation with increasing scale [63]. Therefore, the community is composed of more distantly related species on small scales and more closely related species on large scales. This is probably because of the low variation in habitat conditions on a small scale, combined with the harsh environmental conditions in the study area, where competition for a particular resource between species makes competitive exclusion the dominant community-building process. Therefore, the genealogical structure tends to diverge. At large scales, habitat heterogeneity increases, habitat-filtering ecological processes begin to dominate [64], and the genealogical structure gradually tends to agglomerate. Studies in the evergreen broadleaf forests of the Gutian Mountains and in the tropical rainforests of Panama [61,65] have also shown that competitive exclusion at small scales has a negative effect on the coexistence of closely related species, and that there is a tendency for the genealogical structure of communities to change from divergence to aggregation with increasing scale, but when the spatial scale exceeds a certain area, there is no correlation between genealogical structure and scale, and the genealogical structure becomes aggregated [66].

4.2. Ecosystem Multifunctionality at Different Scales

Early studies of ecosystem multifunctionality focused on the effects of species diversity on a single ecosystem function on the same scale [50,67–70]. As research has progressed, researchers have realized that differences in the choice of scale of study influence the expression of ecosystem multifunctionality [20]. However, the results of this study showed that ecosystem multifunctionality did not diverge significantly among the three scales. This is probably because variations in ecosystem multifunctionality in this study area are more related to changes in species composition, environmental conditions, or temporal scales than to spatial scales [33,71,72]. Our study area is located in an arid desert region where plant species composition is highly dependent on soil water and salinity conditions, which has led to a high degree of similarity in the overall community species composition of the region; however, significant differences in ecosystem multifunctionality between scales depend not only on significant differences in species diversity between scales but also on diverse community structure [73]. In addition, this study did not consider climatic conditions and time scales when discussing ecosystem multifunctionality, but numerous studies have shown that ecosystem multifunctionality varies significantly depending on the climatic conditions or time span of the study [12,74]. Finally, although three scales—large, medium, and small—were chosen for this study, the largest scale was only 50 m × 50 m. There is still much potential to expand the area of the sampling scale; thus, work on larger scales should be carried out in the future.

4.3. Relationship between Plant Diversity and Ecosystem Multifunctionality at Different Scales

The relationship between plant diversity and ecosystem multifunctionality varies with scale. Our study found that the highest levels of species diversity at large scales explained the greatest amount of variation in ecosystem multifunctionality, suggesting that the maintenance of multiple ecosystem functions simultaneously in this study area required a greater number of species and higher levels of species diversity to support them and that high species diversity could effectively support the maintenance of ecosystem multifunctionality. This is probably due to ecological niche differences on large scales when different species or functional groups coexist in resource-limited communities [75], where larger scales have more species and therefore have complementary advantages in resource use and, thus, the greatest explanatory power for ecosystem multifunctionality. The relationship between species diversity indices and multifunctionality also varied between scales, with the Margalef index being significantly correlated with ecosystem multifunctionality at the meso- and macro-scales but not at the small-scale, and the Pielou index was significantly correlated with ecosystem multifunctionality at the macro-scale. This demonstrated that under the dimension of species diversity, species richness and species evenness played a dominant role in driving multifunctionality. This likely results from the fact that the maintenance of ecosystem multifunctionality requires a high level of species diversity [72,76], and a smaller quantity of species cannot support all ecosystem functions [77]. Domestic and international studies have shown that larger vegetation communities and a larger number of species increase the use of habitat resources by plants and play an important role in maintaining high levels of ecosystem multifunctionality [69,78–81]. Species are relatively sparse in desert areas, and when the scale is small, the number of species in the sample is relatively infrequent, the Margalef index is comparatively low, the level of species diversity is not high, and the impact on ecosystem multifunctionality is limited. When the scale increases, the spatial ecosystem heterogeneity of the sample increases, it can accommodate a larger number and more species of plant samples, the Margalef index and Pielou indexes increase, and the explanatory power of species diversity on multifunctionality is strengthened, which is consistent with the hypothesis that species diversity is positively correlated with ecosystem multifunctionality [82].

Studies have shown that species diversity is no substitute for functional diversity to simply quantify multifunctionality [70], and in recent years, many studies have demonstrated that functional diversity has a stronger explanatory role for multifunctionality [83–85]. In our study, functional diversity contributes most to multifunctionality at a small scale, which is determined by a combination of species composition and structure. At small scales, species richness is low to allow for an increase in ecological niche space, which facilitates the expansion of the range of functional plant traits [86], so functional richness is at a higher level at small scales, which indicates a higher proportion of available resources [79], thus allowing functional diversity to dominate variations in ecosystem multifunctionality at small scales. In addition, the FRic index of functional diversity was significantly correlated with ecosystem multifunctionality at all three scales ($p < 0.05$), and the FDiv index was highly significantly correlated with ecosystem multifunctionality at large scales. The FRic index is a dominant driver of ecosystem multifunctionality at different scales, probably because plant species are limited in desert areas, and changes in species community composition and structure directly alter functional richness [87], which in turn affects single ecosystem functions and thus ecosystem multifunctionality. In addition, functional dispersion is significantly correlated with ecosystem multifunctionality at large scales, probably because as habitat heterogeneity increases on larger scales, variation in functional traits increases, competition among species decreases, complementarity of functional traits for resource use increases, overlap of species' ecological niches decreases, and the overall degree of resource use within scales increases, which also has a positive effect on the maintenance of ecosystem multifunctionality [86]. The results are also similar to the findings of Huang et al. (2019) [79] in Yunnan.

Genealogical diversity reflects the historical course of species evolution [88], can express community information not represented by species diversity and functional diversity, and is an important component in the study of the relationship between plant diversity and ecosystem multifunctionality [89–91]. In our study, genealogical diversity played a dominant role in ecosystem multifunctionality at the mesoscale, probably because habitat heterogeneity has an important influence on the maintenance of genealogical diversity. The results of the study on typical areas of karst landscapes showed that phylogenetic diversity had a single-peaked change with increasing habitat heterogeneity and was greatest in areas of moderate habitat heterogeneity [59]. In addition, a study by Li et al. (2021) [92] in the Daiyun Mountains also showed that phylogenetic diversity showed an intermediate peak with altitude, which was similar to the results of our study. At the mesoscale, habitat heterogeneity is at a moderate level compared to the large and small scales, which contributes more to genealogical diversity, and higher levels of genealogical diversity increase the rate of explanation for ecosystem multifunctionality. Our study also showed that the MPD index was significantly correlated with ecosystem multifunctionality on the large scale, and the PD index was not significantly correlated with ecosystem multifunctionality on any scale, indicating that the proximity of affinities among species plays a major role in maintaining ecosystem multifunctionality at different scales. Although the PD index increases with the number of species, the total number of plant species in the desert area is limited, and when the PD index reaches a certain value, it will not continue to promote the maintenance of ecosystem multifunctionality. In our study area, species relatedness tended to change from distant to close, based on scale expansion. This is likely because the habitat filtering effect in the study area increases with scale, the extreme arid habitat conditions make it easier for species with similar life history strategies to survive [93,94], and similar survival strategies and close affinities use resources in similar ways, allowing limited resources to be fully utilized [95]. Genealogical structure has a strong influence on ecosystem multifunctionality, so the correlation between MNTD and MPD indexes and ecosystem multifunctionality is also stronger than that between PD indices.

5. Conclusions

We uncovered a significant scale effect in the relationship between plant diversity and ecosystem multifunctionality. Differences in sampling scales altered the composition and distribution of species in plant communities. The relative contribution of plant diversity to ecosystem multifunctionality also changed consequently. This highlights the close coupling between scale effects and ecosystem multifunctionality in the plant communities of this study area. Therefore, we will continue to expand the sampling scale in the next research work, aiming to better enhance plant diversity and maintain ecosystem function.

Supplementary Materials: The following supporting information can be downloaded at: https:// www.mdpi.com/article/10.3390/f13091505/s1, Table S1: Calculation formula of plant diversity indexes.

Author Contributions: Conceptualization, J.L. and G.L.; methodology, J.L.; software, J.L., H.W. and D.H.; writing—original draft preparation, J.L.; writing—review and editing, J.L., L.J. and D.H.; supervision, G.L. funding acquisition, G.L. All authors have read and agreed to the published version of the manuscript.

Funding: This research was supported by the National Natural Science Foundation of China (42171026) and Xinjiang Uygur Autonomous Region innovation environment Construction special project & Science and technology innovation base construction project (PT2107).

Institutional Review Board Statement: Not applicable.

Informed Consent Statement: Not applicable.

Data Availability Statement: Not available.

Conflicts of Interest: The authors declare that they have no conflict of interest.

References

1. Cardinale, B.J. Biodiversity loss and its impact on humanity. *Nature* **2012**, *489*, 59. [CrossRef] [PubMed]
2. Srivastava, D.S.; Vellend, M. Biodiversity-Ecosystem Function Research: Is It Relevant to Conservation? *Annu. Rev. Ecol. Evol. Syst.* **2005**, *36*, 267–294. [CrossRef]
3. Swenson, N.G. The role of evolutionary processes in producing biodiversity patterns, and the interrelationships between taxonomic, functional and phylogenetic biodiversity. *Am. J. Bot.* **2011**, *98*, 472–480. [CrossRef] [PubMed]
4. Grace, J.B.; Anderson, T.M.; Smith, M.D.; Seabloom, E.; Andelman, S.J.; Meche, G.; Weiher, E.; Allain, L.K.; Jutila, H.; Sankaran, M. Does species diversity limit productivity in natural grassland communities? *Ecol. Lett.* **2007**, *10*, 680–689. [CrossRef] [PubMed]
5. Ma, W.J.; Zhang, Q.; Niu, J.M.; Kang, S.; Liu, P.T.; He, X.; Yan, Y.; Zhang, Y.N.; Wu, J.G. Relationship of ecosystem primary productivity to species diversity and functional group diversity: Evidence from *Stipa breviflora* grassland in Nei Mongol. *Chin. J. Plant Ecol.* **2013**, *37*, 620–630. [CrossRef]
6. Manning, P.; van der Plas, F.; Soliveres, S.; Allan, E.; Maestre, F.T.; Mace, G.; Whittingham, M.J.; Fischer, M. Redefining ecosystem multifunctionality. *Nat. Ecol. Evol.* **2018**, *2*, 427–436. [CrossRef]
7. Garland, G.; Banerjee, S.; Edlinger, A.; Oliveira, E.M.; Herzog, C.; Wittwer, R.; Philippot, L.; Maestre, F.T.; Van Der Heijden, M.G.A. A closer look at the functions behind ecosystem multifunctionality: A review. *J. Ecol.* **2021**, *109*, 600–613. [CrossRef]
8. Wang, K.; Wang, C.; Feng, X.M.; Wu, X.; Fu, B.J. Research progress on the relationship between biodiversity and ecosystem versatility. *Acta Ecol. Sin.* **2022**, *42*, 11–23.
9. Zhang, L. Influence Mechanism of Typical Desert Plant Diversity on Ecosystem Function. Master's Thesis, Xinjiang University, Ürümqi, China, 2019.

10. Meng, B.; Wang, J. A Review on the Methodology of Scaling with Geo-Data. *Acta Geogr. Sin.* **2005**, *60*, 12.
11. Stürck, J.; Verburg, P.H. Multifunctionality at what scale? A landscape multifunctionality assessment for the European Union under conditions of land use change. *Landsc. Ecol.* **2017**, *32*, 481–500. [CrossRef]
12. Pasari, J.R.; Levi, T.; Zavaleta, E.S.; Tilman, D. Several scales of biodiversity affect ecosystem multifunctionality. *Proc. Natl. Acad. Sci. USA* **2013**, *110*, 10219–10222. [CrossRef] [PubMed]
13. Ye, W.H.; Wang, Z.G.; Cao, H.L.; Lian, J.Y. Spatial distribution of species diversity indices in a monsoon evergreen broadleaved forest at Dinghushan Mountain. *Biodivers. Sci.* **2008**, *16*, 454–461.
14. Deng, H.J.; Li, W.Z.; Cao, Z.; Wang, Q.; Wang, G.R. Species diversity based on sample size in a mixed forest. *J. Zhejiang A F Univ.* **2015**, *32*, 67–75.
15. Pedro, M.S.; Rammer, W.; Seidl, R. A disturbance-induced increase in tree species diversity facilitates forest productivity. *Landsc. Ecol.* **2016**, *31*, 989–1004. [CrossRef]
16. Dang, H. Research on Multi-Scale Construction Mechanism of Desert Communities Based on Traits in Arid Region. Master's Thesis, Northwest University, Shanxi, China, 2018.
17. Wang, S.Y.; Lv, G.H.; Jiang, L.M.; Wang, H.F.; Li, Y.; Wang, J.L. Functional diversity and phylogenetic diversity of typical plant communities in Ebi Lake at different scales. *J. Ecol. Environ.* **2020**, *29*, 889–900.
18. Huang, Z.P. Patterns and Mechanisms of Phylogenetic Diversity of Limestone Cave Plant Communities in Guangxi. Master's Thesis, Guangxi Normal University, Guilin, China, 2019.
19. Grman, E.; Zirbel, C.R.; Bassett, T.; Brudvig, L.A. Ecosystem multifunctionality increases with beta diversity in restored prairies. *Oecologia* **2018**, *188*, 837–848. [CrossRef]
20. Isbell, F.; Calcagno, V.; Hector, A.; Connolly, J.; Harpole, W.S.; Reich, P.B.; Scherer-Lorenzen, M.; Schmid, B.; Tilman, D.; Van Ruijven, J.; et al. High plant diversity is needed to maintain ecosystem services. *Nature* **2011**, *477*, 199–202. [CrossRef]
21. Van der Plas, F.; Manning, P.; Soliveres, S.; Allan, E.; Scherer-Lorenzen, M.; Verheyen, K.; Wirth, C.; Zavala, M.A.; Ampoorter, E.; Baeten, L.; et al. Biotic homogenization can decrease landscape-scale forest multifunctionality. *Proc. Natl. Acad. Sci. USA* **2016**, *113*, 3557–3562. [CrossRef]
22. Hautier, Y.; Isbell, F.; Borer, E.T.; Seabloom, E.W.; Harpole, W.S.; Lind, E.M.; MacDougall, A.S.; Stevens, C.J.; Adler, P.B.; Alberti, J.; et al. Local loss and spatial homogenization of plant diversity reduce ecosystem multifunctionality. *Nat. Ecol. Evol.* **2018**, *2*, 50–56. [CrossRef]
23. Whitford, W.G. *Ecology of Desert Systems*; Academic Press: New York, NY, USA, 2002.
24. Zou, C.; Fan, Z.A. Effects of wind and sand disasters on Photosynthetic Characteristics of cotton in Qira Oasis. *J. Chuzhou Univ.* **2019**, *21*, 10–14.
25. Mao, D.; Lei, J.; Li, S.; Zeng, F.; Zhou, J. Characteristics of Meteorological Factors over Different Landscape Types During Dust Storm Events in Cele, Xinjiang, China. *J. Meteorol. Res.* **2014**, *28*, 576–591. [CrossRef]
26. Qie, Y.D.; Jiang, L.M.; Lv, G.H.; Yang, X.D.; Wang, H.F.; Teng, D.X. Response ofPlant Leaf Functional Traits to Soil Aridity and Salinity in Temperate Desert Ecosystem. *Ecol. Environ. Sci.* **2018**, *27*, 24–34.
27. Bao, S.D. *Soil Agrochemical Analysis*, 3rd ed.; China Agricultural Press: Beijing, China, 2000.
28. Zhang, J.T. *Quantitative Ecology*, 2nd ed.; Science Press: Beijing, China, 2011.
29. Ricotta, C. A note on functional diversity measures. *Basic Appl. Ecol.* **2005**, *6*, 479–486. [CrossRef]
30. Song, Y.T.; Wang, P.; Zhou, D.W. Methods of measuring plant community functional diversity. *Chin. J. Ecol.* **2011**, *30*, 2053–2059.
31. Faith, D.P. *Phylogenetic Diversity: A General Framework for the Prediction of Feature Diversity*; Clarendon Press: Oxford, UK, 1994; Volume 50.
32. Balvanera, P.; Pfisterer, A.B.; Buchmann, N.; He, J.S.; Nakashizuka, T.; Raffaelli, D.; Schmid, B. Quantifying the evidence for biodiversity effects on ecosystem functioning and services. *Ecol. Lett.* **2006**, *9*, 1146–1156. [CrossRef]
33. Maestre, F.T.; Castillo-Monroy, A.P.; Bowker, M.A.; Ochoa-Hueso, R. Species richness effects on ecosystem multifunctionality depend on evenness, composition and spatial pattern. *J. Ecol.* **2012**, *100*, 317–330. [CrossRef]
34. Bowker, M.A.; Maestre, F.T.; Mau, R.L. Diversity and Patch-Size Distributions of Biological Soil Crusts Regulate Dryland Ecosystem Multifunctionality. *Ecosystems* **2013**, *16*, 923–933. [CrossRef]
35. R Team; R Core Team. *R: A Language and Environment for Statistical Computing 2014*; R Foundation for Statistical Computing: Vienna, Austria, 2008.
36. Bartoń, K. *MuMIn: Multi-Model Inference, Version 1.9*; R Package; R Foundation for Statistical Computing: Vienna, Austria, 2013.
37. Burnham, K.P.; Anderson, D.R. *Formal Inference from More Than One Model: Multimodel Inference (MMT) 4.1*; Springer: New York, NY, USA, 2002.
38. Hou, J.H.; Mi, X.C.; Liu, C.R.; Ma, K.P. Spatial patterns and associations in a *Quercus-Betula* forest in northern China. *J. Veg. Sci.* **2004**, *15*, 407–414.
39. Wiegand, T.; Moloney, K.A. *A Handbook of Spatial Point Pattern Analysis in Ecology*, 5th ed.; CRC Press: Boca Raton, FL, USA, 2013.
40. He, Z.B.; Zhao, W.Z.; Chang, X.X.; Chang, X.L. Scale dependence in desert plant biodiversity. *Acta Ecol. Sin.* **2004**, *24*, 4.
41. Jia, Z.T.; Yang, J.Y.; Sun, Y.X.; Chen, Q.; Yan, R.Q.; Li, N.N. Species diversity and its correlation with environmental factors in the Alxa Plateau, China. *Chin. J. Grassl.* **2021**, *43*, 1–9.
42. Huang, Z.L.; Kong, G.H.; He, D.Q. Plant community diversity in Dinghushan Nature Reserve. *Acta Ecol. Sin.* **2000**, *20*, 193–198.

43. Violle, C.; Enquist, B.J.; Mcgill, B.J.; Jiang, L.; Cécile, H.A.; Hulshof, C.; Jung, V.; Messier, J. The return of the variance: Intraspecific variability in community ecology. *Trends Ecol. Evol.* **2012**, *27*, 244–252. [CrossRef] [PubMed]
44. Violle, C.; Castro, H.; Richarte, J.; Navas, M.L. Intraspecific seed trait variations and competition: Passive or adaptive response? *Funct. Ecol.* **2010**, *23*, 612–620. [CrossRef]
45. Gubsch, M.; Buchmann, N.; Schmid, B.; Schulze, E.D.; Lipowsky, A.; Roscher, C. Differential effects of plant diversity on functional trait variation of grass species. *Ann. Bot.* **2011**, *107*, 157–169. [CrossRef]
46. García-Cervigón, A.I.; Linares, J.C.; Aibar, P.; Olano, J.M. Facilitation promotes changes in leaf economics traits of a perennial forb. *Oecologia* **2015**, *179*, 103–116. [CrossRef]
47. Vesk, P.A.; Leishman, M.R.; Westoby, M. Simple traits do not predict grazing response in Australian dry shrublands and woodlands. *J. Appl. Ecol.* **2010**, *41*, 22–31. [CrossRef]
48. Díaz, S.; Cabido, M. Vive la difference: Plant functional diversity matters to ecosystem processes. *Trend. Ecol Evolut.* **2001**, *16*, 646–655. [CrossRef]
49. Saugier, B. Plant Strategies, Vegetation Processes, and Ecosystem Properties. *Plant Sci.* **2001**, *161*, 813. [CrossRef]
50. Hooper, D.U.; Chapin, F.S.; Ewel, J.J.; Hector, A.; Setala, H. Effects of biodiversity on ecosystem functioning: A consensus of current knowledge: Ecological Monog. *Ecol. Monogr.* **2004**, *75*, 95–105.
51. Li, X.; Tian, P.; Cheng, X.Q. The species diversity and functional diversity of the typical forest of Taiyue Mountain Shanxi, China. *J. Shanxi Agric. Univ. Nat. Sci. Ed.* **2017**, *9*, 1671–8151.
52. Clark, C.M.; Flynn, D.; Butterfield, B.J.; Reich, P.B.; Cahill, J.F. Testing the Link between Functional Diversity and Ecosystem Functioning in a Minnesota Grassland Experiment. *PLoS ONE* **2012**, *7*, e52821. [CrossRef] [PubMed]
53. Zuo, X.; Jing, Z.; Lv, P.; Wang, S.; Yang, Y.; Yue, X.; Zhou, X.; Li, Y.; Chen, M.; Lian, J. Effects of plant functional diversity induced by grazing and soil properties on above- and belowground biomass in a semiarid grassland. *Ecol. Indic.* **2018**, *93*, 555–561. [CrossRef]
54. Zhang, Q.D.; Duan, X.M.; Bai, Y.F.; Wang, W.L.; Zhang, J.T. Functional diversity of *Ulmus lamellosa* community in the Taiyue Mountain of Shanxi. *Chin. Bull. Bot.* **2016**, *51*, 218–225.
55. Sasaki, T.; Okubo, S.; Okayasu, T.; Jamsran, U.; Ohkuro, T.; Takeuchi, K. Two-phase functional redundancy in plant communities along a grazing gradient in Mongolian rangelands. *Ecology* **2009**, *90*, 2598–2608. [CrossRef]
56. Chen, C.; Zhu, Z.; Yingnian, L.I.; Yao, T.; Pan, S.; Wei, X.; Kong, B.; Jiali, D.U. Effects of interspecific trait dissimilarity and species evenness on the relationship between species diversity and functional diversity in an alpine meadow. *Acta Ecol. Sin.* **2016**, *36*, 661–674.
57. Dong, S.K.; Tang, L.; Liu, S.L.; Liu, Q.R.; Sun, X.K.; Zhang, Y.; Wu, X.Y.; Zhao, Z.Z.; Li, Y.; Sha, W. Relationship between plant species diversity and functional diversity in alpine grasslands. *Acta Ecol. Sin.* **2018**, *37*, 1472–1483.
58. Wu, H.; Xiao, N.N.; Lin, T.T. Coupling relationship between functional diversity, species diversity and environmental heterogeneity of Pinus oak forest in Qinling Mountains. *Eco-Environ. Sci.* **2020**, *29*, 1090–1100.
59. Tan, Q.Y. Distribution Patterns of Species and Lineage Diversity of Karst Plant Communities Based on Habitat Heterogeneity. Master's Thesis, Guizhou University, Guiyang, China, 2021.
60. Cavender-Bares, J.; Keen, A.; Miles, B. Phylogenetic structure of Floridian plant communities depends on taxonomic and spatial scale. *Ecology* **2006**, *87*, 109–122. [CrossRef]
61. Kembel, S.W.; Hubbell, S.P. The phylogenetic structure of a neotropical forest tree community. *Ecology* **2006**, *87*, 86–99. [CrossRef]
62. Sweson, N.G.; Enquist, B.J.; Pither, J.; Thompson, J.; Zimmerman, J.K. The Problem and Promise of Scale Dependency in Community Phylogenetics. *Ecology* **2006**, *87*, 2418–2424. [CrossRef]
63. Swenson, N.G.; Enquist, B.J.; Thompson, J.; Zimmerman, J.K. The influence of spatial and size scale on phylogenetic relatedness in tropical forest communities. *Ecology* **2007**, *88*, 1770–1780. [CrossRef] [PubMed]
64. Willis, C.G.; Halina, M.; Lehman, C.; Reich, P.B.; Keen, A.; McCarthy, S.; Cavender-Bares, J. Phylogenetic community structure in Minnesota oak savanna is influenced by spatial extent and environmental variation. *Ecography* **2010**, *33*, 565–577. [CrossRef]
65. Huang, J.X.; Zheng, F.Y.; Mi, X.C. Influence of environmental factors on phylogenetic structure at multiple spatial scales in an evergreen broad-leaved forest of China. *Chin. J. Plant Ecol.* **2010**, *34*, 309–315.
66. Vamosi, S.M.; Heard, S.B.; Vamosi, J.C.; Webb, C.O. Emerging patterns in the comparative analysis of phylogenetic community structure. *Mol. Ecol.* **2010**, *18*, 572–592. [CrossRef] [PubMed]
67. Loreau, M.; Naeem, S.; Inchausti, P.; Bengtsson, J.; Grime, J.P.; Hector, A.; Hooper, D.U.; Huston, M.A.; Raffaelli, D.; Schmid, B.; et al. Biodiversity and ecosystem functioning: Current knowledge and future challenges. *Science* **2001**, *294*, 804–808. [CrossRef]
68. Cardinale, B.J.; Srivastava, D.S.; Emmett Duffy, J.; Wright, J.P.; Downing, A.L.; Sankaran, M.; Jouseau, C. Effects of biodiversity on the functioning of trophic groups and ecosystems. *Nature* **2006**, *443*, 989–992. [CrossRef]
69. Hector, A.; Bagchi, R. Biodiversity and ecosystem multifunctionality. *Nature* **2007**, *448*, 188. [CrossRef]
70. Tilman, D.; Knops, J.; Wedin, D.; Reich, P.; Ritchie, M.; Siemann, E. The Influence of Functional Diversity and Composition on Ecosystem Processes. *Science* **1997**, *277*, 1300–1302. [CrossRef]
71. Hillebrand, H.; Matthiessen, B. Biodiversity in a complex world: Consolidation and progress in functional biodiversity research. *Ecol. Lett.* **2010**, *12*, 1405–1419. [CrossRef]
72. Byrnes, J.E.K.; Gamfeldt, L.; Isbell, F.; Lefcheck, J.S.; Griffin, J.N.; Hector, A.; Cardinale, B.J.; Hooper, D.U.; Dee, L.E.; Duffy, J.E. Investigating the relationship between biodiversity and ecosystem multifunctionality: Challenges and solutions. *Methods Ecol. Evol.* **2014**, *5*, 111–124. [CrossRef]

73. Zavaleta, E.S.; Pasari, J.R.; Hulvey, K.B.; Tilman, G.D. Sustaining multiple ecosystem functions in grassland communities requires higher biodiversity. *Proc. Natl. Acad. Sci. USA* **2010**, *107*, 1443–1446. [CrossRef] [PubMed]
74. Perkins, D.M.; Bailey, R.A.; Dossena, M.; Gamfeldt, L.; Reiss, J.; Trimmer, M.; Woodward, G. Higher biodiversity is required to sustain multiple ecosystem processes across temperature regimes. *Glob. Change Biol.* **2015**, *21*, 396–406. [CrossRef] [PubMed]
75. Xie, H.; Wang, G.G.; Yu, M. Ecosystem multifunctionality is highly related to the shelterbelt structure and plant species diversity in mixed shelterbelts of eastern China. *Glob. Ecol. Conserv.* **2018**, *16*, e00470. [CrossRef]
76. Gamfeldt, L.; Hillebrand, H.; Jonsson, P.R. Multiple functions increase the importance of biodiversity for overall ecosystem functioning. *Ecology* **2008**, *89*, 1223–1231. [CrossRef]
77. Lohbeck, M.; Bongers, F.; Martinez-Ramos, M.; Poorter, L. The importance of biodiversity and dominance for multiple ecosystem functions in a human-modified tropical landscape. *Ecology* **2016**, *97*, 2772–2779. [CrossRef]
78. Soliveres, S.; Maestre, F.T.; Eldridge, D.J.; Delgado-Baquerizo, M.; Quero, J.L.; Bowker, M.A.; Gallardo, A. Plant diversity and ecosystem multifunctionality peak at intermediate levels of woody cover in global drylands. *Glob. Ecol. Biogeogr. J. Macroecology* **2014**, *23*, 1408–1416. [CrossRef]
79. Huang, X.B.; Su, J.R.; Li, S.F.; Liu, W.D.; Lang, X.D. Functional diversity drives ecosystem multifunctionality in a *Pinus yunnanensis* natural secondary forest. *Sci. Rep.* **2019**, *9*, 6979. [CrossRef]
80. Xiong, D.P.; Zhao, G.S.; Wu, J.S.; Shi, P.L.; Zhang, X.Z. Relationship between species diversity and ecosystem multifunction in Qiangtang Alpine grassland. *Acta Ecol. Sin.* **2016**, *36*, 3362–3371.
81. Chen, T.; Zhang, Z.W.; Liang, S.M.; Lv, Y.H.; Zhang, Y.; Zhao, J.J. Relationship between plant species diversity and ecosystem function in water-level-fluctuation zone of Xiaolangdi Reservoir area. *J. Shanxi Norm. Univ. Nat. Sci. Ed.* **2022**, *36*, 80–88.
82. Maestre, F.T.; Quero, J.L.; Gotelli, N.J.; Escudero, A.; Ochoa, V.; Delgado-Baquerizo, M.; García-Gómez, M.; Bowker, M.A.; Soliveres, S.; Escolar, C.; et al. Plant species richness and ecosystem multifunctionality in global drylands. *Science* **2012**, *335*, 214–218. [CrossRef]
83. Roscher, C.; Schumacher, J.; Gubsch, M.; Lipowsky, A.; Weigelt, A.; Buchmann, N.; Schmid, B.; Schulze, E.D. Using plant functional traits to explain diversity-productivity relationships. *PLoS ONE* **2012**, *7*, e36760. [CrossRef] [PubMed]
84. Chanteloup, P.; Bonis, A. Functional diversity in root and above-ground traits in a fertile grassland shows a detrimental effect on productivity. *Basic Appl. Ecol.* **2013**, *14*, 208–216. [CrossRef]
85. Fu, H.; Zhong, J.Y.; Yuan, G.X.; Ni, L.Y.; Xie, P.; Cao, T. Functional traits composition predict macrophytes community productivity along a water depth gradient in a freshwater lake. *Ecol. Evol.* **2014**, *4*, 1516–1523. [CrossRef] [PubMed]
86. Song, Y.T.; Wang, P.; Li, G.D.; Zhou, D.W. Relationships between functional diversity and ecosystem functioning: A review. *Acta Ecol. Sin.* **2014**, *34*, 85–91. [CrossRef]
87. Liu, M.X.; Zhang, G.J.; Li, L.; Mu, R.L.; Xu, L.; Yu, R.X. Relationship between functional diversity and ecosystem multifunctionality of alpine meadow along an altitude gradient in Gannan, China. *Chin. J. Appl. Ecol.* **2022**, *33*, 1291–1299.
88. Knapp, S.; Kühn, I.; Schweiger, O.; Klotz, S. Challenging urban species diversity: Contrasting phylogenetic patterns across plant functional groups in Germany. *Ecol. Lett.* **2008**, *11*, 1054–1064. [CrossRef]
89. Cadotte, M.W.; Cardinale, B.J.; Oakley, T.H. Evolutionary history and the effect of biodiversity on plant productivity. *Proc. Natl. Acad. Sci. USA* **2008**, *105*, 17012–17017. [CrossRef]
90. Yuan, Z.; Wang, S.; Gazol, A.; Mellard, J.; Lin, F.; Ye, J.; Hao, Z.; Wang, X.; Loreau, M. Multiple metrics of diversity have different effects on temperate forest functioning over succession. *Oecologia* **2016**, *182*, 1175–1185. [CrossRef]
91. Cadotte, M.W. Phylogenetic diversity and productivity: Gauging interpretations from experiments that do not manipulate phylogenetic diversity. *Funct. Ecol.* **2015**, *29*, 1603–1606. [CrossRef]
92. Li, M.J.; He, Z.S.; Jiang, L.; Gu, X.G.; Jin, M.R.; Chen, B.; Liu, J.F. Distribution patterns and driving factors of altitudinal gradients of species diversity and phylogenetic diversity in Daiyun Mountain. *Acta Ecol. Sin.* **2021**, *41*, 1148–1157.
93. Meng, Q.X. Distribution Pattern of Plant Community DIversity and Its Response to Environmental Factors in Taihang Mountain. Master's Thesis, Shanxi University, Taiyuan, China, 2020.
94. Xiao, Y.M.; Yang, L.C.; Nie, X.Q.; Chang-Bin, L.I.; Xiong, F.; Zhou, G.Y. Phylogenetic patterns of shrub communities along the longitudinal and latitudinal gradients on the northeastern Qinghai–Tibetan Plateau. *J. Mt. Sci. Engl. Ed.* **2020**, *17*, 9. [CrossRef]
95. Lin, H.L. Analysis on Community Structure and Pedigree Structure of Deciduous Broadleaved Forest in Karst Stone Mountain of Guilin. Master's Thesis, Guangxi Normal University, Guilin, China, 2021.

 *forests*

Article

α Diversity of Desert Shrub Communities and Its Relationship with Climatic Factors in Xinjiang

Yan Luo [1] and Yanming Gong [2,3,*]

1   College of Ecology and Environment, Xinjiang University, Urumqi 830017, China
2   Bayinbuluk Grassland Ecosystem Research Station, Xinjiang Institute of Ecology and Geography, Chinese Academy of Sciences, Bayingolin Mongolian Autonomous Prefecture 841314, China
3   CAS Research Center for Ecology and Environment of Central Asia, Urumqi 830011, China
*   Correspondence: gongym@ms.xjb.ac.cn

Abstract: In the past 30 years, Northwest China has experienced a warm and humid climate increase trend. How this climate change will affect the species diversity of plant communities is a hot issue in ecological research. In this study, four α diversity indexes were applied in 29 shrub communities at desert sites in Xinjiang, including the Margalef index, Simpson index, Shannon–Wiener index, and Pielou index, to explore the relationship between the α diversity of the desert shrub communities and climate factors (mean annual temperature (MAT) and mean annual precipitation (MAP)). The species diversity indexes varied across these different desert shrub communities. *Tamarix ramosissima* communities had the highest Margalef index, while the *Krascheninnikovia ewersmannia* communities had the lowest Margalef index; *T. ramosissima* communities also showed the highest Simpson index and Shannon–Wiener index, but *Alhagi sparsifolia* communities showed the lowest Simpson index and Shannon–Wiener index. The *Ephedra przewalskii* communities and *Karelinia caspica* communities showed the highest and the lowest Pielou index, respectively. The α diversity indexes (except the Pielou index) of desert shrub communities had a significantly positive correlation with MAP ($p < 0.05$) but a non-significantly correlation with MAT ($p > 0.05$). These results indicate that, compared with temperature, water conditions are still a more vital climatic factor affecting the species diversity of desert shrub communities in Xinjiang, and thus, the recent "warm and humid" climate trend in Xinjiang affects the α diversity of desert shrub communities.

Keywords: desert plants; shrub community; biodiversity; climate change; arid region

Citation: Luo, Y.; Gong, Y. α Diversity of Desert Shrub Communities and Its Relationship with Climatic Factors in Xinjiang. *Forests* **2023**, *14*, 178. https://doi.org/10.3390/f14020178

Academic Editor: Cate Macinnis-Ng

Received: 22 November 2022
Revised: 14 January 2023
Accepted: 17 January 2023
Published: 18 January 2023

1. Introduction

As the basis of ecosystem biodiversity, plant diversity plays an important role in maintaining ecosystem versatility, productivity, stability, and anti-interference ability [1,2]. Climate change has a significant impact on the water and heat dynamics of ecosystems, resulting in significant changes in the species composition and community structure of ecosystems [3]. Some research has found that climate change restricts the growth and distribution of plants, drives changes in plant diversity, affects interspecific relationships and community productivity, and seriously threatens the biodiversity of desert ecosystems [4,5]. Plant diversity will also alleviate the impact of climate change [4]. Therefore, the impact of climate change on plant diversity is a mutual process. In addition, to explain the distribution pattern of species diversity, a variety of hypotheses have been proposed, the most discussed of which is the energy hypothesis, which argues that changes in species diversity are controlled by energy according to the different forms of energy and their impact on the species diversity mechanism. The water-energy dynamic hypothesis holds that the large-scale pattern of species diversity is determined by water and energy. In this hypothesis, energy refers to thermal kinetic energy (or thermal energy), usually expressed in terms of potential evapotranspiration or temperature [5]. In addition, some scholars have studied the species richness model of 3637 vascular plants in arid areas of northern China. It was found that water had a greater impact on species richness than energy, which indicated that the species richness pattern of plants in arid areas was mainly limited by water [6]. Therefore, under extreme climatic conditions, the influence of limiting factors on species diversity should be considered first. Desert ecosystems are the most sensitive areas to climate change. Desert plant diversity is one of the key factors determining the structure and functional diversity of desert plant communities [4,7]. Under the background of the rapid loss of biodiversity caused by global change, it is of great significance to study the maintenance mechanism of desert plant diversity for the maintenance and management of biodiversity in this area and the protection of biodiversity in the future.

Due to different regions and ecological types, the effects of temperature and precipitation on plant diversity will also show different response results. On a global scale, rising surface temperatures increase plant growth in the mid- to high latitudes of the Southern and Northern Hemispheres [8]; in inland Asia, temperature is also a major environmental factor for vegetation growth [9]. Other studies have shown that warming reduces the diversity and richness of plant species. For example, in the Arctic tundra ecosystem, short-term warming reduces community species diversity [10]. In the Tibetan Plateau, the increase in temperature accelerates the decrease in species richness [11]. However, studies from Inner Mongolia grassland found that warming had no significant effect on plant diversity [12]. In addition to temperature, the precipitation distribution pattern also affects plant species diversity. Studies have shown that a warm and humid climate is conducive to the increase in perennial plant diversity, while a warm and dry climate is conducive to the improvement of annual plant diversity [13]. In the warm and humid climate, the vegetation coverage in most arid areas of inland China increased significantly [14]. With the increase in annual precipitation, plant diversity changes along the steppe types (from desert to desert steppe-typical to steppe-meadow steppe) [15], which is consistent with Lopez-Angulo research results on the correlation between vegetation species richness and annual precipitation in different grassland types in Xinjiang. However, studies have shown that increased annual precipitation has no significant effect on plant diversity [16,17]. In addition, under the influence of climate change, species composition and vegetation patterns are changing. Kazakis et al. [18] and Pickering et al. [19] believed that climate warming caused the distribution of thermophilic shrubs, herbs, and invasive weeds to tend to occur at higher altitudes. The increase in temperature caused the dominant dwarf shrub, *Dryas octopetala*, in the alpine zone of southern Norway to be replaced by herbs, and the response of shrubs to global changes was more intense [20]. However, there are still few studies on the diversity of desert shrubs.

In recent years, against the background of global warming, the climate in Xinjiang has become more unstable, precipitation variability has increased, and the characteristics of "warm and humid" have become more obvious. Strong human activities have greatly increased the possibility of sudden climate change in Xinjiang, and extreme climate events have increased [21]. This makes the response of the Xinjiang desert ecosystem to global climate change unique and complex. Desert shrubland plants are an important part of biodiversity in the desert ecosystem of Xinjiang. Climate change has reduced plant diversity in originally extremely fragile desert ecosystems. The effects of climate change on the maintenance mechanism of plant diversity and other important scientific questions need to be answered. The study of species diversity is a key issue in desert ecosystem ecology, because the research content involves many aspects, such as the sustainable development of desert ecosystems and the protection and reconstruction of ecosystems. Species diversity is the degree of diversity at the level of species organization within a biological community. Margalef index, Simpson index, Shannon–Wiener index, and Pielou index are mostly used for community species diversity research. In this paper, the shrub community of the desert ecosystem in Xinjiang was taken as the research object. The four diversity indexes of the shrub community of desert plants were systematically studied, and the correlation between species diversity and climatic factors (temperature and precipitation) was discussed to provide a theoretical basis for the optimal management of desert ecosystems and the protection of plant community species diversity.

2. Materials and Methods

2.1. Study Area

Xinjiang is located in the core of the arid region of Central Asia, deep inland, and is the farthest province from the sea in China. The region has a typical warm temperate continental climate, with dry summers, less rain, cold winters, and large temperature differences between day and night. Sandstorm disasters are serious, precipitation is scarce, and evaporation is strong. The study area spans 80°39′–91°19′ E and 40°14′–46°14′ N. The average annual temperature is 9.37 °C, and the average annual precipitation is 107.93 mm (Figure 1; Table 1).

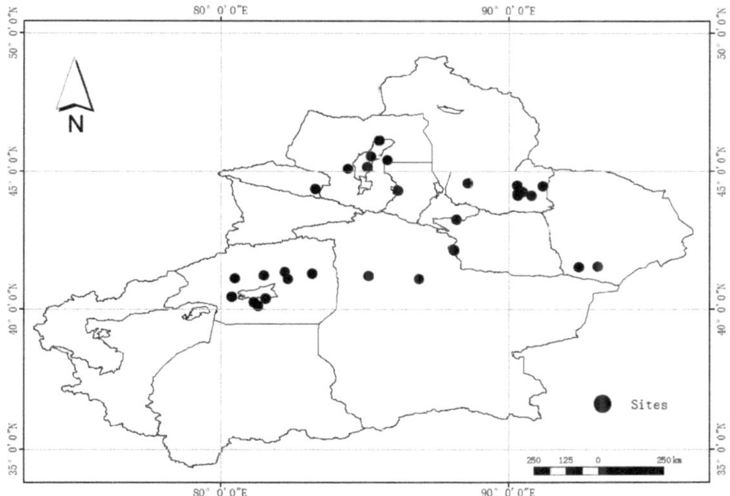

Figure 1. Locations of the 29 sampling sites in Xinjiang, China. The figure was drawn based on the map of Xinjiang at a scale of 1:5,500,000 (Xinjiang Bureau of Surveying Mapping and Geoinformation, No. Xin S (2016) 250).

Table 1. Characteristics of sites and climatic factors.

Sites	Longitude (°)	Latitude (°)	Altitude (m)	MAT (°C)	MAP (mm)
1	86.27	44.48	365	7.3	141
2	85.03	45.29	270	8.9	114
3	85.86	46.24	526	7.9	159
4	85.15	45.72	366	9.1	122
5	85.39	45.92	307	9.2	128
6	84.72	45.18	270	8.7	118
7	83.49	44.57	365	9.4	159
8	83.02	41.95	1451	9.1	169
9	82.78	41.91	1110	9.5	167
10	82.59	41.67	1031	11.2	135
11	82.44	41.61	997	11.1	138
12	81.09	41.40	1024	11.0	94
13	80.82	41.25	984	10.7	83
14	81.24	40.43	975	11.5	50
15	81.05	40.24	1025	11.9	45

Table 1. *Cont.*

Sites	Longitude (°)	Latitude (°)	Altitude (m)	MAT (°C)	MAP (mm)
16	80.66	40.75	1024	11.4	53
17	81.94	40.68	980	11.6	70
18	83.30	41.52	958	11.6	118
19	85.23	41.36	916	11.5	64
20	86.16	41.18	878	11.7	52
21	88.19	42.26	892	10.0	75
22	88.34	43.40	1196	6.0	173
23	90.54	44.19	870	6.6	91
24	90.56	44.05	958	5.9	94
25	91.33	44.76	766	8.1	70
26	90.82	44.41	905	7.4	83
27	90.14	44.18	756	6.9	110
28	90.06	44.81	563	8.7	109
29	89.00	44.94	571	8.1	146

2.2. Field Sampling

Sampling was conducted in the Xinjiang desert area in August 2018. According to the vegetation characteristics and environmental characteristics, a total of 29 survey plots were selected in the study area. The typical plot method was used to investigate 29 plant communities in Xinjiang desert areas. The basic information of 39 desert species was investigated. The sample design area of the plot is as follows: the shrub sample size is 15×15 m^2. In each quadrat, plant species and coverage, plant height, number of plants, and other information of each community were counted to calculate community species diversity, abundance, frequency, dominance, and other indicators.

2.3. Meteorological Data Acquisition

In this study, annual average temperature (MAT, °C) and annual precipitation (MAP, mm) were selected as climatic factors. Annual mean temperature and annual precipitation data are from WorldClim Version 2.0 (http://worldclim.org/version2 (accessed on 7 March 2020)), Fick& Hijmans, 2017). Selecting these two indicators, we can investigate the changes in the diversity characteristics of desert shrub communities with temperature and water factors.

2.4. Data Processing

SPSS (Version 26.0, Armonk, NY, USA) and Origin (Version 2019, Northampton, MA, USA) were used to analyze the experimental data. One-way ANOVA was used to test the significance of the diversity indexes of different plant communities, and the difference was significant at the 0.05 level. The one-way ANOVA method was used to analyze the variance of the Margalef index, Simpson index, Shannon–Wiener index, and Pielou index of different plant communities. Using linear regression model to analyze the relationship between Margalef index, Simpson index, Shannon–Wiener index, and Pielou index and climatic factors.

The following four diversity indexes were selected as community diversity indexes. Margalef index represents the richness of species; the Simpson index represents the sum of probabilities of all species in a community or sample plot, the Shannon–Wiener index indicates the number of species in a community or sample plot and the evenness of individual distribution among species, and the Pielou index reflects the evenness of individual number distribution of each species, the calculation formula of each index is:

$$\text{Simpson index } (C): \quad C = \sum_{i=1}^{n} N_i^2 \tag{1}$$

$$\text{Shannon–Wiener index } (H): \quad H' = -\sum_{i=1}^{n} Pi \ln Pi \tag{2}$$

$$\text{Margalef index } (Ma) \; Ma = (S - 1)/\ln N \tag{3}$$

$$\text{Pielou index } (E) \; E = H'/\ln(S) \tag{4}$$

In the above formula, S is the total number of species in the quadrat, Pi is the proportion of the abundance of the i-th species in the total abundance, and N is the total abundance. Gleason index is adopted for plant abundance (D). The calculation formula is:

$$D = S/\ln A \tag{5}$$

In the above formula, S and A are the total number of plant species and total area (m^2) in the sample plot.

3. Results and Analysis

3.1. Species Composition and Structural Characteristics of the Desert Shrub Community

Through the investigation of the study area, a total of 39 species, 27 genera, and 12 families were recorded, represented by Chenopodiaceae, Polygonaceae, Leguminosae, Tamaricaceae, and Zygophyllaceae. Common species include *Tamarix ramosissima, Haloxylon ammodendron, Halostachys caspica, Calligonum mongolicum, Alhagi sparsifolia, Reaumuria soongarica, Ephedra przewalskii*, etc. (Table 2).

Table 2. Species classification and characteristics of the desert shrub community.

Species	Family	Genus	Life Form
Alhagi sparsifolia	Leguminosae	Alhagi	Subshrub
Anabasis aphylla	Chenopodiaceae	Anabasis	Subshrub
Anabasis brevifolia	Chenopodiaceae	Anabasis	Subshrub
Anabasis cretacea	Chenopodiaceae	Anabasis	Subshrub
Apocynum venetum	Apocynaceae	Apocynum	Subshrub
Artemisia ordosica	Compositae	Artemisia	Subshrub
Calligonum ebinuricum	Polygonaceae	Calligonum	Shrub
Calligonum junceum	Polygonaceae	Calligonum	Shrub
Calligonum leucocladum	Polygonaceae	Calligonum	Shrub
Calligonum mongolicum	Polygonaceae	Calligonum	Shrub
Caragana acanthophylla	Leguminosae	Caragana	Shrub
Caragana dasyphylla	Leguminosae	Caragana	Shrub
Cynanchum sibiricum	Asclepiadaceae	Cynanchum	Herb
Ephedra intermedia	Ephedraceae	Ephedra	Shrub
Ephedra przewalskii	Ephedraceae	Ephedra	Shrub
Gymnocarpos przewalskii	Caryophyllaceae	Gymnocarpos	Subshrub
Halimodendron halodendron	Leguminosae	Halimodendron	Shrub
Halocnemum strobilaceum	Chenopodiaceae	Halocnemum	Subshrub
Halostachys caspica	Chenopodiaceae	Halostachys	Shrub
Haloxylon ammodendron	Chenopodiaceae	Haloxylon	Shrub
Iljinia regelii	Chenopodiaceae	Iljinia	Subshrub
Kalidium caspicum	Chenopodiaceae	Kalidium	Subshrub
Kalidium foliatum	Chenopodiaceae	Kalidium	Shrub
Karelinia caspica	Compositae	Karelinia	Herb
Krascheninnikovia ewersmannia	Chenopodiaceae	Krascheninnikovia	Shrub
Leymus racemosus	Gramineae	Leymus	Herb
Lycium ruthenicum	Solanaceae	Lycium	Shrub
Nitraria sphaerocarpa	Zygophyllaceae	Nitraria	Shrub
Nitraria tangutorum	Zygophyllaceae	Nitraria	Shrub
Poacynum pictum	Apocynaceae	Poacynum	Subshrub
Reaumuria soongarica	Tamaricaceae	Reaumuria	Subshrub
Salsola arbuscula	Chenopodiaceae	Salsola	Subshrub
Suaeda microphylla	Chenopodiaceae	Suaeda	Subshrub
Sympegma regelii	Chenopodiaceae	Sympegma	Subshrub
Tamarix arceuthoides	Tamaricaceae	Tamarix	Shrub
Tamarix hispida	Tamaricaceae	Tamarix	Shrub
Tamarix ramosissima	Tamaricaceae	Tamarix	Shrub
Zygophyllum fabago	Zygophyllaceae	Sarcozygium Bunge	Herb
Zygophyllum xanthoxylon	Zygophyllaceae	Sarcozygium	Shrub

The plant composition and structural characteristics of 29 plant communities were investigated (Table 3). The overall plant with extremely low coverage (ranging from 6% to 26%), most of the communities contain shrub and subshrub layers, indicating that the distribution of plants is relatively scattered and the coverage is relatively low. Most of them are *T. ramosissima*, *H. ammodendron*, and *A. sparsifolia*, which can grow under extremely dry conditions. In 29 shrub communities, *T. ramosissima* (in S4, S5, S21) and *H. amodendron* (in S2, S25, S26, S27, S29) has the highest frequency with 0.38 and 0.34, respectively. The plant height ranges from 0.07 m to 6.43 m, and the crown width ranges from 0.11 m to 6.48 m in different habitats vary greatly.

Table 3. Plant composition and structural characteristics of different community types.

Sites	Community Types	Vegetation Composition	Number of Seedlings	Average Height (m)	Average Crown (m)	Coverage (%)	Frequency
S1	*Karelinia caspica*	*H. caspica*	30	1.28	1.33	24	0.10
		T. ramosissima	15	3.55	3.60	10	0.38
		K. caspica	45	0.81	0.78	21	0.14
		S. microphylla	20	0.26	0.28	6	0.07
S2	*Haloxylon ammodendron*	*H. ammodendron*	45	2.85	2.89	26	0.34
		C. mongolicum	10	0.72	0.69	8	0.07
		S. arbuscula	30	0.18	0.21	24	0.07
S3	*Ephedra intermedia*	*E. intermedia*	60	0.66	0.65	10	0.17
		T. arceuthoides	10	3.11	3.16	22	0.03
		C. sibiricum	15	0.35	0.31	7	0.03
		A. sparsifolia	20	0.45	0.40	24	0.24
S4	*Tamarix ramosissima*	*T. ramosissima*	45	2.96	3.01	10	0.38
		H. ammodendron	15	2.06	2.10	23	0.34
		L. ruthenicum	10	0.97	0.94	10	0.14
		I. regelii	20	0.42	0.38	22	0.03
		K. caspica	20	0.68	0.63	8	0.14
S5	*Tamarix ramosissima*	*T. ramosissima*	50	3.82	3.84	26	0.38
		H. ammodendron	10	1.12	1.10	7	0.34
		E. przewalskii	20	0.79	0.76	23	0.14
		H. halodendron	30	1.70	1.68	9	0.03
		A. ordosica	30	0.37	0.38	22	0.07
		S. regelii	35	0.41	0.42	7	0.07
		A. sparsifolia	30	0.62	0.60	23	0.17
S6	*Suaeda microphylla*	*H. ammodendron*	10	1.83	1.82	11	0.34
		R. soongarica	30	0.57	0.58	24	0.24
		S. microphylla	40	0.07	0.11	8	0.07
S7	*Tamarix hispida*	*T. hispida*	30	4.30	4.33	26	0.10
		R. soongarica	25	0.45	0.44	8	0.24
		K. caspicum	15	0.78	0.72	23	0.03
		N. tangutorum	26	0.50	0.49	9	0.03
S8	*Reaumuria soongarica*	*Z. xanthoxylon*	10	0.63	0.59	25	0.07
		R. soongarica	25	0.29	0.32	6	0.24
		C. dasyphylla	8	1.24	1.24	22	0.03
		G. przewalskii	10	0.51	0.49	7	0.03
S9	*Kalidium foliatum*	*E. przewalskii*	10	0.5	0.48	24	0.24
		K. foliatum	30	0.32	0.33	7	0.03
		R. soongarica	20	0.16	0.18	22	0.24
		L. racemosus	10	0.36	0.34	6	0.03
		A. ordosica	15	0.28	0.30	21	0.07
S10	*Anabasis cretacea*	*A. cretacea*	20	0.41	0.39	8	0.07
		E. przewalskii	5	0.36	0.37	22	0.34

Table 3. *Cont.*

Sites	Community Types	Vegetation Composition	Number of Seedlings	Average Height (m)	Average Crown (m)	Coverage (%)	Frequency
S11	*Tamarix hispida*	*T. hispida*	30	4.93	4.98	9	0.10
		C. junceum	10	0.55	0.51	25	0.07
		A. sparsifolia	10	0.66	0.62	6	0.20
S12	*Ephedra przewalskii*	*E. przewalskii*	20	0.57	0.56	22	0.45
		C. junceum	15	0.79	0.79	8	0.07
S13	*Lycium ruthenicum*	*T. ramosissima*	10	1.97	1.97	24	0.38
		L. ruthenicum	15	0.84	0.86	7	0.14
		A. sparsifolia	10	0.41	0.42	21	0.23
S14	*Alhagi sparsifolia*	*A. sparsifolia*	50	0.89	0.81	7	0.26
		T. ramosissima	10	2.53	2.58	23	0.38
S15	*Poacynum pictum*	*T. ramosissima*	10	3.21	3.26	9	0.38
		P. pictum	20	1.06	1.06	22	0.07
S16	*Lycium ruthenicum*	*H. caspica*	15	1.60	1.62	9	0.10
		T. hispida	10	2.76	2.78	25	0.10
		L. ruthenicum	30	0.76	0.74	6	0.14
S17	*Alhagi sparsifolia*	*P. pictum*	10	1.22	1.20	22	0.07
		T. ramosissima	20	4.03	4.03	9	0.38
		A. sparsifolia	30	1.24	1.20	21	0.29
		S. regelii	10	0.79	0.78	6	0.07
S18	*Karelinia caspica*	*T. ramosissima*	10	3.48	3.50	25	0.38
		H. caspica	15	1.21	1.20	8	0.10
		K. caspica	30	0.81	0.80	22	0.14
S19	*Karelinia caspica*	*T. ramosissima*	10	3.08	3.10	9	0.38
		K. caspica	70	0.93	0.94	22	0.14
		Z. fabago	5	0.44	0.48	7	0.03
S20	*Alhagi sparsifolia*	*T. ramosissima*	10	2.82	2.81	25	0.38
		L. ruthenicum	20	0.88	0.89	7	0.14
		A. venetum	25	1.09	1.08	21	0.03
		A. sparsifolia	30	0.44	0.52	6	0.32
S21	*Tamarix ramosissima*	*T. ramosissima*	10	6.43	6.48	24	0.38
		H. strobilaceum	5	1.65	1.69	8	0.03
S22	*Caragana acanthophylla*	*C. acanthophylla*	45	0.84	0.86	25	0.03
		Z. xanthoxylon	15	0.85	0.82	8	0.07
		A. aphylla	20	0.14	0.19	22	0.03
		A. brevifolia	10	0.44	0.45	6	0.03
S23	*Krascheninnikovia ewersmannia*	*H. ammodendron*	40	1.81	1.86	25	0.34
		K. ewersmannia	80	0.34	0.32	7	0.07
S24	*Anabasis cretacea*	*A. cretacea*	80	0.51	0.49	22	0.07
		R. soongarica	25	0.28	0.31	8	0.24
		N. sphaerocarpa	7	0.74	0.73	24	0.07
S25	*Haloxylon ammodendron*	*H. ammodendron*	65	1.25	1.25	8	0.34
		C. leucocladum	35	1.50	1.53	25	0.10
		S. arbuscula	10	0.18	0.21	6	0.07
		E. intermedia	5	0.71	0.71	21	0.17
		N. sphaerocarpa	20	1.07	1.06	8	0.07
S26	*Haloxylon ammodendron*	*H. ammodendron*	55	2.35	2.4	24	0.34
		E. intermedia	25	0.66	0.67	8	0.17
		R. soongarica	15	0.43	0.48	22	0.24

Table 3. *Cont.*

Sites	Community Types	Vegetation Composition	Number of Seedlings	Average Height (m)	Average Crown (m)	Coverage (%)	Frequency
S27	*Haloxylon ammodendron*	*H. ammodendron*	76	1.85	1.89	10	0.34
		C. leucocladum	23	0.82	0.82	23	0.10
		K. ewersmannia	65	0.83	0.83	6	0.07
S28	*Krascheninnikovia ewersmannia*	*C. leucocladum*	13	0.58	0.58	23	0.10
		E. intermedia	35	0.57	0.57	7	0.17
		H. ammodendron	21	1.13	1.18	24	0.34
S29	*Haloxylon ammodendron*	*H. ammodendron*	65	1.7	1.7	10	0.34
		E. intermedia	10	0.74	0.74	21	0.17
		R. soongarica	20	0.6	0.63	7	0.24

Note: S1 to S29 indicated sites with serial numbers from 1 to 29.

3.2. Relationship between Diversity Index Characteristics of the Desert Shrub Community and Climatic Factors

The Margalef index is related to the number of species and the total number of plants in the sample community. The Shannon–Wiener index is an important index reflecting the diversity of the community, which is used to explain the richness of species in the sample community. The Simpson index is also called the Simpson dominance index, which reflects the change in the number of species in the community. The larger the Pielou evenness index, the more uniform the distribution of individual species in the community. Table 4 shows the different variation characteristics among the species diversity indexes of the different plant communities. The Margalef species richness index ranged from 0.21 to 1.13. Among them, the *T. ramosissima* community (in S5) was the highest, and the *K. wersmannia* community (in S23) was the lowest, indicating that the plant species in the *T. ramosissima* community were the most species number and *K. wersmannia* community was the least species number. The Shannon–Wiener index ranged from 0.53 to 1.86, among which the *T. ramosissima* community (in S5) was the highest and the *A. sparssifolia* community (in S14) was the lowest, indicating that the *T. ramosissima* community contained a large amount of plant information and that the complexity of the community was higher than that in the other communities. The Simpson index ranged from 0.28 to 0.84 in this study, with the highest in the *T. ramosissima* community (in S5) and the lowest in the *A. sparssifolia* community (in S14), indicating that *T. ramosissima* plays an important role in Xinjiang desert ecosystem. The Pielou index ranged from 0.27 to 0.99, with the highest in the *E. przewalskii* community (in S12) and the lowest in the *K. caspica* community (in S19), indicating that the distribution of *E. przewalskii* community was the most uniform.

The relationship between the diversity characteristics of desert shrub communities and climatic factors was obtained by linear regression model analysis. Figure 2 shows that the Margalef index, Simpson index, Shannon–Wiener index, and Pielou index were negatively correlated with MAT but did not reach a significant level. Except for the Pielou index, the Margalef index ($R^2 = 0.18$, $p < 0.05$), Simpson index ($R^2 = 0.19$, $p < 0.05$), and Shannon–Wiener index ($R^2 = 0.21$, $p < 0.05$) were significantly positively correlated with MAP. This indicates that community characteristics are closely related to precipitation conditions and that precipitation conditions have a significant effect on the structure and composition of flora.

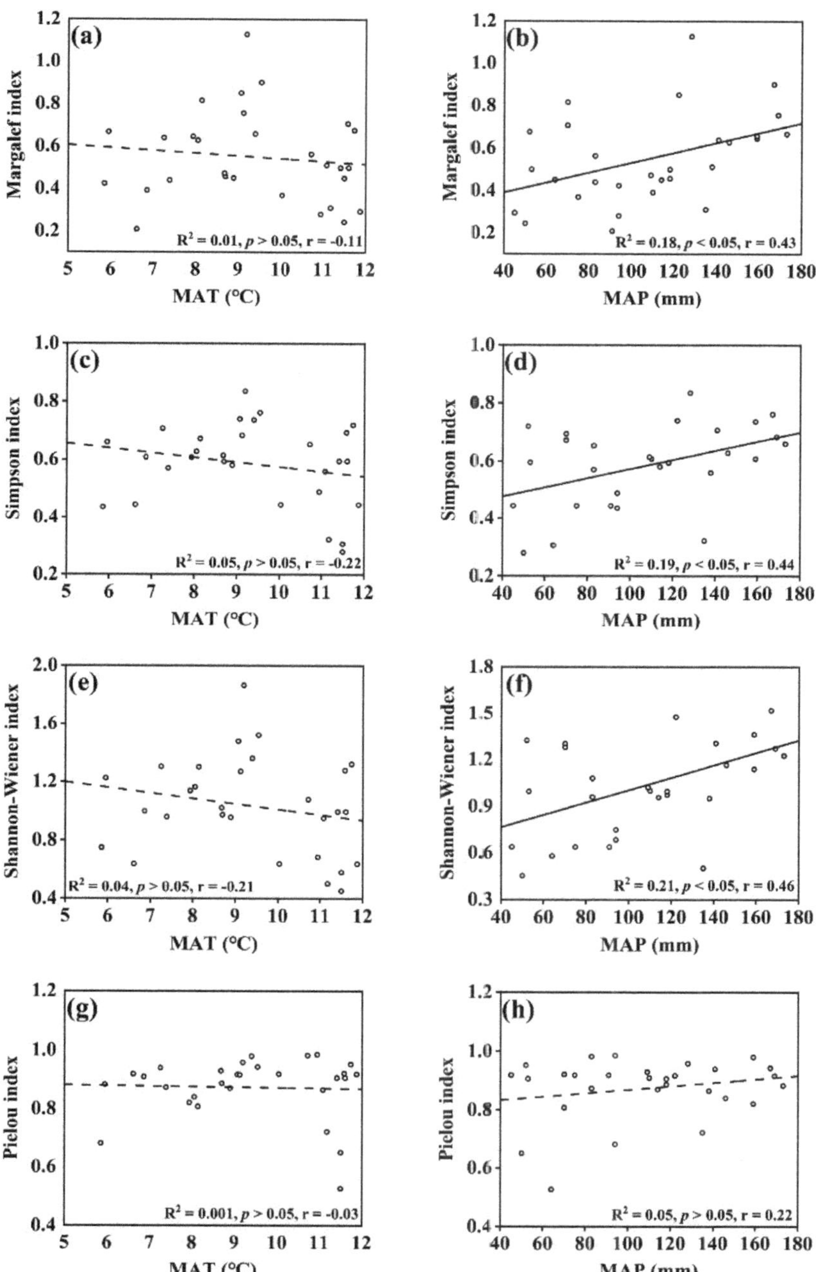

Figure 2. Relationship between plant community diversity index and climate factors. Note: (**a,c,e,g**) diagrams represent the relationship between Margalef index, Simpson index, Shannon–Wiener index, Pielou index and MAT respectively; (**b,d,f,h**) diagrams represent the relationship between Margalef index, Simpson index, Shannon–Wiener index, Pielou index and MAP respectively.

Table 4. Diversity index in different plant communities.

Sites	Community Types	Margalef Index	Simpson Index	Shannon–Wiener Index	Pielou Index
S1	*K. caspia*	0.64	0.71	1.30	0.94
S2	*H. ammodendron*	0.45	0.58	0.96	0.87
S3	*E. intermedia*	0.64	0.61	1.14	0.82
S4	*T. ramosissima*	0.85	0.74	1.48	0.92
S5	*T. ramosissima*	1.13	0.84	1.86	0.96
S6	*S. microphylla*	0.46	0.59	0.97	0.89
S7	*T. hispida*	0.66	0.74	1.36	0.98
S8	*R. songarica*	0.76	0.68	1.27	0.92
S9	*K. foliatum*	0.90	0.76	1.52	0.94
S10	*A. cretacea*	0.31	0.32	0.50	0.72
S11	*T. hispida*	0.51	0.56	0.95	0.86
S12	*E. przewalskii*	0.28	0.49	0.68	0.99
S13	*L. ruthenicum*	0.56	0.65	1.08	0.98
S14	*A. sparsifolia*	0.24	0.28	0.45	0.65
S15	*P. hendersonii*	0.29	0.44	0.64	0.92
S16	*L. ruthenicum*	0.50	0.60	0.99	0.91
S17	*A. sparsifolia*	0.71	0.69	1.28	0.92
S18	*K. caspia*	0.50	0.60	0.99	0.91
S19	*K. caspia*	0.45	0.30	0.58	0.53
S20	*A. sparsifolia*	0.68	0.72	1.32	0.95
S21	*T. ramosissima*	0.37	0.44	0.64	0.92
S22	*C. acanthophylla*	0.67	0.66	1.22	0.88
S23	*K. eversmanniana*	0.21	0.44	0.64	0.92
S24	*A. cretacea*	0.42	0.44	0.75	0.68
S25	*H. ammodendron*	0.82	0.67	1.30	0.81
S26	*H. ammodendron*	0.44	0.57	0.96	0.87
S27	*H. ammodendron*	0.39	0.61	1.00	0.91
S28	*K. eversmanniana*	0.47	0.61	1.02	0.93
S29	*H. ammodendron*	0.63	0.63	1.16	0.84

Note: S1 to S29 indicated sites with serial numbers from 1 to 29.

4. Discussion

4.1. Diversity Characteristics of the Desert Shrub Community

The species and number of natural vegetation in desert shrub communities are relatively small, dominated by xerophytic, halophytic, or ultra-xerophytic small shrubs and perennial herbs, reflecting the characteristics of desert, semi-desert, and steppe desert plant communities [22,23]. In this study, the dominant layer of the community is the shrub layer, and its species composition is dominated by dwarf subshrubs. The dominant plants are mainly shrubs or subshrubs of Chenopodiaceae, Tamaricaceae, and Zygophyllaceae. The proportion of herbaceous plants is lower than that of shrubs and subshrubs, reflecting the harsh desert environment in Xinjiang. Under the conditions of a low degree of heterogeneity, from the perspective of layers, the life form of vegetation composition species gradually tends to be simple or even single.

The community species diversity index can reflect community structure characteristics. The larger the value of the Margalef species richness index, the higher the species number reflecting the community or habitat, and the greater the number of species [22–24]. Among the 29 communities in the Xinjiang desert, the Margalef index is the highest in the *T. ramosissima* community. *T. ramosissima*, as the constructive species, was widely distributed in Xinjiang. Because of its large number of species, large size, and strong ability to adapt to barren habitats, *T. ramosissima* has significant control over the formation of its community structure and community environment. The constructive species of *T. ramosissima* is more likely to form tall shrubs than other species in the process of the barren desert environment, but it will also form significant differences in height, coverage, and other aspects due to environmental changes [25]. The Shannon–Wiener diversity index is used to describe the amount of information contained in a community. The larger the index, the higher the complexity of the community [26]. In this study, the Shannon–Wiener index was the highest in the *T. ramosissima* community. *T. ramosissima* is the most salt-tolerant species in *Tamarix*, which has the characteristics of strong adaptability, resistance to wind, salt and alkali, drought and moisture, barren soil, sand burial, and developed roots [27]. Therefore, the associated species of the *T. ramosissima* community are abundant. The Simpson index, also known as the ecological dominance index, comprehensively reflects the richness of species in the community. It is one of the more commonly used diversity indexes. The higher the index is, the higher the ecological dominance of dominant species and the higher the probability of species occurrence [24]. In different communities, the Simpson index was still the highest in the *T. ramosissima* community, indicating that the *T. ramosissima* community plays an important role in the desert ecosystem, has good adaptability to the extreme environment, and has a great impact on the improvement of the desert environment. The Pielou index is an index of species evenness, which is generally used to characterize the uniformity of the spatial distribution of species in a community. The larger the value of the index is, the more uniform the distribution of plants [24]. In this study, the Pielou index was the highest in the *E. pseudacorus* community and the lowest in the *K. caspica* community, indicating that the species in the *E. pseudacorus* community had the highest uniformity of spatial distribution, while the *K. caspica* community had the lowest. In desert ecosystems, water, salinity, and other factors have certain effects on the distribution of species, so they will also affect the diversity characteristics of plant communities. However, more samples need to be collected to confirm the specific impact.

4.2. Relationship between Desert Shrub Community Diversity and Climatic Factors

The results of this study confirmed that climatic factors, especially water conditions, were the main factors leading to differences in plant community types and species composition in the Xinjiang desert. The Margalef index, Simpson index, Shannon–Wiener index, and Pielou index had no significant correlation with MAT but had a significant positive correlation with MAP. This shows that water is still the main limiting factor affecting the diversity of desert plant communities in the Xinjiang desert ecosystem. The frequent occurrence of extreme events in the context of global warming has a particularly significant impact on Xinjiang, which is located in arid and semi-arid regions, resulting in increased risks of extreme precipitation events, storm floods, and snowmelt floods. The increase in precipitation will help to improve the α diversity characteristics of desert shrub communities.

In this study, most diversity indexes had a significant positive correlation with the MAP. This is because the Xinjiang desert area is located in the temperate desert area of China. The environment is harsh, the soil is barren, the community type is poor, and the species' life type is single. It is mainly composed of temperate desert shrubs and semi-shrub communities, and temperate shrubs and meadows are formed in some areas [22,23,26]. In terms of community species composition, shrubs and subshrubs account for a higher proportion, less precipitation in the habitat, simpler community composition, and higher dominance of dominant communities. Environmental factors have important effects on the distribution and diversity of plant species [26]. Extremely arid climatic conditions in the study area are the basis for the formation of desert plants in Xinjiang. At the regional scale, temperature and precipitation are the determinants of vegetation type distribution and species life form and are the basis for the formation of zonal vegetation [16,28]. The growth and development of various plants require precipitation. Precipitation will affect the distribution of different plant species and the species diversity of communities. Precipitation is a comprehensive environmental factor affected by many factors. Especially in recent years, due to the over-exploitation of the earth's resources and the rapid development of industrialization it has accelerated the process of global warming and has also had a huge impact on global plant species diversity [28]. Global warming can not only affect plant growth and productivity by prolonging the seasonal cycle and changing phenological conditions but also directly affect plant photosynthesis by changing precipitation [29]. In addition, precipitation will also affect the change in soil moisture. The water absorbed by plants mainly comes from soil water, while soil water mainly comes from precipitation, so most of the water available to vegetation comes from natural precipitation [28,29].

Plants absorb water mainly from soil moisture, and soil moisture mainly comes from precipitation, so most of the water that vegetation can use comes from natural precipitation [28]. The effect of precipitation on plant community growth is a cumulative effect, and the plant community structure will change with changes in precipitation [27–29]. Precipitation changes have a significant impact on ecosystem structure and processes, such as community composition and dynamics, species diversity, and species competition. In the desert area of northwest China, water is the main limiting factor controlling plant growth and plant community diversity [29]. The results of this study show that precipitation can significantly improve the Margalef index, Simpson index, and Shannon–Wiener index. Yuan et al. found that the species diversity of alpine grassland in the northern Tibetan Plateau showed an exponential growth relationship with precipitation, and the increase in precipitation led to the optimization of community structure, which was significantly promoted [30]. Soliveres et al. [31] found that there is a good linear relationship between species richness and seasonal precipitation changes. Precipitation pattern changes directly affect species diversity and ecosystem versatility. Therefore, in the future, this paper also needs to consider investigating more data and further revealing the impact of rainfall changes on plant community diversity on a longer time scale.

5. Conclusions

In a word, this study analyzed the characteristics of different diversity indexes among desert plant communities and their responses to climate factors. Our results show that the *T. ramosissima* community had the highest Margalef index, and the *K. ewersmannia* community had the lowest. The Simpson index and Shannon–Wiener index were the highest in *T. ramosissima* and the lowest in *A. sparsifolia*. The Pielou index was the highest in the *E. przewalskii* community and the lowest in the *K. caspica* community. These results showed that there are significant differences among different plant communities in different diversity indexes. In addition, the α diversity index (except the Pielou index) of desert shrub community species was significantly positively correlated with the annual average precipitation but not significantly negatively correlated with the annual average temperature. These results indicated that water conditions are still an important factor affecting the species diversity of desert shrub communities in Xinjiang. These results provide an important reference for understanding the characteristics of plant community diversity in desert ecosystems and the relationship between desert plants and climate change.

Author Contributions: Y.L. carried out the fieldwork and wrote the first draft of the manuscript and Y.G. assisted with revising and editing the draft manuscript. All authors have read and agreed to the published version of the manuscript.

Funding: Third Xinjiang Scientific Expedition Program (Grant No. 2022xjkk040301), China Postdoctoral Science Foundation (Grant No. 2022M722667), and the Department of Education of Xinjiang Uygur Autonomous Region, Dr. Tianchi Program Project (Grant No. TCBS202123).

Data Availability Statement: Anyone who needs available data can directly consult and contact the first author of this article.

References

1. Eisenhauer, N. Plant diversity effects on soil microorganisms: Spatial and temporal heterogeneity of plant inputs increase soil biodiversity. *Pedobiologia* **2016**, *59*, 175–177. [CrossRef]
2. Vellend, M.; Geber, M.A. Connections between species diversity and genetic diversity. *Ecol. Lett.* **2005**, *8*, 767–781. [CrossRef]
3. van Oijen, M.; Bellocchi, G.; Hoglind, M. Effects of Climate Change on Grassland Biodiversity and Productivity: The Need for a Diversity of Models. *Agronomy* **2018**, *8*, 14. [CrossRef]

4. Parra-Tabla, V.; Albor-Pinto, C.; Tun-Garrido, J.; Angulo-Perez, D.; Barajas, C.; Silveira, R.; Ortiz-Diaz, J.J.; Arceo-Gomez, G. Spatial patterns of species diversity in sand dune plant communities in Yucatan, Mexico: Importance of invasive species for species dominance patterns. *Plant Ecol. Divers.* **2018**, *11*, 157–172. [CrossRef]
5. Wright, D.H. Species-energy theory: An extension of species-area theory. *Oikos* **1983**, *41*, 496–506. [CrossRef]
6. Banerjee, A.K.; Mukherjee, A.; Guo, W.X.; Liu, Y.; Huang, Y.L. Spatio-temporal patterns of climatic niche dynamics of an invasive plant *Mikania micrantha* kunth and its potential distribution under projected climate change. *Front. Ecol. Evol.* **2019**, *7*, 291. [CrossRef]
7. Zuo, X.A.; Zhao, S.L.; Cheng, H.; Hu, Y.; Wang, S.K.; Yue, P.; Liu, R.T.; Knapp, A.K.; Smith, M.D.; Yu, Q.; et al. Functional diversity response to geographic and experimental precipitation gradients varies with plant community type. *Funct. Ecol.* **2021**, *35*, 2119–2132. [CrossRef]
8. Mao, J.F.; Shi, X.Y.; Thornton, P.E.; Hoffman, F.M.; Zhu, Z.C.; Myneni, R.B. Global latitudinal-asymmetric vegetation growth trends and their driving mechanisms: 1982–2009. *Remote Sens.* **2013**, *5*, 1484–1497. [CrossRef]
9. Anwar, M.; Wang, X.H.; Xu, X.T.; Peng, L.Q.; Yang, Y.; Zhang, X.P.; Myneni, R.B.; Piao, S.L. Drought and spring cooling induced recent decrease in vegetation growth in Inner Asia. *Agric. For. Meteorol.* **2013**, *178*, 21–30. [CrossRef]
10. Walker, M.D.; Wahren, C.H.; Hollister, R.D.; Henry, G.H.R.; Ahlquist, L.E.; Alatalo, J.M.; Bret-Harte, M.S.; Calef, M.P.; Callaghan, T.V.; Carroll, A.B.; et al. Plant community responses to experimental warming across the tundra biome. *Proc. Natl. Acad. Sci. USA* **2006**, *103*, 1342–1346. [CrossRef]
11. Klein, J.A.; Harte, J.; Zhao, X.Q. Experimental warming causes large and rapid species loss, dampened by simulated grazing on the Tibetan Plateau. *Ecol. Lett.* **2004**, *7*, 1170–1179. [CrossRef]
12. Wu, Q.; Ren, H.Y.; Wang, Z.W.; Li, Z.G.; Liu, Y.H.; Wang, Z.; Li, Y.H.; Zhang, R.Y.; Zhao, M.L.; Chang, S.X.; et al. Additive negative effects of decadal warming and nitrogen addition on grassland community stability. *J. Ecol.* **2020**, *108*, 1442–1452. [CrossRef]
13. Zhang, G.G.; Huang, J.; Jia, M.Q.; Liu, F.H.; Yang, Y.H.; Wang, Z.W.; Han, G.D. Ammonia-Oxidizing Bacteria and Archaea: Response to Simulated Climate Warming and Nitrogen Supplementation. *Soil Sci. Soc. Am. Merica J.* **2019**, *83*, 1683–1695. [CrossRef]
14. Li, W.; Cheng, J.M.; Yu, K.L.; Epstein, H.E.; Guo, L.; Jing, G.H.; Zhao, J.; Du, G.Z. Plant functional diversity can be independent of species diversity: Observation based on the impact of 4-yrs of nitrogen and phosphorus additions in an alpine meadow. *PLoS ONE* **2015**, *10*, e0136040. [CrossRef] [PubMed]
15. Bai, Y.F.; Wu, J.G.; Xing, Q.; Pan, Q.M.; Huang, J.H.; Yang, D.L.; Han, X.G. Primary production and rain use efficiency across a precipitation gradient on the Mongolia plateau. *Ecology* **2008**, *89*, 2140–2153. [CrossRef]
16. Lopez-Angulo, J.; Pescador, D.S.; Sanchez, A.M.; Luzuriaga, A.L.; Cavieres, L.A.; Escudero, A. Impacts of climate, soil and biotic interactions on the interplay of the different facets of alpine plant diversity. *Sci. Total Environ.* **2019**, *698*, 133960. [CrossRef]
17. Grime, J.P.; Fridley, J.D.; Askew, A.P.; Thompson, K.; Hodgson, J.G.; Bennett, C.R. Long-term resistance to simulated climate change in an infertile grassland. *Proc. Natl. Acad. Sci. USA* **2008**, *105*, 10028–10032. [CrossRef]
18. Kazakis, G.; Ghosn, D.; Vogiatzakis, I.N.; Papanastasis, V.P. Vascular plant diversity and climate change in the alpine zone of the Lefka Ori, Crete. *Biodivers. Conserv.* **2007**, *16*, 1603–1615. [CrossRef]
19. Pickering, C.; Hill, W.; Green, K. Vascular plant diversity and climate change in the alpine zone of the Snowy Mountains. *Aust. Biodivers. Conserv.* **2008**, *17*, 1627–1644. [CrossRef]
20. Ni, J. A biome classification of China based on plant functional types and the BIOME3 model. *Folia Geobot.* **2001**, *36*, 113–129. [CrossRef]
21. Wang, Q.; Zhai, P.M.; Qin, D.H. New perspectives on 'warming-wetting' trend in Xinjiang, China. *Adv. Clim. Chang. Res.* **2010**, *11*, 252–260. [CrossRef]
22. Pekas, K.M.; Schupp, E.W. Influence of aboveground vegetation on seed bank composition and distribution in a Great Basin Desert sagebrush community. *J. Arid Environ.* **2013**, *88*, 113–120. [CrossRef]
23. Li, S.J.; Su, P.X.; Zhang, H.N.; Zhou, Z.J.; Xie, T.T.; Shi, R.; Gou, W. Distribution patterns of desert plant diversity and relationship to soil properties in the Heihe River Basin, China. *Ecosphere* **2018**, *9*, e02355. [CrossRef]
24. Therriault, T.W.; Kolasa, J. Physical determinants of richness, diversity, evenness and abundance in natural aquatic microcosms. *Hydrobiologia* **1999**, *412*, 123–130. [CrossRef]
25. Zhang, Z.; Ullah, I.; Wang, Z.; Ma, P.; Zhao, Y.; Xia, X.; Li, Y. Reconstruction of temperature for the past 400 years in the southern margin of the Taklimakan desert based on carbon isotope fractionation of *Tamarix* leaves. *Appl. Ecol. Environ. Res.* **2019**, *17*, 271–284. [CrossRef]
26. Garland, J.L.; Lehman, R.M. Dilution/extinction of community phenotypic characters to estimate relative structural diversity in mixed communities. *Fems Microbiol. Ecol.* **1999**, *30*, 333–343. [CrossRef]
27. Chen, Y.; Wang, H.; Zhang, F.S.; Xi, J.B.; He, Y.K. The characteristic of salt excretion and its affected factors on *Tamarix ramosissima* Leded under desert saline-alkali habitat in Xinjiang Province. *Acta Ecol. Sin.* **2010**, *30*, 511–518, (Abstract in Chinese).
28. Mokany, K.; Ferrier, S. Predicting impacts of climate change on biodiversity: A role for semimechanistic community-level modeling. *Divers. Distrib.* **2015**, *17*, 374–380. [CrossRef]
29. Sun, Y.; He, M.Z.; Wang, L. Effects of precipitation control on plant diversity and biomass in a desert region. *Acta Ecol. Sin.* **2018**, *38*, 2425–2433, (Abstract in Chinese).

30. Yuan, Q.; Yuan, Q.Z.; Ren, P. Coupled effect of climate change and human activities on the restoration/degradation of the Qinghai-Tibet Plateau grassland. *J. Geogr. Sci.* **2021**, *31*, 1299–1327. [CrossRef]
31. Soliveres, S.; Maestre, F.T.; Eldridge, D.J.; Delgado-Baquerizo, M.; Quero, J.L.; Bowker, M.A.; Gallardo, A. Plant diversity and ecosystem multifunctionality peak at intermediate levels of woody cover in global drylands. *Glob. Ecol. Biogeogr.* **2015**, *23*, 1408–1416. [CrossRef] [PubMed]

Article

Spatial Distribution Pattern and Genetic Diversity of *Quercus wutaishanica* Mayr Population in Loess Plateau of China

Dong Hu [1,2], Yao Xu [1,2], Yongfu Chai [1,2], Tingting Tian [1], Kefeng Wang [1,2], Peiliang Liu [1,2], Mingjie Wang [3], Jiangang Zhu [3], Dafu Hou [4] and Ming Yue [1,2,5,*]

[1] Key Laboratory of Resource Biology and Biotechnology in Western China, Ministry of Education, Northwest University, Xi'an 710069, China
[2] College of Life Science, Northwest University, Xi'an 710069, China
[3] Shuanglong State-Owned Ecological Experimental Forest Farm of Qiaoshan State-Owned Forestry Administration of Yan'an City, Yan'an 727306, China
[4] Shaanxi Micangshan National Nature Reserve Administration, Hanzhong 723000, China
[5] Xi'an Botanical Garden of Shaanxi Province, Institute of Botany of Shaanxi Province, Xi'an 710069, China
* Correspondence: yueming@nwu.edu.cn

Abstract: The *Quercus wutaishanica* forest influences the ecological environment and climate characteristics and plays an important ecological role in the Loess Plateau region. However, we still know relatively little about the genetic diversity and spatial distribution of *Q. wutaishanica*. Here, we assessed the genetic diversity of *Q. wutaishanica* using simple sequence repeats and used the point pattern method to analyze the spatial distribution patterns as well as intraspecific relationships. Our results indicate that the diameter structure of the *Q. wutaishanica* population was inverted J-type, showing a growing population. In addition, the population maintained high genetic diversity on a small scale. Due to dispersal constraints, the spatial distribution pattern of *Q. wutaishanica* seedlings (DBH < 1 cm) tended to aggregate at small scales and the degree of aggregation decreased with increasing spatial scale. However, trees (DBH > 5 cm) and saplings (1 cm ≤ DBH < 5 cm) showed more random distribution at the scale, indicating that *Q. wutaishanica* individuals shift from aggregation to random distribution at the spatial scale. In addition, although individuals of different diameter classes showed facilitative (trees vs. saplings, 5–6.5 m) and competitive effects (trees vs. seedlings, 13.5–16 m) on some scales, they showed no correlation on other scales, especially for saplings and seedlings, where they were not correlated on any scale. The results contribute to revealing the status and dynamics of *Q. wutaishanica* in the Loess Plateau, thereby providing a theoretical basis for further study on the maintenance mechanism of the population.

Keywords: *Quercus wutaishanica*; simple sequence repeats (SSR); point pattern; diameter classes structure; intraspecific association

Citation: Hu, D.; Xu, Y.; Chai, Y.; Tian, T.; Wang, K.; Liu, P.; Wang, M.; Zhu, J.; Hou, D.; Yue, M. Spatial Distribution Pattern and Genetic Diversity of *Quercus wutaishanica* Mayr Population in Loess Plateau of China. *Forests* **2022**, *13*, 1375. https://doi.org/10.3390/f13091375

Academic Editors: Xiao-Dong Yang, Nai-Cheng Wu and Xue-Wei Gong

Received: 27 July 2022
Accepted: 26 August 2022
Published: 28 August 2022

1. Introduction

The spatial distribution patterns of populations, which are the basic constituent units of plant communities, have been one of the trending topics of ecological research [1–3]. Different distribution patterns and spatial correlations can directly reflect the processes of individual biology, intra-species competition, and population–environment interactions [4], and the mechanism of population formation can be deduced from the distribution patterns and spatial correlations of populations [5]. Therefore, studying the spatial distribution patterns of populations and their correlations is of great importance to real successional trends, intra-species relationships, environmental adaptation mechanisms, and the formation of forest community structure [6–8].

One of the main drivers of species coexistence and community dynamics is species competition for scarce resources [9,10]. Competition can occur through density-dependent mortality events, which affect the survival and growth of species by robbing each other of living space and resources [11,12]. When larger trees disproportionately affect the growth and survival of smaller trees, asymmetric competition between individuals occurs [13]. For example, adult trees may inhibit seedling formation through competition, allelopathy, or increased populations of herbivores and pathogens [14]. Thus, tree size is a key factor influencing the spatial pattern and structure of forest populations.

Biological polymorphism and species diversity are based on genetic variation. Ecological processes such as interspecific interactions and community structure are significantly influenced by genetic diversity [15]. In recent years, the development of polymorphic markers and statistical analysis tools has allowed us to understand population maintenance mechanisms from a molecular perspective. For example, researchers have used inter-simple sequence repeat analysis (ISSR) to determine that genetic variation can explain intraspecific variation in plant–soil biotic interactions [16]. In addition, genetic diversity can also be analyzed by other molecular makers, such as random amplified polymorphic DNA (RAPD), restriction fragment length polymorphisms (RFLPs), amplified fragment length polymorphisms (AFLPs), single nucleotide polymorphisms (SNPs) [17,18] and simple sequence repeats (SSRs) [19]. In the last two decades, SSR markers have become a powerful tool for such studies because they are highly informative, polymorphic, and co-dominant, and present transferability between closely related species [20,21]. Researchers have studied the genetic diversity of some *Quercus* spp. by using SSR markers [22–26], but there are fewer experiments on the study of *Quercus wutaishanica* as a target, especially in the Loess Plateau region, and there are almost no studies on the genetic diversity of *Q. wutaishanica* for this region. Therefore, the study of the genetic diversity of *Q. wutaishanica* is important for understanding the level of species diversity and population genetic structure as well as the development of species conservation strategies and measures.

Q. wutaishanica is a deciduous tree belonging to the *Quercus* subgenus *Quercus* in the Fagaceae family. It is closely related to *Quercus mongolica* and considered by some scholars to be a synonym of the *Q. mongolica*. However, some researchers used SSR and AFLP analysis methods to analyze 15 separately distributed and mixed populations of *Q. wutaishanica* and *Q. mogolica* in China and found that *Q. wutaishanica* and *Q. mogolica* have clearly identifiable independent gene pools not only in separately distributed populations, but also in mixed populations, so they should remain as independent taxonomic units [27]. The Loess Plateau region, with its severe erosion and intense soil erosion, is a key area for ecological restoration in China, and the dynamics of its ecological environment have traditionally received widespread attention. The *Q. wutaishanica* forest is a relatively stable terminal forest community in the Loess Plateau region, which has an important influence on the ecological environment and climate characteristics of the region and plays an important ecological role [28]. At present, research on *Q. wutaishanica* in the Loess Plateau region has focused on taxonomy, morphological, and physiological characteristics [29,30], seed dispersal, and population renewal [31], but less on genetic diversity and spatial distribution patterns. Therefore, in this paper, we investigated populations' genetic diversity and spatial distribution patterns and their associations using SSR analysis and scale-based point pattern analysis, respectively, with populations of *Q. wutaishanica* in the Loess Plateau region as the research object. Specifically, our primary aims were to: (1) assess the population structure and genetic diversity of *Q. wutaishanica*; and (2) determine the distribution pattern and compare differences in the spatial association between trees of different diameter classes. The findings are discussed in the context of conservation and restoration strategies of *Q. wutaishanica* in the Loess Plateau region.

2. Materials and Methods

2.1. Study Area

The research was conducted in the Shuanglong Forest Farm (35°32′06″–35°45′55″ N, 108°33′40″–109°19′41″ E), Yan'an City, Shaanxi Province, China, which belongs to the Ziwuling Mountain region in the middle of the Loess Plateau. This area has a warm-temperate continental monsoon climate with an average annual temperature of 9–11 °C [32]. The mean annual precipitation is 550–650 mm [33]. The elevation ranges from 800 to 1700 m.

Historically, the natural forests in the Ziwuling Mountain region of the Loess Plateau were widely distributed, but the forests were severely damaged due to the interference of historical human activities [34]. Since the vegetation began to recover naturally, it has been 150 years, and now the Ziwuling Mountain has formed a large and continuous secondary forest landscape [34–36], with a denseness of 0.7–0.9. Typical secondary forest species in the forest area include *Q. wutaishanica*, *Betula platyphylla* Sulk., *Populus davidiana* Dode, etc. Among them, the *Q. wutaishanica* forest was the most widely distributed as the natural climax vegetation [37], and its mixed plants mainly include *Quercus acutissim* var. *acutissima*, *Quercus aliena* Blume, *B. platyphylla*, *Carpinus turczaninowii* Hance, etc. The forest area is currently growing well and playing an important role in the regional ecosystem.

2.2. Data Collection

In July 2018, a 50 m × 50 m plot was set up at the Qiaoshan forest ecosystem positioning and research station in Shaanxi Province, and this plot was divided into 25 quadrats of 10 m × 10 m (Figure 1). We recorded the coordinate position of all *Q. wutaishanica* individuals in the plot, measured diameter at breast height (DBH, defined as 1.3 m above the ground) or basal diameter (for seedlings), and numbered each individual. Then, we divided stems into separate sizes classed based on DBH. We defined seedlings as individuals <1 cm DBH, saplings as ≥1 cm and <5 cm DBH, and trees as ≥5 cm DBH [38,39]. For all saplings and trees, about three fresh and healthy leaves were collected from each plant, placed in a self-sealing bag, and dried with silica gel for genetic analysis.

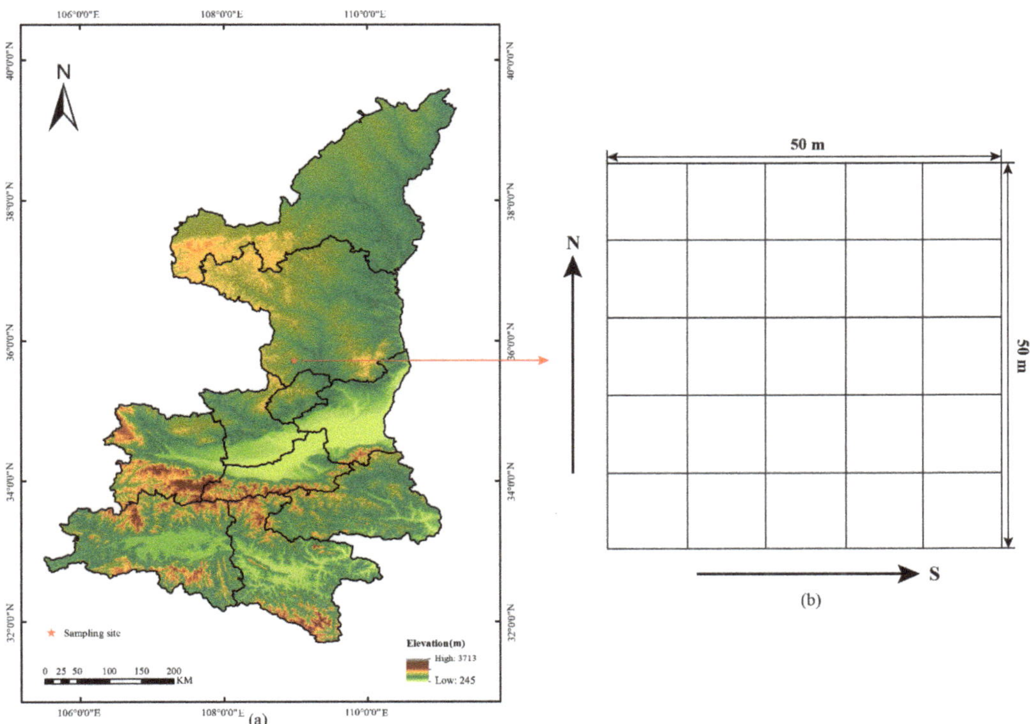

Figure 1. Map of the study area showing (**a**) the sampling site in an elevation map of the administrative regions of the Shaanxi Province, and (**b**) the layout of the plot and quadrats.

2.3. DNA Extraction and SSR Analysis

The dried leaves were ground on a TissueLysser (Scientz-48) before the DNA extraction. Genomic DNA was extracted following the method of Hormaza [40] and using the Speedtools Plant DNA Extraction Kit (Bioteke, Beijing, China) according to the manufacturer's instructions [41–43]. The extracted DNA was measured by a nucleic acid analyzer and quality was checked by 0.8% agarose gel electrophoresis.

A preliminary screening of primers for *Q. wutaishanica* was carried out by reviewing the literature [25,26,44–46], and a total of 20 pairs of primers were finally selected. Forty samples were randomly selected from the extracted DNA samples and subsequent experiments were carried out with the initially screened pairs of primers. The PCR reaction system was 25 μL: 1 μL template DNA, 1 μL each upstream and downstream primers, 12.5 μL PCR Premix (*Taq* DNA polymerase, dNTPs, $MgCl_2$, KCl reaction buffer, other stabilizers and enhancers), 9.5 μL ddH_2O. The PCR reaction procedure was: pre-denaturation at 94 °C for 5 min, denaturation at 94 °C for 45 s, denaturation at 46–59 °C for 40 s, extension at 72 °C for 45 s, and final extension at 72 °C for 7 min. The PCR products were detected by 1% agarose gel electrophoresis, and the primers with bright and clear bands were selected.

Of the 20 primer pairs selected, 12 were amplified successfully and the amplified products were obtained. Five pairs of primers with clear bands, good polymorphism, and high stability were selected for SSR analysis, and the primer sequences are shown in Table 1. The DNA samples were sent to the Tsingke Biotechnology Co., Ltd. (Bejing, China) to complete SSR analysis.

Table 1. Sequences and proper annealing temperature of 12 primer pairs.

Locus	Primer Sequence (5′–3′)	Annealing Temperature (°C)
PL123-124	(F) GCTTGAGAGTTGAGATTTGT (R) GCAACACCCTTTAACTACCA	55
PL127-128	(F) GCAATTACAGGCTAGGCTGG (R) GTCTGGACCTAGCCCTCATG	55
PL125-126	(F) CTTCACTGGCTTTTCCTCCT (R) TGAAGCCCTTGTCAACATGC	58
E79	**(F) CCATTAAAAGAAGCAGTATTTTGT** **(R) GCAACACTCAGCCTATATCTAGAA**	**52**
E71	**(F) CGTCTATAAGTTCTTGGGTGA** **(R) GTAACTATGATGTGATTCTTACTTCA**	**46**
Qden 05011	(F) CCCACTCCCTGTCCATTGT (R) CACTGTGTGCTGCGACTTG	59
ssrQrZAG 96	(F) CCCAGTCACATCCACTACTGTCC (R) GGTTGGGAAAAGGAGATCAGA	59
ssrQrZAG 7	(F) CAACTTGGTGTTCGGATCAA (R) GTGCATTTCTTTTATAGCATTCAC	55
01b	(F) GTTCAACAATTTTATTAGGGTGC (R) GCCTATTACACACAACAAGCC	56
02b	**(F) ATGTCAATATGGTCACCTACCG** **(R) TTTTTGTAGATTTTTAAGCACGC**	**53**
04b	**(F) TTCCTTTTCCTCAGTTTGGG** **(R) CCCGCATCAAAGAACTATTG**	**52**
10b	**(F) GAATGGATCTTCATTTATCGTTG** **(R) TCTGCATATTTTCAACATACATTTAG**	**55**

Note: Bold are 5 pairs of primers with clear bands, good polymorphism, and high stability selected in this study.

Popgene32 software was used to calculate the genetic diversity of trees and saplings, including the effective number of alleles (Ne), Shannon's information index (I), observed number of alleles (Na), observed heterozygosity (Ho), expected heterozygosity (He), and fixation index (Fis). In addition, we tested for differences in the number of alleles, genetic diversity, and fixation index between trees and saplings using the Wilcoxon Matched Pairs Test.

It is worth noting that due to data deficiencies (only two size groups' data), we did not calculate genetic differentiation indices (allele frequency differentiation, AFD) [47] for the three stages, as in previous studies [48,49].

2.4. Point Pattern Analysis and Null Model

The pairwise correlation function $g(r)$ was used to study the distribution patterns of populations at different scales. The pairwise correlation function $g(r)$ is derived from the K function, which is a non-accumulative distribution function compared with the K function, and can more sensitively discern the extent to which the actual distribution of points at a given scale deviates from the expected value, and is not affected by the cumulative effects at small scales when analyzing patterns at large scales [50]. We analyzed the spatial distribution patterns within tree species using the univariate $g(r)$ function and the spatial association of size classes within species using the bivariate $g(r)$ function.

The choice of a null model that has a clear ecological meaning and accurately describes the extent to which the data deviate from the theory is important for the analysis of spatial point patterns [38,51]. In our study, there were two different patterns (univariate pairwise correlation and bivariate pairwise correlation) using two different null model groups, and both of these two null model groups were based on heterogeneous Poisson. The heterogeneous Poisson model defines the distribution of individuals based on the density function $\lambda(x, y)$, which can exclude the effect of environmental heterogeneity at large scales. We used sigma = 30 m for this analysis [50].

The univariate pairwise correlation function was used to analyze the spatial distribution patterns of tree species as a whole and populations of different diameter classes and the bivariate pairwise correlation function was used to analyze the spatial relationships among individuals at different diameter classes. We performed 99 Monte-Carlo simulations, and the maximum and minimum values of the results were used to generate 99% confidence intervals formed by the upper and lower Poisson distribution envelopes. When the observed values were above the upper limits of the envelope line, within the intervals and below the lower limits, they corresponded to the aggregated or positive correlation, random or uncorrelated and uniform distribution or negative correlation, respectively [51]. All analyses were conducted using the 'statstat' package in R 4.1.2 (R Core Team, 2021).

3. Results

3.1. Individual Distribution and Population Structure

In this study, a total of 2964 individuals of *Q. wutaishanica* were counted (Figure 2a), of which 53 trees with DBH more than 5 cm, 53 saplings with DBH between 1 cm and 5 cm, and 2858 seedlings (DBH <1 cm). The large trees in the community were mainly distributed evenly in the plot, with only a few individual clusters. In addition, there were almost no saplings around the trees, but a large number of seedlings were distributed. However, the aggregation of seedlings around the saplings was not apparent.

The DBH size structure of the *Q. wutaishanica* showed an obvious inverted J-type distribution, in other words, the number of individuals decreased with increasing DBH size classes (Figure 2b). Moreover, the age structure population was dominated by small-sized individuals, and the age structure of the population was pyramidal-type and showed a growing population.

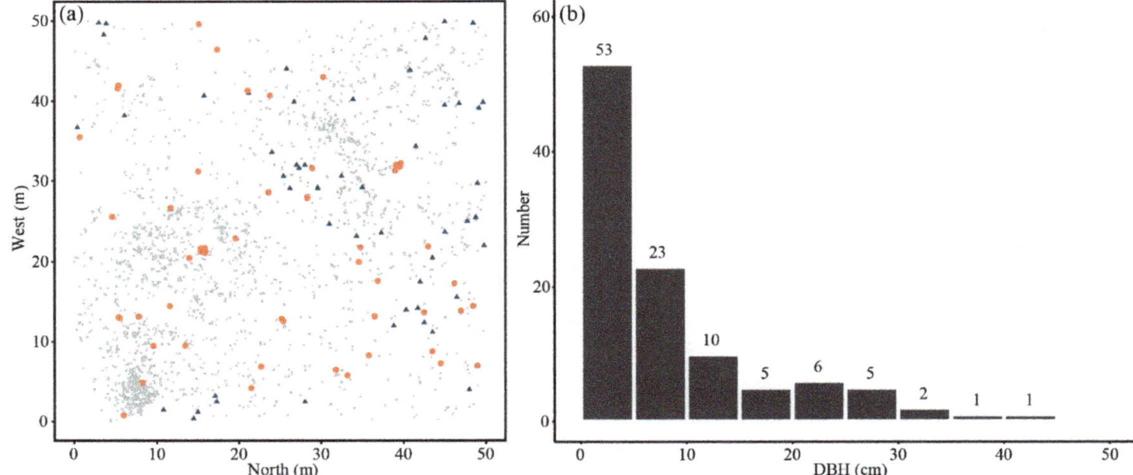

Figure 2. Spatial distribution of the sample plot (**a**), and diameter at breast height (DBH) structure of the *Quercus wutaishanica* (**b**). Trees, red circles; saplings, blue triangle; seedlings, gray rectangle.

3.2. Genetic Diversity

A total of 66 alleles were amplified using 5 SSR primers across 106 *Q. wutaishanica* (53 trees and 53 saplings). The number of alleles per locus (Na) ranged from 7 to 22, with an average value of 13.2 (Table 2). The lowest observed heterozygosity (Ho) ranged from 0.7980 (10b) to 0.8371 (E79), and the mean value was 0.8371 for all accessions. The expected heterozygosity (He) values ranged between 0.7419 for E79 and 0.8973 for 04b, with an average of 0.8159 per locus. In addition, the lowest Shannon's information index (I) was 1.5015 and the highest was 2.5257, with an average of 1.9410. The F-statistics showed moderate population differentiations for each locus and fixation index (Fis) varied from −0.2125 to 0.0810.

Although trees and saplings had high genetic diversity, the populations showed significant inbreeding in both life stages and total population (Table 2). The Wilcoxon paired test showed that trees and saplings were not significantly different in number of alleles (Na, $P = 0.68$), genetic diversity (He, $P = 0.58$), and fixation index (Fis, $P = 0.22$) ($P > 0.05$ for all comparisons).

Table 2. Genetic diversity in the different life stages in a population of *Quercus wutaishanica*.

Life Stage	Locus	N	Na	Ne	PIC	I	He	Ho	Fis
Trees and saplings	02b	105	7	3.8677	0.5525	1.5015	0.7450	0.8000	−0.0790
	04b	106	22	9.3555	0.4253	2.5257	0.8973	0.8208	0.0810
	10b	99	13	7.3060	0.5260	2.1323	0.8675	0.7980	0.0755
	E71	101	11	5.6719	0.5156	1.9138	0.8278	0.8713	−0.0578
	E79	105	13	3.8215	0.2940	1.6315	0.7419	0.8952	−0.2125
	Mean	—	13.2	6.0045	0.4627	1.9410	0.8159	0.8371	—
Trees	02b	53	7	3.6985	0.5284	1.4758	0.7736	0.7366	0.0370
	04b	53	18	9.3478	0.5193	2.4809	0.8302	0.9015	−0.0713
	10b	49	10	6.7067	0.6707	2.0250	0.7143	0.8597	−0.1454
	E71	50	10	5.7670	0.5767	1.9331	0.8800	0.8349	0.0451
	E79	53	11	4.3652	0.3968	1.7378	0.8302	0.7783	0.0519
	Mean	—	11.2	5.9770	0.5384	1.9305	0.8057	0.8222	—

Table 2. *Cont.*

Life Stage	Locus	N	Na	Ne	PIC	I	He	Ho	Fis
Saplings	02b	52	6	3.9132	0.6522	1.4981	0.7517	0.8269	−0.0752
	04b	53	16	7.9575	0.4973	2.2968	0.8827	0.8113	−0.0714
	10b	50	13	7.8125	0.6010	2.1930	0.8808	0.8800	0.0008
	E71	51	9	5.2229	0.5803	1.8111	0.8165	0.8627	−0.0462
	E79	52	8	3.1333	0.3917	1.3315	0.6875	0.9615	−0.2740
	Mean	—	10.4	5.6079	0.5445	1.8261	0.8038	0.8685	—

Note: N, number of sampled individuals; Na, observed number of alleles; Ne, effective number of alleles; PIC, Polymorphism Information Content; I, Shannon's information index; He, expected heterozygosity; Ho, observed heterozygosity; Fis, fixation index.

3.3. Spatial Distributions Patterns and Intraspecific Association

We found differences in the spatial distribution of three life stages of *Q. wutaishanica* in this study (Figure 3a–c). The trees were almost randomly distributed at the whole scale. For saplings of *Q. wutaishanica*, the spatial distribution was aggregated at some medium and small scales (0–1.5 m, 2.5–4.5 m and 10–11 m). In other scales, the saplings' distribution was random. However, *Q. wutaishanica* seedlings showed a regular distribution at large scale (>21 m) and aggregated distribution between 0 and 20 m. The degree of aggregation then decreased with increasing scale.

There was a positive correlation between individuals of different diameter classes of *Q. wutaishanica* at a small interval (Figure 3d–f). Trees and seedlings showed a positive association at 5–6.5 m, but there was no association at other scales. For trees and saplings, they showed a negative association at some medium scale (13.5–16 m). At small and large scales, they were not spatially correlated. Meanwhile, no obvious associations were observed between saplings and seedlings of *Q. wutaishanica*.

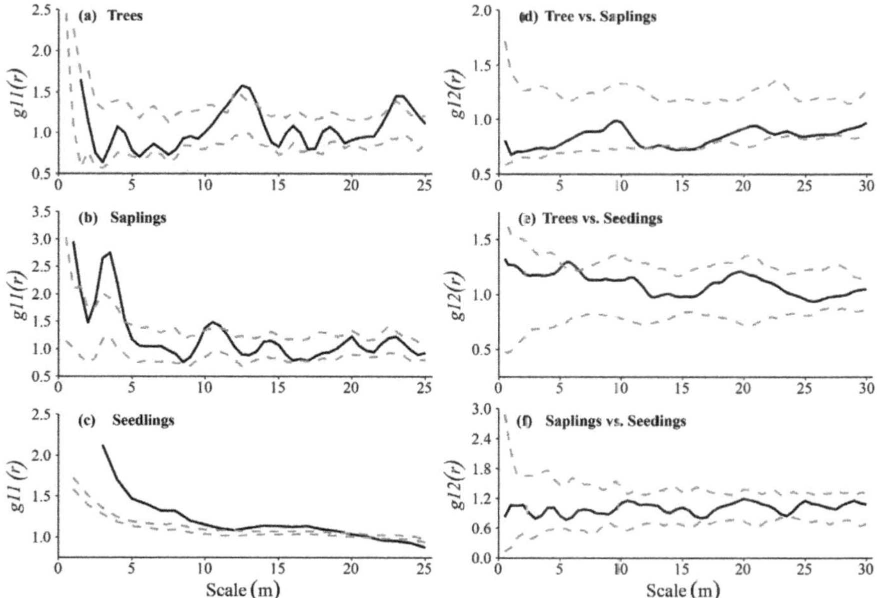

Figure 3. Univariate point pattern analyses show the spatial pattern using the pair-correlation function (**a–c**), and bivariate point pattern analyses show intra-species spatial associations among three size classes (**d–f**) of *Q. wutaishanica*. Black lines indicated the g11(r)/g12(r) function, dotted lines indicated the upper and lower limits of the 99% confidence interval.

4. Discussion

4.1. Diameter Classes Structure of Q. wutaishanica

The diameter class structure of plant populations is the result of the interaction between population viability and the external environment, and to a certain extent, it can reflect the current structure and renewal strategy of the population [52,53]. In the present study, the diameter class structure of the *Q. wutaishanica* population showed a typical inverted *J*-type distribution, with an abundance of individuals in seedlings and a gradual decrease in the number of individuals as the diameter class increased, and no individuals missing at any diameter class. In addition, a large number of renewed seedings were present and growing well in this area. The possible reason is that *Q. wutaishanica* will produce a large number of seeds, but its tannin content is high, bitter, and toxic, resulting in *Q. wutaishanica* seeds becoming an alternative food for animals [54], which in turn leads to a high number of saved seeds and the formation of a large number of seedlings. However, with the growth of individual plants, the demand for resources in the environment will become greater [55], the degree of nutrient specialization will increase [56], and strong competition will be formed within and between populations. Therefore, only a small number of seedlings will grow into saplings, and this will be more likely to occur in an area far from the adult trees, as predicted by the Janzen–Connell hypothesis. This also explains why only a small number of large-diameter individuals are in the plot. It can be seen that seed number and environmental adaptability are important reasons for the good regeneration of *Q. wutaishanica*.

4.2. Gene Diversity of Q. wutaishanica

Genetic diversity is a condition for the long-term survival and evolutionary basis of species [57], and the higher the genetic diversity, the better the species' ability to adapt to its environment [58]. During the survey and collection in the field, we speculated that the *Q. wutaishanica* population might have been propagated from the largest individual at DBH in the sample plot, and that the *Q. wutaishanica* individuals in the sample might be extremely close in kinship, or even have the possibility of identical SSR analysis results. However, the SSR analysis results do not correspond to our speculation. A total of 66 alleles were amplified by the set of SSR markers, emphasizing their high degree of polymorphism. In addition, the genetic diversity of *Q. wutaishanica* in this study was at a high level, with an average expected heterozygosity of 0.816, compared to 0.678 for *Castanea mollissima*, which is also a member of the Fagaceae family [59]. Furthermore, the expected heterozygosity of other families such as *Abies ziyuanensis* is 0.337 [60] and poplar is 0.552 [61]. This suggests that *Q. wutaishanica* at small scales can still maintain rich genetic diversity, which is very beneficial for the conservation of populations. Some researchers suggested that mating patterns have an important influence on the genetic diversity of plants [62]. In general, an increase in the proportion of self or inbred in plants leads to a decrease in genetic diversity, while plant outbreeding can increase the genetic diversity [63]. In addition, hybridization between closely related species of the same genus can also increase genetic diversity, and natural hybridization is very common in *Quercus* spp. Moreover, *Q. wutaishanica* is a anemophilous plant, and can be outbred [64]. Therefore, we guess that natural hybridization between *Q. wutaishanica* and *Q. aliena* leads to a high genetic diversity of *Q. wutaishanica*.

The level of genetic diversity of a species is the result of a combination of factors, including the biology of the species itself, its biological characteristics (life type, breeding system), external natural environmental changes, interactions with other species, and anthropogenic disturbances, all of which have an impact on the level of genetic diversity of the species [65]. In this study, the correlation between genetic distance and geographical distance among individuals of *Q. wutaishanica* was weak (Mantel test, $r = 0.09$, $P = 0.044$), indicating that at a small scale, spatial distribution pattern was not the main reason for influencing genetic diversity. In addition, because the *Q. wutaishanica* community is the climax community in the Loess Plateau region, its community species richness is higher and the intensity of interactions between species is stronger, which also promotes the higher genetic diversity of *Q. wutaishanica* individuals to some extent.

4.3. Spatial Distribution Pattern of Q. wutaishanica at Different Life Stages

Spatial distribution patterns of plant populations reflect the survival strategies and adaptive mechanisms by which populations adapt to their environment [66]. Under natural conditions, biotic and abiotic factors and processes such as habitat heterogeneity, dispersal limitation, interactions between organisms, and disturbance are considered potential ecological mechanisms that influence the formation and variation of species patterns at different spatial scales [53,67–69]. In general, at smaller scales, the distribution patterns of populations are mainly determined by biological characteristics such as seed dispersal mechanisms, individual reproductive characteristics, and intraspecific competition; while at larger scales, they may be influenced by environmental factors such as topography, soil, moisture, and light [70,71]. In this study, seedlings showed clustered, random, and regular distributions in order of increasing spatial scale, i.e., the degree of aggregation decreased as the spatial scale increased. This generally differs from the results observed in the tropics but is consistently observed in the temperate zones. While both tropical and temperate studies showed species aggregation at certain scales, tropical studies did not find a significant decrease in species aggregation with increasing scale, and temperate forests showed a significant decrease in species aggregation [1,72,73]. The possible reasons are that the intensity and scale of spatial aggregation of populations are related to the way seeds disperse [74]. Seed dispersal is an extremely important ecological process in population dynamics and an important mechanism for explaining changes in the spatial pattern of populations [75,76]. In addition to environmental factors such as wind, water, animals, and topography, the factors affecting seed dispersal are also closely related to the reproductive and biological characteristics of plants [77,78]. The initial distribution pattern of the *Q. wutaishanica* population in this study may be related to seed dispersal and environmental factors [79]. Because the main way of seed dispersal is gravity and animal dispersal [80], and the seed can easily fall off after maturity, thus forming a cluster distribution pattern centered on the mother tree, while cluster distribution is conducive to mutual shelter between individuals of small diameter wood and improving interspecific competition [81]. In addition, as the spatial scale increases, it is also influenced by environmental heterogeneity, thus showing a random or regular distribution at large scales.

At smaller scales, the spatial distribution of adult trees is more dispersed than that of juveniles and saplings, mainly due to extreme competition among individuals [38]. In our study, sapling shows a random distribution at intermediate and large scales and trees at 0–25 m scales, a distribution pattern that is also found in other regional forest ecosystems [7,53,82,83]. As predicted by the classical Janzen–Connell hypothesis, large individuals tend to over-disperse in small areas because of intense competition among individuals of the same species. These results indicated that individuals of *Q. wutaishanica* changed from aggregated distribution (seedlings) to random distribution (trees) at a certain spatial scale. This finding confirms previous studies that density-dependent mortality leads to the observed increase in overdispersion from small to large sized trees [12,71]. This is because *Q. wutaishanica* seedlings have good shade tolerance and the shade of the forests provides good environmental conditions for the seedings to survive. However, as the plant grows, individuals of *Q. wutaishanica* need more light to promote their growth. This leads to rapid growth of plants at the edge of the forests or forest gaps, while plants in shaded conditions do not receive enough light and die [84]. In other words, when the population enters the middle and late life history, the plants' demand for light, water, nutrients, space, and other resources increases, intra- and interspecific competition intensifies, and the self-thinning and alien-thinning effects are enhanced, leading to a large number of plant deaths and eventually tending to a random distribution [85]. In summary, *Q. wutaishanica* communities showed different spatial patterns at different life stages, which is not only beneficial for individual plants to obtain sufficient water, growing space, light, soil, and other environmental resources under different spatial distribution patterns, but also a survival strategy and adaptation mechanism for the population [86].

4.4. Spatial Correlation of Q. wutaishanica at Different Life Stages

Spatial correlations between different developmental stages of the same population can reveal the interactions between individuals within a population over some time and help to describe the status and dynamics of the population [6]. In addition, spatial relatedness can reflect the results of plants' mutual facilitation under unfavorable conditions or competing for limited resources [87]. Spatial associations between plants depend to some extent on the spatial and temporal distribution patterns of habitat resources, and habitat heterogeneity spatially constrains the distribution of individuals at different diameter classes, leading to different spatial associations between individuals at different diameter classes within a population. Many previous studies have proposed the existence of competition and facilitation in temperate forests [88–91]. Competition regulates the abundance and distribution of tree populations through intra- or interspecific density-dependent effects [11,92]. Facilitation may occur mainly among trees of different sizes or life history characteristics, such as sheltering of small trees by large trees and sheltering of newly established seedlings by shrubs [93,94]. In this study, *Q. wutaishanica* populations showed both facilitative and competitive effects between individuals of different diameter classes. Specifically, competitive effects occurred mainly between trees and saplings and only at intermediate scales, while trees and seedlings showed facilitative effects at small scales. The positive association indicates that individuals of different diameter classes of plants have similar needs for environmental resources. trees with larger canopy width can create a suitable renewal and growth microenvironment for seedlings, including protection from strong light, reduction of ground evaporation, and maintenance of soil moisture, providing shelter for the growth of small diameter classes, resulting in the higher survival rate of small diameter classes and thus dominance in the plots [95], thus creating a facilitating effect between trees and seedlings. In turn, when seedlings develop into young trees, there is competition for resources with adults [85], resulting in a negative correlation between adult and young trees at some scales. In addition, saplings and seedlings showed no correlation at all scales, due to the relatively low competition for resources such as nutrients and water between individuals at small diameter levels, and similar resource preferences that allow them to harmonize their relationships with each other and to jointly resist disturbance by herbivores [39].

4.5. Suggestions for Conservation and Management

The findings on the population structure and genetic diversity of *Q. wutaishanica* forests can be used as one of the bases for the management, can provide theoretical references for the conservation and management of populations, and have certain guiding significance for the sustainable management of Quercus forests in the Loess Plateau region. Given the population structure of this *Q. wutaishanica* forest, it is suggested that more efforts should be made to nurture *Q. wutaishanica* forests in this area, selectively adopt disturbance and carry out artificial thinning, which is not only conducive to the renewal and development of populations, but also of great significance to the conservation of forest ecosystems and biodiversity in the Loess Plateau region.

In addition, this study has some limitations. The scale of this study was small, and only a 50 m × 50 m sample plot was investigated. Subsequent studies can expand the scale and the research object and continue the survey sampling analysis to make the results more credible and scientific, and provide a theoretical basis for the conservation of this natural secondary forest.

Author Contributions: Conceptualization, D.H. (Dong Hu), M.Y. and Y.C.; investigation, Y.X., T.T., K.W., P.L., M.W., J.Z. and D.H. (Dafu Hou); formal analysis, D.H. (Dong Hu), Y.X., K.W. and P.L.; writing—original draft, D.H. (Dong Hu); writing—review and editing, M.Y., Y.C. and D.H. (Dafu Hou); data curation, M.W. and J.Z.; supervision, M.Y. All authors have read and agreed to the published version of the manuscript.

Funding: This research was financially supported by the National Natural Science Foundation of China (41571500, 41871036).

Institutional Review Board Statement Not applicable.

Informed Consent Statement: Not applicable.

Data Availability Statement: Not available.

Conflicts of Interest: The authors declare that they have no conflict of interest.

References

1. Condit, R.; Ashton, P.S.; Baker, P.; Bunyavejchewin, S.; Gunatilleke, S.; Gunatilleke, N.; Hubbell, S.P.; Foster, R.B.; Itoh, A.; LaFrankie, J.V.; et al. Spatial Patterns in the Distribution of Tropical Tree Species. *Science* **2000**, *288*, 1414–1418. [CrossRef] [PubMed]
2. Lydersen, J.M.; North, M.P.; Knapp, E.E.; Collins, B.M. Quantifying Spatial Patterns of Tree Groups and Gaps in Mixed-Conifer Forests: Reference Conditions and Long-Term Changes Following Fire Suppression and Logging. *For. Ecol. Manag.* **2013**, *304*, 370–382. [CrossRef]
3. May, F.; Huth, A.; Wiegand, T. Moving beyond Abundance Distributions: Neutral Theory and Spatial Patterns in a Tropical Forest. *Proc. Royal Soc. B* **2015**, *282*, 20141657. [CrossRef]
4. Lin, Y.; Chang, L.; Yang, K.; Wang, H.; Sun, I. Point Patterns of Tree Distribution Determined by Habitat Heterogeneity and Dispersal Limitation. *Oecologia* **2011**, *165*, 175–184. [CrossRef]
5. Tuo, F.; Liu, X.; Liu, R.; Zhao, W.; Jing, W.; Ma, J.; Wu, X.; Zhao, J.; Ma, X. Spatial distribution patterns and association of *Picea crassifolia* population in Dayekou Basin of Qilian Mountains, northwestern China. *Chin. J. Plant Ecol.* **2020**, *44*, 1172–1183. [CrossRef]
6. Huang, X.; Li, S.; Su, J.; Liu, W.; Lang, X. Distribution of *Pinus yunnanensis* Natural Population in Yunlong Tianchi National Nature Reserve. *For. Res.* **2018**, *31*, 47–52. [CrossRef]
7. Liu, H.; Chen, Q.; Xu, Z.; Wu, C.; Chen, Y. Natural population structure and spatial distribution pattern of rare and endangered species *Dacrydium pectinatum*. *Acta Ecol. Sin.* **2020**, *40*, 2985–2995. [CrossRef]
8. Martínez, I.; Wiegand, T.; González-Taboada, F.; Obeso, J.R. Spatial Associations among Tree Species in a Temperate Forest Community in North-Western Spain. *For. Ecol. Manag.* **2010**, *260*, 456–465. [CrossRef]
9. Tilman, D. Resource Competition between Plankton Algae: An Experimental and Theoretical Approach. *Ecology* **1977**, *58*, 338–348. [CrossRef]
10. Wright, A.; Schnitzer, S.A.; Reich, P.B. Living Close to Your Neighbors: The Importance of Both Competition and Facilitation in Plant Communities. *Ecology* **2014**, *95*, 2213–2223. [CrossRef]
11. Peters, H.A. Neighbour-Regulated Mortality: The Influence of Positive and Negative Density Dependence on Tree Populations in Species-Rich Tropical Forests. *Ecol. Lett.* **2003**, *6*, 757–765. [CrossRef]
12. Getzin, S.; Dean, C.; He, F.A.; Trofymow, J.; Wiegand, K.; Wiegand, T. Spatial Patterns and Competition of Tree Species in a Douglas-Fir Chronosequence on Vancouver Island. *Ecography* **2006**, *29*, 671–682. [CrossRef]
13. Zhang, C.; Jin, W.; Gao, L.; Zhao, X. Scale Dependent Structuring of Spatial Diversity in Two Temperate Forest Communities. *For. Ecol. Manag.* **2014**, *316*, 110–116. [CrossRef]
14. Tilman, D. *Resource Competition and Community Structure*; Princeton University Press: Princeton, NJ, USA, 1982; ISBN 978-0-691-08302-5.
15. Hughes, A.R.; Inouye, B.D.; Johnson, M.T.J.; Underwood, N.; Vellend, M. Ecological Consequences of Genetic Diversity. *Ecol Lett* **2008**, *11*, 609–623. [CrossRef] [PubMed]
16. Liu, X.; Etienne, R.S.; Liang, M.; Wang, Y.; Yu, S. Experimental Evidence for an Intraspecific Janzen-Connell Effect Mediated by Soil Biota. *Ecology* **2015**, *96*, 662–671. [CrossRef] [PubMed]
17. Campoy, J.A.; Lerigoleur-Balsemin, E.; Christmann, H.; Beauvieux, R.; Girollet, N.; Quero-García, J.; Dirlewanger, E.; Barreneche, T. Genetic Diversity, Linkage Disequilibrium, Population Structure and Construction of a Core Collection of Prunus Avium L. Landraces and Bred Cultivars. *BMC Plant Biol.* **2016**, *16*, 49. [CrossRef]
18. Mas-Gómez, J.; Cantín, C.M.; Moreno, M.Á.; Prudencio, Á.S.; Gómez-Abajo, M.; Bianco, L.; Troggio, M.; Martínez-Gómez, P.; Rubio, M.; Martínez-García, P.J. Exploring Genome-Wide Diversity in the National Peach (*Prunus persica*) Germplasm Collection at CITA (Zaragoza, Spain). *Agronomy* **2021**, *11*, 481. [CrossRef]
19. Topp, B.L.; Russell, D.M.; Neumüller, M.; Dalbó, M.A.; Liu, W. Plum. In *Fruit Breeding*; Badenes, M.L., Byrne, D.H., Eds.; Springer: Boston, MA, USA, 2012; pp. 571–621. ISBN 978-1-4419-0762-2.
20. Mason, A.S. SSR Genotyping. *Methods Mol. Biol.* **2015**, *1245*, 77–89. [CrossRef]
21. Vieira, M.L.C.; Santini, L.; Diniz, A.L.; Munhoz, C.d.F. Microsatellite Markers: What They Mean and Why They Are so Useful. *Genet. Mol. Biol.* **2016**, *39*, 312–328. [CrossRef]
22. Dow, B.D.; Ashley, M.V. Factors Influencing Male Mating Success in Bur Oak, Quercus Macrocarpa. *New For.* **1998**, *15*, 161–180. [CrossRef]
23. Lefort, F.; Echt, C.; Streiff, R.; Giovanni Giuseppe, V. Microsatellite Sequences: A New Generation of Molecular Markers for Forest Genetics. *For. Genet.* **1999**, *6*, 15–20.
24. Lefort, F.; Lally, M.; Thompson, D.; Douglas, G. Morphological Traits, Microsatellite Fingerprinting and Genetic Relatedness of a Stand of Elite Oaks (Q. Robur L.) at Tullynally, Ireland. *Silvae Genet.* **1998**, *47*, 257–262.

25. Liu, M. The Research on Genetic Evolution Relationship of *Quercus. mongolia* and *Quercus. wutaishanica*. Master's Thesis, Northeast Forestry University, Haerbin, China, 2012.

26. Qin, Y. Study on the Genetic Diversity of *Quercus liaotungensis* Natural Population in Shanxi Province. Master's Thesis, Beijing Forestry University, Beijing, China, 2012.

27. Zeng, Y.; Liao, W.; Petit, R.J.; Zhang, D. Exploring Species Limits in Two Closely Related Chinese Oaks. *PLoS ONE* **2010**, *5*, e15529. [CrossRef]

28. Wang, M.; Xu, J.; Chai, Y.; Guo, Y.; Liu, X.; Yue, M. Differentiation of Environmental Conditions Promotes Variation of Two Quercus Wutaishanica Community Assembly Patterns. *Forests* **2020**, *11*, 43. [CrossRef]

29. Chen, C.; Liu, D.; Wu, J.; Kang, M.; Zhang, J.; Liu, Q.; Liang, Y. Leaf traits of *Quercus wutaishanica* and their relationship with topographic factors in Mount Dongling. *Chin. J. Ecol.* **2015**, *34*, 2131–2139. [CrossRef]

30. Yang, J.; Lv, J.; He, Q.; Yan, M.; Li, G.; Du, S. Time lag of stem sap flow and its relationships with transpiration characteristics in *Quercus liaotungensis* and *Robina pseudoacacia* in the loess hilly region, China. *Chin. J. Appl. Ecol.* **2019**, *30*, 2607–2613. [CrossRef]

31. Ou, R.; Yan, X.; Ma, H.; Jiang, Y. Predation and removal of *Quercus wutaishanica* and *Prunus davidiana* seeds of different size by rodents. *Seed* **2017**, *36*, 76–80. [CrossRef]

32. Chai, Y.; Yue, M.; Liu, X.; Guo, Y.; Wang, M.; Xu, J.; Zhang, C.; Chen, Y.; Zhang, L.; Zhang, R. Patterns of Taxonomic, Phylogenetic Diversity during a Long-Term Succession of Forest on the Loess Plateau, China: Insights into Assembly Process. *Sci. Rep.* **2016**, *6*, 27087. [CrossRef] [PubMed]

33. Chai, Y.; Yue, M.; Wang, M.; Xu, J.; Liu, X.; Zhang, R.; Wan, P. Plant Functional Traits Suggest a Change in Novel Ecological Strategies for Dominant Species in the Stages of Forest Succession. *Oecologia* **2016**, *180*, 771–783. [CrossRef]

34. Li, Y.; Shao, M. The change of plant diversity during natural recovery process of vegetation in Ziwuling area. *Acta Ecol. Sin.* **2004**, *24*, 252–260. [CrossRef]

35. Wang, J.; Chen, Y.; Cao, Y.; Zhou, J.; Hou, L. Carbon concentration and carbon storage in different components of natural Quercus wutaishanica forest in Ziwuling of Loess Plateau, Northwest China. *Chin. J. Ecol.* **2012**, *31*, 3058–3063. [CrossRef]

36. Zou, H.; Liu, G.; Wang, H. The vegetation development in North Ziwulin forest region in last fifty years. *Acta Bot. Boreali Occident. Sin.* **2002**, *22*, 1–8.

37. Wang, S.; Wang, X.; Guo, H.; Fan, W.; Lv, H.; Duan, R. Distinguishing the Importance between Habitat Specialization and Dispersal Limitation on Species Turnover. *Ecol. Evol.* **2013**, *3*, 3545–3553. [CrossRef]

38. Liu, P.; Wang, W.; Bai, Z.; Guo, Z.; Ren, W.; Huang, J.; Xu, Y.; Yao, J.; Ding, Y.; Zang, R. Competition and Facilitation Co-Regulate the Spatial Patterns of Boreal Tree Species in Kanas of Xinjiang, Northwest China. *For. Ecol. Manag.* **2020**, *467*, 118167. [CrossRef]

39. Qiu, J.; Han, A.; He, C.; Yin, Q.; Jia, S.; Luo, Y.; Li, C.; Hao, Z. Spatial distribution pattern and intraspecific association of the dominant species *Quercus aliena* var. *acutiserrata* in Qinling Mountains, China. *Chin. J. Appl. Ecol.* **2022**, *33*, 1–9. [CrossRef]

40. Hormaza, J.I. Molecular Characterization and Similarity Relationships among Apricot (*Prunus armeniaca* L.) Genotypes Using Simple Sequence Repeats. *Theor. Appl. Genet.* **2002**, *104*, 321–328. [CrossRef] [PubMed]

41. Guerra, M.E.; Guerrero, B.I.; Casadomet, C.; Rodrigo, J. Self-(in)Compatibility, S-RNase Allele Identification, and Selection of Pollinizers in New Japanese Plum-Type Cultivars. *Sci. Hortic.* **2020**, *261*, 109022. [CrossRef]

42. Guerra, M.E.; López-Corrales, M.; Wünsch, A. Improved S-Genotyping and New Incompatibility Groups in Japanese Plum. *Euphytica* **2012**, *186*, 445–452. [CrossRef]

43. Guerrero, B.I.; Guerra, M.E.; Rodrigo, J. Establishing Pollination Requirements in Japanese Plum by Phenological Monitoring, Hand Pollinations, Fluorescence Microscopy and Molecular Genotyping. *JoVE* **2020**, *165*, e61897. [CrossRef]

44. Kampfer, S.; Lexer, C.; Glössl, J.; Steinkellner, H. Characterization of $(GA)_n$ Microsatellite Loci from Quercus Robur. *Hereditas* **1998**, *129*, 183–186. [CrossRef]

45. Steinkellner, H.; Fluch, S.; Turetschek, E.; Lexer, C.; Streiff, R.; Kremer, A.; Burg, K.; Glössl, J. Identification and Characterization of $(GA/CT)_n$- Microsatellite Loci from *Quercus petraea*. *Plant. Mol. Biol.* **1997**, *33*, 1093–1096. [CrossRef] [PubMed]

46. Aldrich, P.R.; Michler, C.H.; Sun, W.; Romero-Severson, J. Microsatellite Markers for Northern Red Oak (Fagaceae: *Quercus rubra*). *Mol. Ecol. Notes* **2002**, *2*, 472–474. [CrossRef]

47. Berner, D. Allele Frequency Difference AFD–An Intuitive Alternative to FST for Quantifying Genetic Population Differentiation. *Genes* **2019**, *10*, 308. [CrossRef] [PubMed]

48. Stölting, K.N.; Paris, M.; Meier, C.; Heinze, B.; Castiglione, S.; Bartha, D.; Lexer, C. Genome-Wide Patterns of Differentiation and Spatially Varying Selection between Postglacial Recolonization Lineages of Populus Alba (Salicaceae), a Widespread Forest Tree. *New Phytol.* **2015**, *207*, 723–734. [CrossRef]

49. Chen, J.; Källman, T.; Ma, X.-F.; Zaina, G.; Morgante, M.; Lascoux, M. Identifying Genetic Signatures of Natural Selection Using Pooled Population Sequencing in Picea Abies. *G3 Genes | Genomes | Genet.* **2016**, *6*, 1979–1989. [CrossRef]

50. Wiegand, T.A.; Moloney, K. Rings, Circles, and Null-Models for Point Pattern Analysis in Ecology. *Oikos* **2004**, *104*, 209–229. [CrossRef]

51. Wiegand, T.; Moloney, K.A. *Handbook of Spatial Point-Pattern Analysis in Ecology*; Chapman and Hall/CRC Press: Boca Raton, FL, USA, 2013; ISBN 978-1-4200-8254-8.

52. Suriguga; Zhang, J.; Cheng, J.; Zhang, B. Population structure and distribution pattern of dominant species in *Tilia mandshurica* forest in Dongling Mountain of Beijing. *Chin. J. Ecol.* **2009**, *28*, 1253–1258. [CrossRef]

53. Zhang, X.; Zhang, X.; Guo, C.; Zhang, Q. Point pattern analysis of *Pteroceltis tatarinowii* population at its different development stages in limestone mountain area of north Anhui, East China. *Chin. J. Ecol.* **2013**, *32*, 542–550. [CrossRef]

54. Shimada, T. Nutrient Compositions of Acorns and Horse Chestnuts in Relation to Seed-Hoarding. *Ecol. Res.* **2001**, *16*, 803–808. [CrossRef]

55. Yang, X.; Anwar, E.; Zhou, J.; He, D.; Gao, Y.; Lv, G.; Cao, Y. Higher Association and Integration among Functional Traits in Small Tree than Shrub in Resisting Drought Stress in an Arid Desert. *Environ. Exp. Bot.* **2022**, *201*, 104993. [CrossRef]

56. Zhu, Y.; Bai, F.; Liu, H.; Li, W.; Li, L.; Li, G.; Wang, S.; Sang, W. Population distribution patterns and interspecific spatial associations in warm temperate secondary forests, Beijing. *Biodivers. Sci.* **2011**, *19*, 252–259. [CrossRef]

57. Gao, J.; Liu, Z.-L.; Zhao, W.; Tomlinson, K.W.; Xia, S.-W.; Zeng, Q.-Y.; Wang, X.-R.; Chen, J. Combined Genotype and Phenotype Analyses Reveal Patterns of Genomic Adaptation to Local Environments in the Subtropical Oak Quercus Acutissima. *J. Syst. Evol.* **2021**, *59*, 541–556. [CrossRef]

58. Yang, M.; Zhang, M.; Shi, S.; Kang, Y.; Liu, J. Analysis of Genetic Structure of *Magnolia sprengeri* Populations Based on ISSR Markers. *Sci. Silvae Sin.* **2014**, *50*, 76–81. [CrossRef]

59. Tian, H.; Kang, M.; Li, L.; Yao, X.; Huang, H. Genetic diversity in natural populations of *Castanea mollissima* inferred from nuclear SSR markers. *Biodivers. Sci.* **2009**, *17*, 296–302. [CrossRef]

60. Tang, S.; Dai, W.; Li, M.; Zhang, Y.; Geng, Y.; Wang, L.; Zhong, Y. Genetic Diversity of Relictual and Endangered Plant Abies Ziyuanensis (Pinaceae) Revealed by AFLP and SSR Markers. *Genetica* **2008**, *133*, 21–30. [CrossRef]

61. Song, Y.; Jiang, X.; Zhang, M.; Wang, Z.; Bo, W.; An, X.; Zhang, Z. Genetic differences revealed by Genomic-SSR and EST-SSR in poplar. *J. Beijing For. Univ.* **2010**, *32*, 1–7. [CrossRef]

62. Wen, Y.; Han, W.; Wu, S. Plant genetic diversity and its influencing factors. *J. Cent. South Univ. For. Technol.* **2010**, *30*, 80–87. [CrossRef]

63. O'Connell, L.M.; Mosseler, A.; Rajora, O.P. Extensive Long-Distance Pollen Dispersal in a Fragmented Landscape Maintains Genetic Diversity in White Spruce. *J. Hered.* **2007**, *98*, 640–645. [CrossRef]

64. Rushton, B.S. Natural Hybridization within the *Genus Quercus* L. *Ann. For. Sci.* **1993**, *50*, 73s–90s. [CrossRef]

65. Souza, I.G.B.; Souza, V.A.B.; Lima, P.S.C. Molecular Characterization of Platonia Insignis Mart. ("Bacurizeiro") Using Inter Simple Sequence Repeat (ISSR) Markers. *Mol. Biol. Rep.* **2013**, *40*, 3835–3845. [CrossRef]

66. Ma, X.; Zhao, C.; Zhang, Q.; Li, Y.; Hou, Z. Spatial pattern and spatial association of *Melica przewalskyi* and *Artemisia frigida* in degraded grassland. *Chin. J. Ecol.* **2013**, *32*, 299–304. [CrossRef]

67. Zhang, J. Analysis of spatial point pattern for plant species. *Acta Phytopathol. Sin.* **1998**, *22*, 344–349.

68. Jin, X.; Zhang, Q.; Xu, Q.; Ji, Y.; Bi, R. Population distribution patterns and interspecific spatial associations of *Acanthopanax senticosus* populations in Lingkong Mountain, Shanxi Province, China. *Plant Sci. J.* **2018**, *36*, 327–335. [CrossRef]

69. Wiegand, T.; Gunatilleke, S.; Gunatilleke, N. Species Associations in a Heterogeneous Sri Lankan Dipterocarp Forest. *Am. Nat.* **2007**, *170*, e77–e95. [CrossRef] [PubMed]

70. Liu, G.; Ding, Y.; Zang, R.; Guo, Z.; Zhang, X.; Cheng, K.; Bai, Z.; Ayoufu, B. Distribution patterns of *Picea schrenkiana* var. tianschanica population in Tianshan Mountains. *Chin. J. Appl. Ecol.* **2011**, *22*, 9–13. [CrossRef]

71. Zhang, Z.; Hu, G.; Zhu, J.; Luo, D.; Ni, J. Spatial Patterns and Interspecific Associations of Dominant Tree Species in Two Old-Growth Karst Forests, SW China. *Ecol. Res.* **2010**, *25*, 1151–1160. [CrossRef]

72. Carrer, M.; Castagneri, D.; Popa, I.; Pividori, M.; Lingua, E. Tree Spatial Patterns and Stand Attributes in Temperate Forests: The Importance of Plot Size, Sampling Design, and Null Model. *For. Ecol. Manag.* **2018**, *407*, 125–134. [CrossRef]

73. Wang, X.; Ye, J.; Li, B.; Zhang, J.; Lin, F.; Hao, Z. Spatial Distributions of Species in an Old-Growth Temperate Forest, Northeastern China. *Can. J. For. Res.* **2010**, *40*, 1011–1019. [CrossRef]

74. Jiang, D.; Tang, Y.; Busso, C.A. Effects of Vegetation Cover on Recruitment of Ulmus Pumila L. in Horqin Sandy Land, Northeastern China. *J. Arid Land* **2014**, *6*, 343–351. [CrossRef]

75. Murrell, L.D.J. The Community-Level Consequence of Seed Dispersal Patterns. *Annu. Rev. Ecol. Evol. Syst.* **2003**, *34*, 549–574. [CrossRef]

76. Ismail, S.A.; Ghazoul, J.; Ravikanth, G.; Kushalappa, C.G.; Uma Shaanker, R.; Kettle, C.J. Evaluating Realized Seed Dispersal across Fragmented Tropical Landscapes: A Two-Fold Approach Using Parentage Analysis and the Neighbourhood Model. *New Phytol.* **2017**, *214*, 1307–1316. [CrossRef] [PubMed]

77. Bohrer, G.; Katul, G.G.; Nathan, R.; Walko, R.L.; Avissar, R. Effects of Canopy Heterogeneity, Seed Abscission and Inertia on Wind-Driven Dispersal Kernels of Tree Seeds. *J. Ecol.* **2008**, *96*, 569–580. [CrossRef]

78. Liu, S.; Jiao, J.; Hu, S.; Wu, D.; Deng, N. Effect of flood runoff on seed dispersal and population regeneration—A case study of *Salix matsudana* in the Loess Hill and Gully Region. *Res. Soil Water Conserv.* **2018**, *25*, 99–103. [CrossRef]

79. Fraver, S.; D'Amato, A.W.; Bradford, J.B.; Jonsson, B.G.; Jönsson, M.; Esseen, P.-A. Tree Growth and Competition in an Old-Growth Picea Abies Forest of Boreal Sweden: Influence of Tree Spatial Patterning. *J. Veg. Sci.* **2014**, *25*, 374–385. [CrossRef]

80. Zhang, Y. Seed Rain Dynamics and Seedling Spatial Pattern of Deciduous Broad-Leaved Forest in Malan Forest Region of Loess Plateau. Master's Thesis, Shaanxi Normal University, Xi'an, China, 2015.

81. Wang, X.; Liang, C.; Wang, W. Balance between Facilitation and Competition Determines Spatial Patterns in a Plant Population. *Chin. Sci. Bull.* **2014**, *59*, 1405–1415. [CrossRef]

82. Yang, X.; Miao, Y.; Zhang, Q.; Zhang, L.; Bi, R. Spatial Pattern Analysis of Individuals in Different Age-classes of *Pinus bungeana* in Wulu Mountain Reserve, Shanxi, China. *Bull. Bot. Res.* **2013**, *33*, 24–30.

83. Shen, Z.; Hua, M.; Dan, Q.; Lu, J.; Fang, J. Spatial pattern analysis and associations of *Quercus aquifolioides* population at different growth stages in Southeast Tibet, China. *Chin. J. Appl. Ecol.* **2016**, *27*, 387–394. [CrossRef]

84. Omelko, A.; Ukhvatkina, O.; Zhmerenetsky, A.; Sibirina, L.; Petrenko, T.; Bobrovsky, M. From Young to Adult Trees: How Spatial Patterns of Plants with Different Life Strategies Change during Age Development in an Old-Growth Korean Pine-Broadleaved Forest. *For. Ecol. Manag.* **2018**, *411*, 46–66. [CrossRef]

85. Cai, F. A stuidy on the structure and dynamics of *Cyclobalanopsis glauca* population at hills around West Lake in Hangzhou. *Sci. Silvae Sin.* **2000**, *36*, 67–72. [CrossRef]

86. Hubbell, S.P. Neutral Theory and the Evolution of Ecological Equivalence. *Ecology* **2006**, *87*, 1387–1398. [CrossRef]

87. Brooker, R.W.; Maestre, F.T.; Callaway, R.M.; Lortie, C.L.; Cavieres, L.A.; Kunstler, G.; Liancourt, P.; Tielbörger, K.; Travis, J.M.J.; Anthelme, F.; et al. Facilitation in Plant Communities: The Past, the Present, and the Future. *J. Ecol.* **2008**, *96*, 18–34. [CrossRef]

88. Metz, J.; Annighöfer, P.; Schall, P.; Zimmermann, J.; Kahl, T.; Schulze, E.-D.; Ammer, C. Site-Adapted Admixed Tree Species Reduce Drought Susceptibility of Mature European Beech. *Glob. Chang. Biol.* **2016**, *22*, 903–920. [CrossRef] [PubMed]

89. Muhamed, H.; Maalouf, J.-P.; Michalet, R. Summer Drought and Canopy Opening Increase the Strength of the Oak Seedlings–Shrub Spatial Association. *Ann. For. Sci.* **2013**, *70*, 345–355. [CrossRef]

90. Zanini, L.; Ganade, G.; Hübel, I. Facilitation and Competition Influence Succession in a Subtropical Old Field. *Plant Ecol.* **2006**, *185*, 179–190. [CrossRef]

91. Kubota, Y. Spatial Pattern and Regeneration Dynamics in a Temperate Abies–Tsuga Forest in Southwestern Japan. *J. For. Res.* **2006**, *11*, 191–201. [CrossRef]

92. Gray, L.; He, F. Spatial Point-Pattern Analysis for Detecting Density-Dependent Competition in a Boreal Chronosequence of Alberta. *For. Ecol. Manag.* **2009**, *259*, 98–106. [CrossRef]

93. Muhamed, H.; Touzard, B.; Le Bagousse-Pinguet, Y.; Michalet, R. The Role of Biotic Interactions for the Early Establishment of Oak Seedlings in Coastal Dune Forest Communities. *For. Ecol. Manag.* **2013**, *297*, 67–74. [CrossRef]

94. Comita, L.S.; Hubbell, S.P. Local Neighborhood and Species' Shade Tolerance Influence Survival in a Diverse Seedling Bank. *Ecology* **2009**, *90*, 328–334. [CrossRef]

95. Han, A.; Qiu, J.; He, C.; Yin, Q.; Jia, S.; Luo, Y.; Li, C.; Hao, Z. Spatial distribution patterns and intraspecific and interspecific associations of the dominant shrub species *Lonicera fragrantissima* var. *lancifolia* in Huangguan of Qinling Mountains, China. *Chin. J. Appl. Ecol.* **2022**, *33*, 1–9. [CrossRef]

Article

Changes in Soil Microbial Communities under Mixed Organic and Inorganic Nitrogen Addition in Temperate Forests

Zhaolong Ding [1,2,3], Lu Gong [1,2,3,*], Haiqiang Zhu [4], Junhu Tang [1,2,3], Xiaochen Li [1,2,3] and Han Zhang [1,2,3]

1 College of Ecology and Environment, Xinjiang University, 777 Huarui Road, Urumqi 830017, China
2 Key Laboratory of Oasis Ecology, Xinjiang University, 777 Huarui Road, Urumqi 830017, China
3 Xinjiang Jinghe Observation and Research Station of Temperate Desert Ecosystem, Ministry of Education, Xinjiang University, 777 Huarui Road, Urumqi 830017, China
4 College of Tourism, Xinjiang University, 777 Huarui Road, Urumqi 830017, China
* Correspondence: gonglu721@163.com

Abstract: Investigating the response of soil microbial communities to nitrogen (N) deposition is critical to understanding biogeochemical processes and the sustainable development of forests. However, whether and to what extent different forms of N deposition affect soil microbial communities in temperate forests is not fully clear. In this work, a field experiment with three years of simulated nitrogen deposition was conducted in temperate forests. The glycine and urea were chosen as organic nitrogen (ON) source, while NH_4NO_3 was chosen as inorganic nitrogen (IN) source. Different ratios of ON to IN (CK = 0:0, Mix-1 = 10:0, Mix-2 = 7:3, Mix-3 = 5:5, Mix-4 = 3:7, Mix-5 = 0:10) were mixed and then used with equal total amounts of $10\ kg \cdot N \cdot ha^{-1} \cdot a^{-1}$. We determined soil microbial diversity and community composition for bacteria and fungi (16S rRNA and ITS), and soil parameters. Different forms of N addition significantly changed the soil bacterial and fungal communities. Mixed N sources had a positive effect on soil bacterial diversity and a negative effect on fungal diversity. Bacterial and fungal community structures were significantly separated under different forms of N addition. Soil pH was the main factor affecting the change in fungal community structure, while bacterial community structure was mainly controlled by STN. We also found that Proteobacteria, Acidobacteriota, Basidiomycota and Ascomycota were the most abundant phyla, regardless of the form of N addition. RDA showed that C/P and NH_4^+ were the main factors driving the change in bacterial community composition, and C/P, pH and C/N were the main factors driving the change in fungal community composition. Our results indicate that different components of N deposition need to be considered when studying the effects of N deposition on soil microorganisms in terrestrial ecosystems.

Keywords: temperate forests; nitrogen addition; bacterial community; fungal community; soil microbial community

Citation: Ding, Z.; Gong, L.; Zhu, H.; Tang, J.; Li, X.; Zhang, H. Changes in Soil Microbial Communities under Mixed Organic and Inorganic Nitrogen Addition in Temperate Forests. *Forests* **2023**, *14*, 21. https://doi.org/10.3390/f14010021

Academic Editor: Artur Alves

Received: 14 October 2022
Revised: 14 December 2022
Accepted: 20 December 2022
Published: 22 December 2022

1. Introduction

Human activities (energy production, industry, agricultural practices, and intensive animal husbandry) have led to a substantial increase in nitrogen (N) deposition in terrestrial ecosystems [1]. The increased N deposition has alleviated the demand for N in terrestrial ecosystems and has led to a shift from N deficiency to N loading in some ecosystems [2,3]. However, most forest ecosystems, particularly northern temperate forests, are still considered N-limited [4]. Therefore, the potential effects of N deposition on temperate forest ecosystems, such as changes in soil pH, nutrients, biodiversity, and primary productivity, have attracted much attention [5–7]. Soil microbes play a critical role in soil functions, ecological processes, and ecosystem functions and may be very sensitive to N deposition [8,9]. Studies on the effect of N deposition on soil microbial communities have been conducted worldwide. However, research on the response of soil microbial communities to N deposition in temperate forests is limited.

N deposition has a profound impact on microbial communities in forest ecosystems, including their diversity, structure, and composition [9–12]. However, there is no consensus opinion on the effects of N addition on microbial communities due to differences in ecosystems, background N deposition, N application rates and experimental periods. N addition has negative, positive and neutral effects on soil microbial communities. In temperate forests, N-saturated subtropical and tropical forests, studies have found that N addition reduces soil microbial biomass and diversity [13–15]. In contrast, studies in N-limited temperate meadows and boreal forests have found that N addition promotes the growth of soil microorganisms [16,17]. In addition, it has also been reported that N addition has a neutral effect on soil microorganisms [18]. In general, N addition mainly affects soil microbial communities through two aspects. On the one hand, N addition alters the nutrient use strategy of soil microorganisms, thus affecting their community structure and composition [19]. On the other hand, N addition also alters abiotic variables that affect soil microbial community composition (e.g., available N, pH, C, and P) [20]. Most studies have linked the microbial response to N addition to soil pH, and argue that soil acidification is the main factor controlling these changes [21,22]. Many studies have shown that soil pH is a key factor in predicting soil microbial activity, and N addition significantly reduces soil pH in most ecosystems [15,23].

Recent studies have shown that plants and microbes display selective absorption characteristics for different forms of N resources [24–26]. Therefore, the forms of the N sources may be an important driving factor leading to soil microorganism community assembly. Previous studies have only examined the effects of different forms of inorganic nitrogen (IN) (e.g., NH_4^+ and NO_3^-) on ecosystems, but rarely considered organic nitrogen (ON) [27–29]. In fact, atmospheric N deposition includes ON components (urea, glycine, etc.) in addition to IN components, accounting for approximately 30–36% and approximately 28% in China, and this proportion may continue to increase in the future [30,31]. ON sources have bioavailability similar to that of IN and are more likely to be bioavailable, especially in N-limited ecosystems [32]. Urea fertilizer increases soil microbial biomass faster than NH_4NO_3 fertilizer, indicating that ON may be the preferred N source for soil microbes [33]. Although ON was used as the N source in some field trials, it was used only as a single N source [25]. This approach may not provide complete information about the effect of atmospheric N deposition on soil microorganisms. Recent studies have also focused on the effects of different types of N sources on ecosystems. For example, in tropical forests, mixed N application enhances the ability of soil microorganisms to secrete enzymes, increases soil enzyme activity, and improves the tolerance of soil microorganisms to pH fluctuations [34]. An investigation in a temperate grassland and coniferous forest showed that mixed N application significantly increased the litter decomposition rate [26,35]. Studies have also pointed out that an increase in the ON ratio under mixed N source fertilization alleviates the inhibitory effect of N input on soil respiration, which may increase forest soil CO_2 emissions [36]. In summary, the mixed N sources had positive effects in these investigations. However, these studies only provide limited information, and it is necessary to study the effects of different forms of N addition on soil microbial communities.

In April 2018, we established a N deposition simulation experiment site in a temperate forest in the Tianshan Mountains and carried out a N addition experiment for three years. Then, we applied high-throughput DNA sequencing techniques to determine the change in soil microbial communities under different forms of N addition. Our purposes were to: (1) compare the alteration of microbial community diversity and community structure under different forms of N addition and (2) determine the main soil parameters that cause microbial community diversity and community structure changes. We hypothesized that: (1) the response of microbial community to N addition is regulated by ON:IN ratio. (2) Mixed nitrogen addition has a positive effect on microbial community diversity and structure. (3) Soil pH is the main factor controlling the effects of different forms of N addition on bacterial and fungal communities.

2. Materials and Methods

2.1. Study Site

This research was carried out in a typical temperate forest located in the Tianshan Mountains, Xinjiang, China (42°25.96′ N, 87°28.17′ E; 1992 m.a.s.l.). Schrenk's spruce (*Picea schrenkiana*) is the dominant tree and the stand mostly comprises pristine forest, with a height of approximately 16 m, an average diameter at breast height (DBH) of 17.6 cm, an average tree age of 78 a and a canopy density of 0.6–0.8. There were almost no associated herbaceous plants within the forest, with only sporadically distributed *Geranium rotundifolium* and *Aegopodium podagraria* [37]. The mean annual temperature is approximately 2.5 °C. The average annual precipitation is approximately 650 mm, with most falling during May to October. The annual total radiation is estimated to be 5.85×105 J cm^{-2} a^{-1}. The background N deposition is 10 kg·N·ha^{-1}·a^{-1} [5,38]. The soil is a gray brown forest soil according to the soil classification of China [39].

2.2. Experimental Design and Sampling

The simulated N deposition experiment was established in April 2018. We referred to the experimental designs in previously published studies [34,36]. The glycine and urea were chosen as ON sources, while NH_4NO_3 was chosen as the IN source. Different ratios of ON to IN were mixed with equal total amounts with 10 kg·N·ha^{-1}·a^{-1} for a total of six treatments: CK (ON:IN = 0:0), ON (ON:IN = 10:0), Mix-1 (ON:IN = 7:3), Mix-2 (ON:IN = 5:5), Mix-3 (ON:IN = 3:7), and IN (ON:IN = 0:10). Eighteen (5 m × 5 m) plots separated from each other by a buffer zone of >5 m were established, including six treatments with three replicates. All plots were equivalent in terms of altitude, slope (<15°), soil and vegetation types. The nitrogen fertilizer weighed in the laboratory was dissolved in 10L of water. During the growing season from May to October each year, the nitrogen fertilizer solution (1.667 kg·N·ha^{-1}·a^{-1}) was everly sprayed on the surface with a backpack sprayer at the beginning of the month, and the CK plot was only sprayed with the same amount of deionized water obtained from the laboratory.

Topsoil (0–20 cm) samples were collected at the end of August 2021. The litter layer was removed before sampling. Five soil samples were collected from five random points using a soil sampler (0–20 cm deep with a 2 cm inner diameter) and mixed to obtain a composite sample for each plot. Fresh soil samples were stored in sterile closed polypropylene bags after removing impurities (vegetation residues and stones) and passage through a 2 mm sterile sieve. Each composite soil sample was divided into two parts: one was transported to the laboratory in liquid N and then stored in a freezer at −80 °C for DNA extraction. The second part was further divided into two subsamples: one subsample was stored at 4 °C, and used for the determination of NH_4^+ and NO_3^- contents within one week. The other subsample was air-dried for pH, soil total N (STN), soil organic carbon (SOC), and soil total phosphorus (STP) analyses.

2.3. Soil Property Determinations

Soil water content (SWC) was obtained by the gravimetric method. Soil pH was measured by a glass electrode (PHS-3E, Leici, Shanghai, China) in a soil-water solution (1:2.5). The SOC concentration was determined by performing potassium dichromate oxidation titration with an Fe_2^+ solution [40]. The STN was determined through acid digestion using the Kjeldahl method [41]. The STP was determined using the molybdenum-antimony colorimetric method after the soil samples were digested with H_2SO_4 [40]. The fresh soil samples were extracted from a 2 M KCl solution and filtered, and then the contents of NH_4^+ and NO_3^- were analyzed with a flow injection autoanalyzer (Bran and Luebbe, Norderstedt, Germany).

2.4. DNA Isolation and Illumina HiSeq Sequencing

Total genomic DNA was extracted from 0.5 g of soil by the CTAB method. A 1% agarose gel was used to verify DNA purity and concentration. The V4 region of the

16S rRNA gene was amplified using the primers 515F (GTGCCAGCMGCCGCGGTAA) and 806R (GGACTACHVGGGTWTCTAAT). The PCR primers ITS5–11737 (GGAAG-TAAAAGTCGTAACAAGG) and ITS2–2043R were used to amplify the fungal ITS1 regions (GCTGCGTTCTTCATCG-ATGC). The PCR products were combined with an equivalent amount of 1× loading buffer (including SYBR Green), and DNA was detected by electrophoresis on a 2% agarose gel. The PCR products were purified using the Qiagen Gel Extraction Kit (Qiagen, Hilden, Germany) after being mixed in equal parts. The NEBNext® UltraTM IIDNA Library Prep Kit was used to create sequencing libraries according to the manufacturer's instructions (Cat No. E7645). A Qubit@ 2.0 Fluorometer (Thermo Scientific, Waltham, Massachusetts, United States) and an Agilent Bioanalyzer 2100 system were used to assess the library's quality. Finally, the Illumina NovaSeq platform was used to sequence the library. The original data were then spliced and filtered to produce useful tags (FLASH, Version 1.2.11, http://ccb.jhu.edu/software/FLASH/ (accessed on 24 November 2021)) [42].

To acquire initial amplicon sequence variants (ASVs), denoising of the effective tags obtained as described above was conducted with DADA2 in QIIME2 software (version QIIME2-202006), and then ASVs with abundances less than 5 were deleted. QIIME2 was used for species annotation. For 16S annotation, the Silva Database https://www.arbsilva.de/ (accessed on 24 November 2021) was utilized, and for ITS annotation, the Unite Database https://unite.ut.ee/ (accessed on 24 November 2021) was employed [43]. The species rarefaction curve reflected the community diversity at different sequencing numbers. With the increase of the number of sequencings, the dilution curve of samples finally tended to be gentle (Figure A1), indicating that the amount of sequencing data was sufficient, and the measured data could reflect the real situation of bacterial and fungal communities.

2.5. Statistical Analysis

Alpha diversity was assessed by calculating the Shannon and Chao1 index values (QIIME2 software. version 1.9.1). The differences in microbial community structures among treatments were determined using nonmetric multidimensional scaling (NMDS) analysis, and permutational multivariate analyses of variances (PERMANOVAs) based on the Bray–Curtis dissimilarity were conducted to test for the significant dissimilarity among different N additions. Both above analyses were processed in the "vegan" package of R software [44]. Random forest plots were constructed to determine soil parameters that play an important role in regulating microbial community structure, which was processed in the "random forest" package of R software [44]. One-way analysis of variance (One-way ANOVA) and least significant difference (LSD, $p < 0.05$) were used to detect the significant differences between treatments with different forms of nitrogen in SPSS v24.0 software (SPSS Inc., Chicago, IL, USA) [45]. In the CANOCO 5.0 software (Wageningen UR, Netherlands), Redundant analysis (RDA) was used to identify important soil parameters affecting soil microbial community composition. Graphs were drawn by Origin 2021b.

3. Results

3.1. Effects of N Addition on Soil Properties

N addition changed the soil parameters (Table 1). The ON treatment significantly increased SOC, STN, C/N, C/P, N/P and NO_3^-. The IN treatment significantly increased SOC, STN, STP, NO_3^-, and NH_4^+, but decreased pH. The Mix-1 treatment significantly increased STP, C/N, and NH_4^+, but decreased STN, C/P, N/P. The Mix-2 treatment significantly increased STP, NH_4^+ and SWC, but decreased C/P and N/P. The Mix-3 treatment significantly reduced SOC, STN, C/P, and N/P, but increased STP.

Table 1. Soil properties in different treatments. Abbreviations: NO_3^-, nitrate nitrogen; NH_4^+, ammonium nitrogen; STN, total nitrogen; SOC, soil organic carbon; STP, total phosphorus; SWC, soil water content. Mean $\pm$ standard error, n = 3. Different lowercase letters indicate significant differences between treatments ($p < 0.05$).

Treatment	CK	ON	Mix-1	Mix-2	Mix-3	IN	One-Way ANOVA	
							F	p
SOC (g kg^{-1})	126.80 (2.2)c	194.90 (23.96)a	101.98 (3.48)cd	117.16 (3.37)cd	91.04 (4.68)d	158.59 (16.81)b	19.82	<0.001
STN (g kg^{-1})	9.20 (0.24)b	11.86 (1.04)a	6.64 (0.66)cd	8.51 (0.41)bc	6.49 (0.42)d	11.86 (1.62)a	15.01	<0.001
STP (g kg^{-1})	0.98 (0.03)d	0.96 (0.02)d	1.18 (0.1)bc	2.26 (0.07)a	1.32 (0.09)b	1.14 (0.08)c	92.16	<0.001
C/N	13.79 (0.31)b	16.39 (0.67)a	15.48 (1.19)a	13.78 (0.35)b	14.04 (0.48)b	13.44 (0.47)b	6.45	0.004
C/P	129.62 (5.49)b	204.03 (29.31)a	86.91 (8.49)c	51.76 (0.84)d	69.23 (6.91)cd	138.54 (5.46)b	35.91	<0.001
N/P	9.41 (0.54)b	12.40 (1.31)a	5.69 (0.94)c	3.76 (0.15)d	4.93 (0.45)cd	10.34 (0.75)b	38.40	<0.001
pH	7.81 (0.04)a	7.29 (0.06)a	7.35 (0.16)a	7.51 (0.21)a	7.60 (0.06)a	6.95 (0.15)b	4.05	0.022
NO_3^- (mg kg^{-1})	8.55 (0.85)c	55.39 (9.43)a	2.70 (0.96)c	1.53 (0.77)c	11.09 (2.21)bc	20.82 (4.93)b	40.65	<0.001
NH_4^+ (mg kg^{-1})	9.88 (0.85)d	10.50 (2.95)d	29.18 (4.42)bc	67.44 (15.23)a	20.55 (3.22)cd	35.40 (1.64)b	20.13	<0.001
SWC (%)	43.55 (1.9)bc	53.75 (7.21)ab	55.58 (12.17)ab	63.32 (8.39)a	35.42 (1.32)c	49.12 (8.84)abc	3.24	0.440

3.2. Soil Bacterial and Fungal Diversity

N addition had no significant effect on the Chao1 index of the bacterial community compared with that in the CK treatment. The Mix-1 and Mix-2 treatments had significantly higher values than the ON treatment ($p < 0.05$; Figure 1a). For fungi, the Chao1 index decreased significantly in the Mix-1, Mix-2 and Mix-3 treatments ($p < 0.05$; Figure 1b). The Mix-1 and Mix-2 treatments significantly increased the bacterial Shannon index ($p < 0.05$; Figure 1c). The Shannon index of fungi decreased significantly in Mix-2 treatment and increased significantly in Mix-3 treatment ($p < 0.05$; Figure 1d). According to Pearson analysis, the bacterial Chao1 index had a significant negative correlation with C/P and N/P (Figure A2; $p < 0.05$). The fungal Chao1 index was significantly positively correlated with soil pH (Figure A2; $p < 0.05$). The fungal Shannon index was significantly negatively correlated with STP (Figure A2; $p < 0.05$).

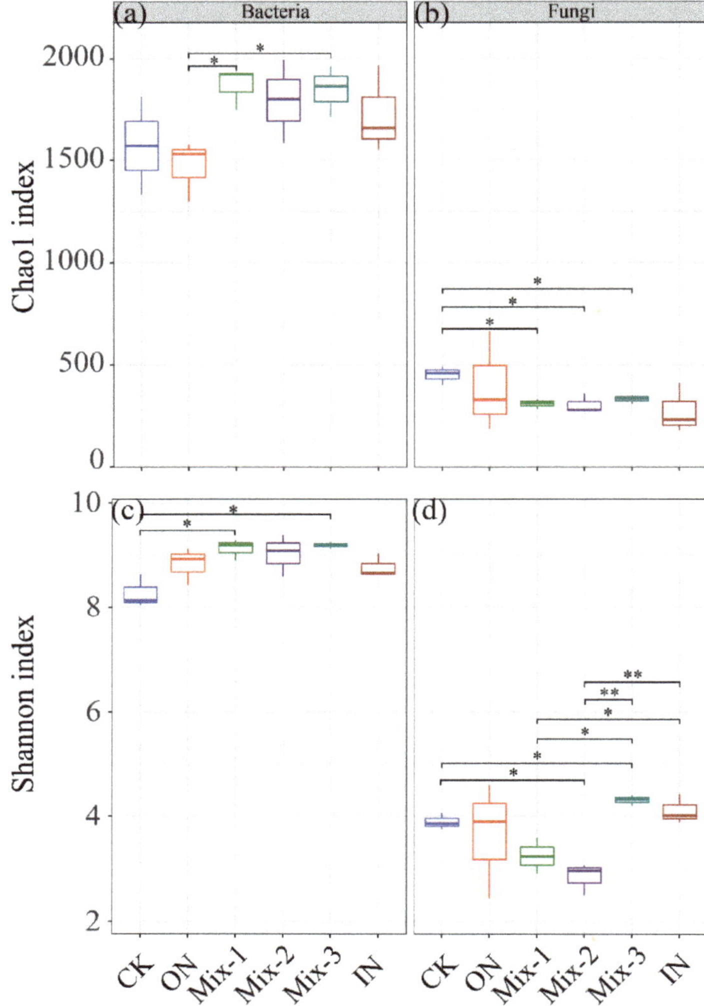

Figure 1. The alpha diversity of bacteria (**a,c**) and fungi (**b,d**) in different treatments. * $p < 0.05$ and ** $p < 0.01$.

3.3. Soil Microbial Species Composition and Community Structure

For the bacterial community, the dominant phyla were Proteobacteria, Acidobacteriota, Gemmatimonadota, Actinobacteriota, and Verrucomicrobiota (Figure 2a). The dominant phyla in the fungal community were Basidiomycota and Ascomycota (Figure 2b). NMDS analysis showed that the community structures of bacteria (PERMANOVA, $R^2 = 0.730$, $p < 0.001$; Figure 3a) and fungi (PERMANOVA, $R^2 = 0.980$, $p < 0.001$; Figure 3b) varied significantly among the treatments (Figure 3).

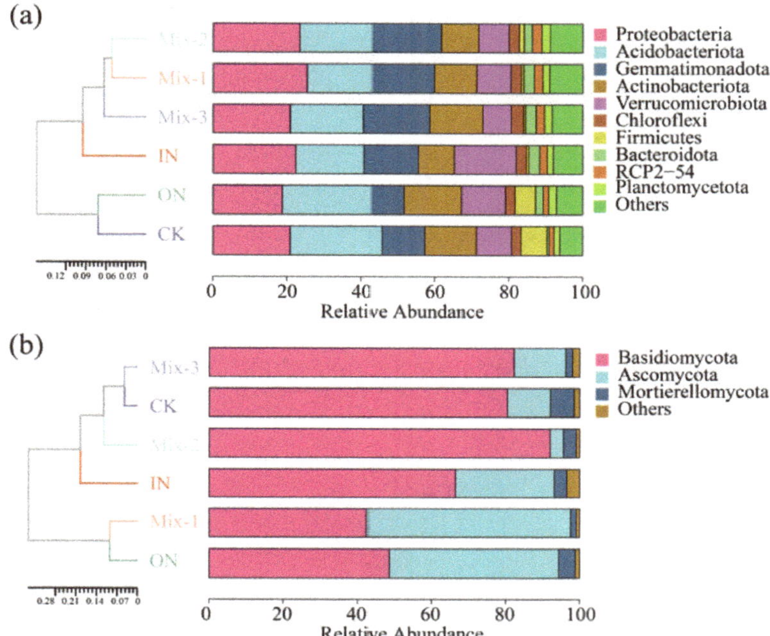

Figure 2. The relative abundance of bacteria (**a**) and fungi (**b**) at the phylum level based on the results of clustering analysis.

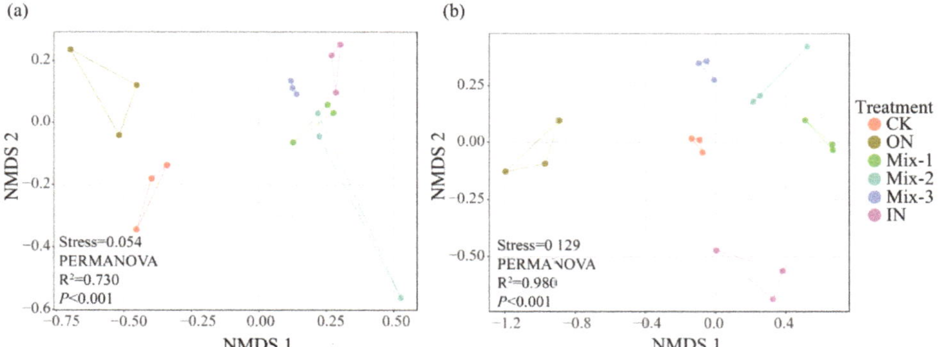

Figure 3. Nonmetric multidimensional scaling (NMDS) ordination plot based on the Bray–Curtis distance of samples for the bacterial (**a**) and fungal communities (**b**) in all treatments.

3.4. Relationship between Soil Factors and Microbial Communities

Random forest analysis revealed that SOC, STN, and N/P were main determinants of bacterial community structure. In contrast, fungal community structure was determined by STN, STP, C/P, C/N, NH_4^+, NO_3^-, and pH (Figure 4). The result of RDA revealed the effects of soil parameters on the relative abundance of microbial taxa (Figure 5). For bacteria, the first two axes associated with the soil parameters together explained 48.74% of the variation in the dominant phyla. The primary variables influencing the soil bacterial community were C/P, and NH_4^+ (23.3%, $p < 0.01$; 10.0%, $p < 0.05$). For fungi, the first two axes associated with the soil parameters together explained 57.82% of the variation in the dominant phyla. C/P, pH, and C/N significantly affected the bacterial community (22.9%, $p < 0.01$; 15.9%, $p < 0.01$; 11.8%, $p < 0.01$).

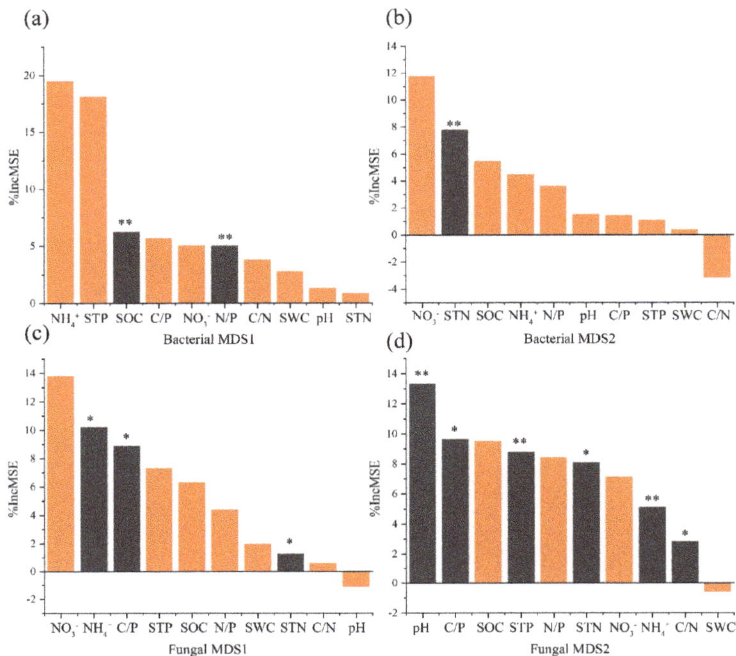

Figure 4. Environmental predictors of the NMDS1 and NMDS2 of bacteria (**a**,**b**) and fungi (**c**,**d**) based on random forest analysis. Only predictors with significant effects are shown in the figures. Abbreviations: NO_3^-, nitrate nitrogen; NH_4^+, ammonium nitrogen; STN, total nitrogen; SOC, soil organic carbon; STP, total phosphorus; SWC, soil water content. * $p < 0.05$ and ** $p < 0.01$.

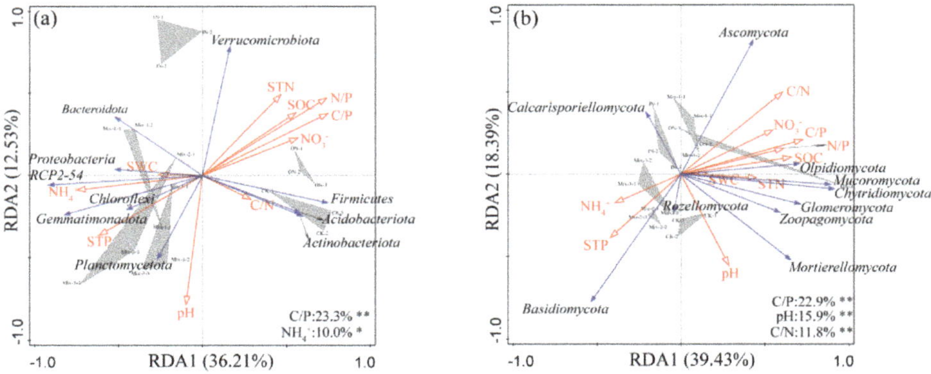

Figure 5. Redundancy analysis (RDA) of soil properties and dominant bacterial phyla (**a**) and dominant fungal phyla (**b**). Abbreviations: NO_3^-, nitrate nitrogen; NH_4^+, ammonium nitrogen; STN, total nitrogen; SOC, soil organic carbon; STP, total phosphorus; SWC, soil water content. Mean ± standard error, n = 3. * $p < 0.05$ and ** $p < 0.01$.

4. Discussion

Consistent with our hypothesis (1), our results suggest that soil fungal and bacterial communities may change with different ON:IN ratios. Inconsistent with our hypothesis (2), our results showed that mixed N addition had a positive effect on bacterial community diversity while showing a negative effect on fungal community diversity. Our results also showed that soil parameters changed under N addition and that the changed soil parameters affected the soil microbial community. For example, as described in our hypothesis (3), soil pH is the main factor controlling the structure and composition of the bacterial community.

4.1. Effects of Different Forms of N Addition on Soil Parameters

Previous studies have shown that N addition significantly alters the physicochemical properties of forest soils [46]. In particular, soil acidification caused by N deposition is of great concern [15,21]. In this study, N addition decreased the soil pH (0.21–0.86) but the decrease was statistically significant in only the IN treatment ($p < 0.05$). Previous studies have reported similar results, suggesting that soil acidification mediated by IN is greater than that mediated by other N forms [21,34,47]. In addition, our study also showed that the soil pH in plots with IN added in isolation was significantly lower than that in plots with mixed N addition. Although there was no significant difference between the mixed N treatments, our data showed an increase trend for pH with a decrease in the organic N ratio. Previous studies have shown that soil pH increases after short-term urea additions but decreases after repeated IN additions [48,49]. Each urea molecule consumes one H^+-ion during the conversion to NH_4^+, so the acidification capacity of urea is lower than that of other forms of N [50]. In addition, we also found that Mix-3 (ON:IN = 3:7) resulted in the smallest decrease in soil pH. This may be caused by the ratios of organic to inorganic N that are closer to those observed under natural conditions, where ON accounts for approximately 30% of total N deposition [30,51]. Therefore, with the increase in the N application cycle, mixed N sources can alleviate soil acidification caused by N enrichment. We found that the SOC, STN, and STP contents responded differently to the different forms of N addition (Table 1). Specifically, SOC and STN increased significantly in the ON and IN treatments but decreased in the mixed N addition plots. Previous studies have shown that mixed N fertilization promotes soil biological activity and thus increases N uptake [36]. Studies have suggested that mixed N increases the soil microbial biomass and soil enzyme activities (e.g., invertase, cellulase, and cellobiohydrolase), which promote litter decomposition [34,52]. Conversely, a single N source may inhibit the decomposition of SOM by reducing the soil pH, leading to an increase in the SOC content [15]. N and P are primary limiting elements in most terrestrial ecosystems [46,53]. N deposition alleviates N limitation while possibly increasing P limitation [54,55]. In general, a large increase in N leads to an increase in soil N/P, which may reduce the availability of soil P and aggravate P limitation [53,56]. However, our results showed that the N/P of the ON plots increased significantly, while those of the Mix-1, Mix-2 and Mix-3 plots decreased significantly. This suggests that the use of ON alone may increase soil P limitation, while the addition of mixed N sources may alleviate soil P limitation. This may be because the mixed N source increased soil microbial biomass and enhances soil microbial P fixation [16].

4.2. Effects of Different Forms of N Addition on the Alpha Diversity of Bacteria and Fungi

Our results provide some evidence that the different forms of N addition influence the diversity and richness of bacteria and fungi. The Mix-1 treatment and Mix-3 treatment significantly increased the bacterial Shannon index. In general, N addition leads to an overall decrease in forest soil bacterial alpha diversity [57–59]. However, the impact of N deposition on soil bacterial diversity may vary by habitat [11]. For example, N addition reduced soil bacterial diversity in N-saturated tropical forests [60], but increased soil microbial diversity in some N-deficient temperate and boreal forests [14,61]. The background value of N deposition in our study area is low compared to that globally and human activity is also low [5,62]. N deposition increased the N content of the soil and alleviated N limitation, thus stimulating the growth of bacterial communities. On the other hand, since most studies investigated only the effects of a single N source (NH_4NO_3, NH_4Cl, or urea), differences in the effects of N addition on bacteria may be due to differences in N sources [28,63]. In contrast, our study examined different forms of N addition (ON:IN), which have rarely been considered. We observed that the Chao1 index of bacteria was significantly higher in the Mix-1-treated and Mix-3-treated plots than in the ON-treated plots. A single N source inhibits microbial activity by disrupting the original balance of ON and IN in the soil [34]. Our results showed that the bacterial Chao1 index was significantly and negatively correlated with C/P and N/P. This is consistent with the findings of previous studies that soil resource stoichiometry is an important driver of bacterial diversity and that low C/P and N/P typically promote bacterial abundance [64,65]. However, the Shannon index was not significantly correlated with the soil parameters. Therefore, we suggest that bacterial alpha diversity in this region may be responding to different forms of N addition through bacterial abundance (the Chao1 index) rather than community diversity (the Shannon index). Our results also indicate that there is no significant correlation between bacterial diversity and soil pH. However, previous studies have pointed out that a decrease in soil pH usually inhibits bacterial activity [66,67]. This may be caused by the small decrease in pH due to N addition that may not reach the threshold required to inhibit bacterial community activity. There is evidence that small changes in soil pH may have little effect on bacterial populations [68].

Here, the Mix-1, Mix-2 and Mix-3 treatments significantly reduced the Chao1 index of the fungi (Figure 2). However, there was no significant difference between a single N source (ON or IN) and the CK. A recent study showed that N addition increases fungal abundance in N-limited boreal forests [17]. Conversely, studies have also shown a significant decrease in fungal species richness due to N addition, but they have not considered the effects of changes in the ON:IN ratio [24]. In addition, Pearson's correlation analysis showed a significant positive correlation between soil pH and the Chao1 index (Figure A1, $p < 0.05$). We hypothesized that changes in soil pH caused by N addition altered fungal abundance, specifically decreasing it with decreasing soil pH. Similar results have been reported in other investigations of fungal communities [11,13]. Some populations with weak acid tolerance may have become small or even disappeared as the soil pH decreased, resulting in a reduction in fungal species. Sufficient evidence may be obtained in the future as the duration of N application increases. As N increases in the soil, bacteria grow faster because they have increased access to resources [69]. Conversely, bacterial competition limits the development of fungi, which may lead to changes in communities dominated by bacterial taxa [70,71].

4.3. Effects of Different Forms of N Addition on Microbial Community Structure and Composition

The structure and species composition of soil microorganisms reflect their role in biogeochemical processes, and N input causes considerable changes in the microbial community structure [11,72]. NMDS analysis revealed that the community structure of bacteria and fungi changed significantly under different treatments, which reflected the response of soil microbial community structure to different ON:IN ratios. Previous studies have shown that soil environment dominate changes in soil microbial community structure [9,73]. Our study yielded the same results that changes in soil parameters caused by N addition affected the community structure of bacteria and fungi. We also found that the differences in fungal community structure among treatments were larger than those for bacteria. This is because fungal populations are more susceptible to changes in the soil environment caused by N addition.

We further investigated changes in the relative abundance of bacterial and fungal phyla. Proteobacteria (21.99%), Acidobacteriota (20.64%), Basidiomycota (68.74%), and Ascomycota (26.13%) were the most abundant phyla, regardless of N addition forms (Table A1). Acidobacteriota were shown to be more sensitive to changes in soil pH, and their relative abundance increased substantially as the soil pH decreased [74–76]. N addition has also been documented to reduce the relative abundance of Acidobacteria [77]. However, our results showed that only the Mix-1 treatment reduced the abundance of these bacteria. This may be because our study investigated different forms of N sources with the same amount of N. A slight decrease in soil pH may not affect such bacteria. Previous studies have indicated that the relative abundance of Acidobacteria may depend on nutrient effectiveness rather than pH [78]. RDA also showed that C/P and NH_4^+ were the main factors controlling the change in bacterial community composition (Figure 4). According to trophic strategies, Acidobacteriota are generally oligotrophic and prefer nutrient-poor environments [79]. N input usually prevents the development of oligotrophic microorganisms [80]. In contrast, Proteobacteria belong to the eutrophic microbiome, which increases in nutrient-rich environments [60]. Our results showed that Ascomycota was significantly more abundant in the ON and Mix-1 plots than in those with other treatments. Large ON inputs (e.g., urea and glycine) have been shown to increase the relative abundance of Ascomycetes and several saprophytic fungi (containing most of the Ascomycete genera) [24,81]. Therefore, the change in ON composition may be the main driving factor for the transformation of fungal community composition. Our study revealed that Basidiomycota was significantly reduced in single-N source treatments, and reached a maximum in the Mix-2 treatment. This may be because a single N source increases the abundance of Ascomycota, thereby reducing Basidiomycota's competitiveness for resources.

5. Conclusions

The goal of this work was to investigate the effects of different forms of N addition on soil microbial communities and soil parameters in temperate forests. We found that the different effects of N addition on soil microbial communities in temperate forests may be attributed to the form of N addition. Compared with the application of a single N source, mixed N addition had a positive effect on bacterial diversity and a negative effect on fungal diversity. NMDS showed that soil microbial community structure changed significantly under different forms of N addition, and soil parameters played a dominant role. In addition, soil pH decreased with increasing IN proportion and was significantly reduced in the IN treatment. Changes in soil pH affected the composition of fungal communities but had no effect on bacteria. In summary, these results may contribute to a better understanding of the mechanisms underlying the effects of N addition on soil microbial community structure and diversity in temperate forests. Our results highlight that studies based on a single N source may underestimate the potential impact of N deposition on soil microbial communities. It is necessary to consider the importance of different components of N deposition.

Author Contributions: Conceptualization, L.G. and Z.D.; methodology, L.G.; investigation, Z.D.; H.Z. (Han Zhang), J.T., X.L. and H.Z. (Haiqiang Zhu); data curation, L.G.; writing—original draft preparation, Z.D.; writing—review and editing, L.G. and H.Z. (Haiqiang Zhu); visualization, Z.D.; funding acquisitional. L.G. and Z.D. All authors have read and agreed to the published version of the manuscript.

Funding: This research was funded by the Third Xinjiang Scientific Expedition Program (2021xjkk0903), the National Natural Science Foundation of China (31760142) and the Xinjiang Uygur Autonomous Region Graduate Research and Innovation Project (XJ2021G045).

Data Availability Statement: Not applicable.

Acknowledgments: The authors greatly acknowledge the help in soil sampling from Ruixi Li and Yuefeng Li. We appreciate the constructive comments and insightful suggestions from three anonymous reviewers that significantly improved this manuscript. We would also like to thank Novogene (Beijing, China) for analyzing high-throughput sequencing and AJE (www.aje.com, accessed on 14 December 2022) for its linguistic assistance during the preparation of this manuscript.

Conflicts of Interest: The authors declare no conflict of interest.

Appendix A

Table A1. Relative abundance of the dominant bacteria phylum and fungi phylum in different treatments. Different lowercase letters indicate significant differences between treatments ($p < 0.05$).

Taxonomic Groups		CK	ON	Mix-1	Mix-2	Mix-3	IN	One-Way ANOVA	
Bacterial	Proteobacteria	0.209 ab	0.188 b	0.255 a	0.234 ab	0.209 ab	0.224 ab	$F = 2.094$	$p = 0.137$
	Acidobacteriota	0.248 a	0.239 a	0.174 b	0.197 ab	0.197 ab	0.184 ab	$F = 2.020$	$p = 0.148$
	Gemmatimonadota	0.116 bc	0.089 c	0.17 a	0.186 a	0.18 a	0.149 ab	$F = 6.527$	$p = 0.004$
	Actinobacteriota	0.14 abc	0.155 a	0.114 abc	0.101 bc	0.144 ab	0.097 c	$F = 2.986$	$p = 0.056$
	Verrucomicrobiota	0.094 b	0.119 b	0.092 b	0.082 b	0.076 b	0.167 a	$F = 4.98$	$p = 0.011$
Fungal	Basidiomycota	0.805 b	0.486 d	0.424 d	0.921 a	0.823 b	0.665 c	$F = 46.004$	$p = <0.001$
	Ascomycota	0.117 de	0.458 b	0.552 a	0.035 e	0.14 d	0.267 c	$F = 54.469$	$p = <0.001$

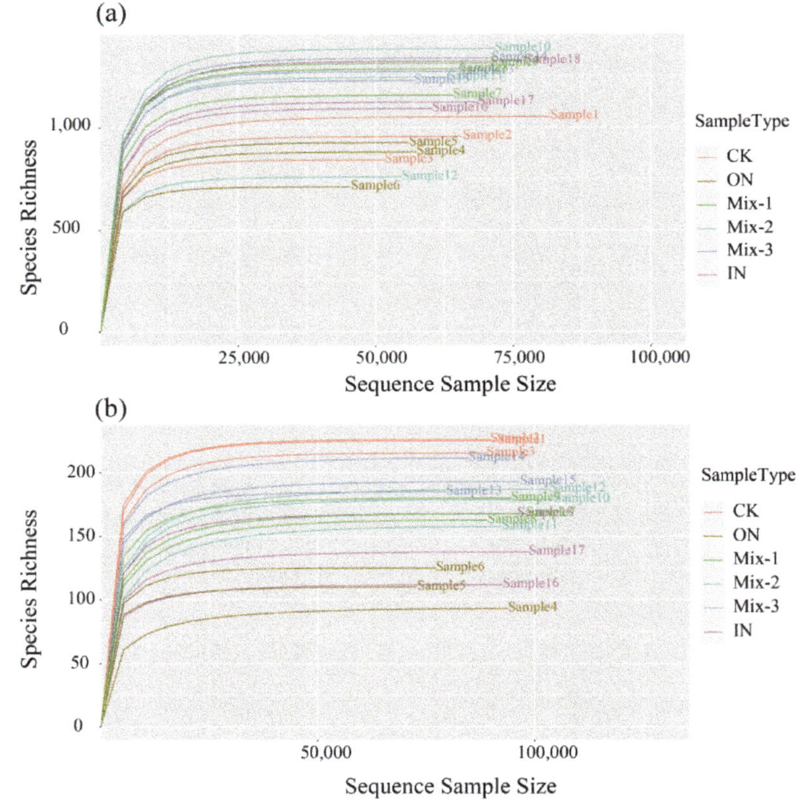

Figure A1. Observed species rarefaction curves of the bacteria (**a**) and fungi (**b**) under different treatments.

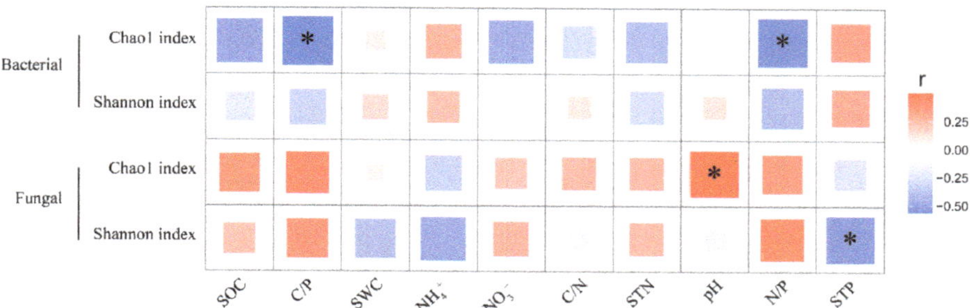

Figure A2. Pearson correlation between alpha diversity and environmental factors. Blue indications a negative correlation, and red a positive correlation, and the strength of color reflects the strength of the correlation. Abbreviations: NO_3^-, nitrate nitrogen; NH_4^+, ammonium nitrogen; STN, total nitrogen; SOC, soil organic carbon; STP, total phosphorus; SWC, soil water content. * $p < 0.05$.

References

1. Kanakidou, M.; Myriokefalitakis, S.; Daskalakis, N.; Fanourgakis, G.; Nenes, A.; Baker, A.R.; Tsigaridis, K.; Mihalopoulos, N. Past, Present and Future Atmospheric Nitrogen Deposition. *J. Atmos. Sci.* **2016**, *73*, 2039–2047. [CrossRef] [PubMed]
2. Mason, R.E.; Craine, J.M.; Lany, N.K.; Jonard, M.; Ollinger, S.V.; Groffman, P.M.; Fulweiler, R.W.; Angerer, J.; Read, Q.D.; Reich, P.B.; et al. Evidence, causes, and consequences of declining nitrogen availability in terrestrial ecosystems. *Science* **2022**, *376*, eabh3767. [CrossRef] [PubMed]
3. Du, E.; Jiang, Y.; Fang, J.Y.; de Vries, W. Inorganic nitrogen deposition in China's forests: Status and characteristics. *Atmos. Env.* **2014**, *98*, 474–482. [CrossRef]
4. Bobbink, R.; Hicks, K.; Galloway, J.; Spranger, T.; Alkemade, R.; Ashmore, M.; Bustamante, M.; Cinderby, S.; Davidson, E.; Dentener, F.; et al. Global assessment of nitrogen deposition effects on terrestrial plant diversity: A synthesis. *Ecol. Appl.* **2010**, *20*, 30–59. [CrossRef] [PubMed]
5. Liu, X.; Duan, L.; Mo, J.; Du, E.; Shen, J.; Lu, X.; Zhang, Y.; Zhou, X.; He, C.; Zhang, F. Nitrogen deposition and its ecological impact in China: An overview. *Env. Pollut.* **2011**, *159*, 2251–2264. [CrossRef] [PubMed]
6. Payne, R.J.; Dise, N.B.; Field, C.D.; Dore, A.J.; Caporn, S.J.M.; Stevens, C.J. Nitrogen deposition and plant biodiversity: Past, present, and future. *Front. Ecol. Env.* **2017**, *15*, 431–436. [CrossRef]
7. Gilliam, F.S. Responses of Forest Ecosystems to Nitrogen Deposition. *Forests* **2021**, *12*, 1190. [CrossRef]
8. Delgado-Baquerizo, M.; Maestre, F.T.; Reich, P.B.; Jeffries, T.C.; Gaitan, J.J.; Encinar, D.; Berdugo, M.; Campbell, C.D.; Singh, B.K. Microbial diversity drives multifunctionality in terrestrial ecosystems. *Nat. Commun.* **2016**, *7*, 10541. [CrossRef]
9. Zhou, Z.H.; Wang, C.K.; Zheng, M.H.; Jiang, L.F.; Luo, Y.Q. Patterns and mechanisms of responses by soil microbial communities to nitrogen addition. *Soil Biol. Biochem.* **2017**, *115*, 433–441. [CrossRef]
10. Waldrop, M.P.; Zak, D.R.; Sinsabaugh, R.L. Microbial community response to nitrogen deposition in northern forest ecosystems. *Soil Biol. Biochem.* **2004**, *36*, 1443–1451. [CrossRef]
11. Zhang, T.; Chen, H.Y.H.; Ruan, H. Global negative effects of nitrogen deposition on soil microbes. *ISME J.* **2018**, *12*, 1817–1825. [CrossRef] [PubMed]
12. Liu, Y.; Tan, X.; Fu, S.; Shen, W. Canopy and Understory Nitrogen Addition Alters Organic Soil Bacterial Communities but Not Fungal Communities in a Temperate Forest. *Front. Microbiol.* **2022**, *13*. 888121. [CrossRef] [PubMed]
13. Moore, J.A.M.; Anthony, M.A.; Pec, G.J.; Trocha, L.K.; Trzebny, A.; Geyer, K.M.; van Diepen, L.T.A.; Frey, S.D. Fungal community structure and function shifts with atmospheric nitrogen deposition. *Glob. Chang. Biol.* **2021**, *27*, 1349–1364. [CrossRef] [PubMed]
14. Zechmeister-Boltenstern, S.; Michel, K.; Pfeffer, M. Soil microbial community structure in European forests in relation to forest type and atmospheric nitrogen deposition. *Plant Soil* **2010**, *343*, 37–50. [CrossRef]
15. Lu, X.; Mao, Q.; Gilliam, F.S.; Luo, Y.; Mo, J. Nitrogen deposition contributes to soil acidification in tropical ecosystems. *Glob. Chang. Biol.* **2014**, *20*, 3790–3801. [CrossRef] [PubMed]
16. Wei, K.; Sun, T.; Tian, J.H.; Chen, Z.H.; Chen, L.J. Soil microbial biomass, phosphatase and their relationships with phosphorus turnover under mixed inorganic and organic nitrogen addition in a *Larix gmelinii* plantation. *For. Ecol. Manag.* **2018**, *422*, 313–322. [CrossRef]
17. Yan, Y.; Sun, X.; Sun, F.; Zhao, Y.; Sun, W.; Guo, J.; Zhang, T. Sensitivity of soil fungal and bacterial community compositions to nitrogen and phosphorus additions in a temperate meadow. *Plant Soil* **2021**, *471*, 477–490. [CrossRef]
18. Allison, S.D.; Czimczik, C.I.; Treseder, K.K. Microbial activity and soil respiration under nitrogen addition in Alaskan boreal forest. *Glob. Chang. Biol.* **2008**, *14*, 1156–1168. [CrossRef]
19. Lilleskov, E.A.; Kuyper, T.W.; Bidartondo, M.I.; Hobbie, E.A. Atmospheric nitrogen deposition impacts on the structure and function of forest mycorrhizal communities: A review. *Env. Pollut.* **2019**, *246*, 148–162. [CrossRef]

20. Li, J.; Sang, C.P.; Yang, J.Y.; Qu, L.R.; Xia, Z.W.; Sun, H.; Jiang, P.; Wang, X.G.; He, H.B.; Wang, C. Stoichiometric imbalance and microbial community regulate microbial elements use efficiencies under nitrogen addition. *Soil Biol. Biochem.* **2021**, *156*, 108207. [CrossRef]

21. Tian, D.S.; Niu, S.L. A global analysis of soil acidification caused by nitrogen addition. *Env. Res. Lett.* **2015**, *10*, 024019. [CrossRef]

22. Zhang, X.M.; Liu, W.; Zhang, G.M.; Jiang, L.; Han, X.G. Mechanisms of soil acidification reducing bacterial diversity. *Soil Biol. Biochem.* **2015**, *81*, 275–281. [CrossRef]

23. Chen, D.M.; Lan, Z.C.; Bai, X.; Grace, J.B.; Bai, Y.F. Evidence that acidification-induced declines in plant diversity and productivity are mediated by changes in below-ground communities and soil properties in a semi-arid steppe. *J. Ecol.* **2013**, *101*, 1322–1334. [CrossRef]

24. Cline, L.C.; Huggins, J.A.; Hobbie, S.E.; Kennedy, P.G. Organic nitrogen addition suppresses fungal richness and alters community composition in temperate forest soils. *Soil Biol. Biochem.* **2018**, *125*, 222–230. [CrossRef]

25. Lim, H.; Jamtgard, S.; Oren, R.; Gruffman, L.; Kunz, S.; Nasholm, T. Organic nitrogen enhances nitrogen nutrition and early growth of *Pinus sylvestris* seedlings. *Tree Physiol.* **2022**, *42*, 513–522. [CrossRef] [PubMed]

26. Tian, J.; Wei, K.; Sun, T.; Jiang, N.; Chen, Z.; Feng, J.; Cai, K.; Chen, L. Different forms of nitrogen deposition show variable effects on soil organic nitrogen turnover in a temperate forest. *Appl. Soil Ecol.* **2022**, *169*, 104212. [CrossRef]

27. Zhang, C.; Zhang, X.Y.; Zou, H.T.; Kou, L.; Yang, Y.; Wen, X.F.; Li, S.G.; Wang, H.M.; Sun, X.M. Contrasting effects of ammonium and nitrate additions on the biomass of soil microbial communities and enzyme activities in subtropical China. *Biogeosciences* **2017**, *14*, 4815–4827. [CrossRef]

28. Qu, W.D.; Han, G.X.; Eller, F.; Xie, B.H.; Wang, J.; Wu, H.T.; Li, J.Y.; Zhao, M.L. Nitrogen input in different chemical forms and levels stimulates soil organic carbon decomposition in a coastal wetland. *Catena* **2020**, *194*, 104672. [CrossRef]

29. Koulympoudi, L.; Papafilippou, A.; Tzanoudaki, M.; Chatzissavvidis, C.; Salamalikis, V. Effect of nitrogen form on trifoliate orange (*Poncirus trifoliata* (L.) Raf.) and sour orange (*Citrus aurantium* L.) plants grown under saline conditions. *J. Plant Nutr.* **2021**, *44*, 2546–2558. [CrossRef]

30. Cornell, S.E. Atmospheric nitrogen deposition: Revisiting the question of the importance of the organic component. *Env. Pollut.* **2011**, *159*, 2214–2222. [CrossRef]

31. Zhang, Y.; Song, L.; Liu, X.J.; Li, W.Q.; Lü, S.H.; Zheng, L.X.; Bai, Z.C.; Cai, G.Y.; Zhang, F.S. Atmospheric organic nitrogen deposition in China. *Atmos. Env.* **2012**, *46*, 195–204. [CrossRef]

32. Moran-Zuloaga, D.; Dippold, M.; Glaser, B.; Kuzyakov, Y. Organic nitrogen uptake by plants: Reevaluation by position-specific labeling of amino acids. *Biogeochemistry* **2015**, *125*, 359–374. [CrossRef]

33. Geisseler, D.; Horwath, W.R.; Joergensen, R.G.; Ludwig, B. Pathways of nitrogen utilization by soil microorganisms—A review. *Soil Biol. Biochem.* **2010**, *42*, 2058–2067. [CrossRef]

34. Guo, P.; Wang, C.Y.; Jia, Y.; Wang, Q.A.; Han, G.M.; Tian, X.J. Responses of soil microbial biomass and enzymatic activities to fertilizations of mixed inorganic and organic nitrogen at a subtropical forest in East China. *Plant Soil* **2011**, *338*, 355–366. [CrossRef]

35. Dong, L.L.; Berg, B.; Sun, T.; Wang, Z.W.; Han, X.G. Response of fine root decomposition to different forms of N deposition in a temperate grassland. *Soil Biol. Biochem.* **2020**, *147*, 107845. [CrossRef]

36. Du, Y.; Guo, P.; Liu, J.; Wang, C.; Yang, N.; Jiao, Z. Different types of nitrogen deposition show variable effects on the soil carbon cycle process of temperate forests. *Glob. Chang. Biol.* **2014**, *20*, 3222–3228. [CrossRef]

37. Zhu, H.; Gong, L.; Luo, Y.; Tang, J.; Ding, Z.; Li, X. Effects of Litter and Root Manipulations on Soil Bacterial and Fungal Community Structure and Function in a Schrenk's Spruce (*Picea schrenkiana*) Forest. *Front. Plant Sci.* **2022**, *13*, 849483. [CrossRef]

38. Zhao, J.J.; Gong, L. Response of Fine Root Carbohydrate Content to Soil Nitrogen Addition and Its Relationship with Soil Factors in a Schrenk (*Picea schrenkiana*) Forest. *J. Plant Growth Regul.* **2021**, *40*, 1210–1221. [CrossRef]

39. Group, C. *China Soil System Classification (Amendment Scheme)*; Agricultural Science and Technology Press of China: Beijing, China, 1995.

40. Bao, S. *Soil Agrochemical Analysis*; China Agricultural Press: Beijing, China, 2000; p. 30. (In Chinese)

41. Zhu, H.; Gong, L.; Ding, Z.; Li, Y. Effects of litter and root manipulations on soil carbon and nitrogen in a Schrenk's spruce (*Picea schrenkiana*) forest. *PLoS ONE* **2021**, *16*, e0247725. [CrossRef]

42. Magoc, T.; Salzberg, S.L. FLASH: Fast length adjustment of short reads to improve genome assemblies. *Bioinformatics* **2011**, *27*, 2957–2963. [CrossRef]

43. Haas, B.J.; Gevers, D.; Earl, A.M.; Feldgarden, M.; Ward, D.V.; Giannoukos, G.; Ciulla, D.; Tabbaa, D.; Highlander, S.K.; Sodergren, E.; et al. Chimeric 16S rRNA sequence formation and detection in Sanger and 454-pyrosequenced PCR amplicons. *Genome Res.* **2011**, *21*, 494–504. [CrossRef] [PubMed]

44. Team, R.C. R: A Language and Environment for Statistical Computing. R Foundation for Statistical Computing: Vienna, Austria. Available online: https://www.R-project.org/ (accessed on 20 September 2022).

45. Yang, X.D.; Anwar, E.; Zhou, J.; He, D.; Gao, Y.C.; Lv, G.H.; Cao, Y.E. Higher association and integration among functional traits in small tree than shrub in resisting drought stress in an arid desert. *Env. Exp. Bot.* **2022**, *201*, 104993. [CrossRef]

46. Zhang, Y.; Guo, Y.; Tang, Z.; Feng, Y.; Zhu, X.; Xu, W.; Bai, Y.; Zhou, G.; Xie, Z.; Fang, J. Patterns of nitrogen and phosphorus pools in terrestrial ecosystems in China. *Earth Syst. Sci. Data* **2021**, *13*, 5337–5351. [CrossRef]

47. Wang, C.Y.; Zhou, J.W.; Liu, J.; Jiang, K.; Du, D.L. Responses of soil N-fixing bacteria communities to *Amaranthus retroflexus* invasion under different forms of N deposition. *Agr. Ecosyst. Env.* **2017**, *247*, 329–336. [CrossRef]
48. Barak, P.; Jobe, B.O.; Krueger, A.R.; Peterson, L.A.; Laird, D.A. Effects of long-term soil acidification due to nitrogen fertilizer inputs in Wisconsin. *Plant Soil* **1997**, *197*, 61–69. [CrossRef]
49. Martikainen, P.J.; Aarnio, T.; Taavitsainen, V.-M.; Päivinen, L.; Salonen, K. Mineralization of carbon and nitrogen in soil samples taken from three fertilized pine stands: Long-term effects. *Plant Soil* **1989**, *114*, 99–106. [CrossRef]
50. Bai, T.S.; Wang, P.; Ye, C.L.; Hu, S.J. Form of nitrogen input dominates N effects on root growth and soil aggregation: A meta-analysis. *Soil Biol. Biochem.* **2021**, *157*, 108251. [CrossRef]
51. Song, L.; Kuang, F.; Skiba, U.; Zhu, B.; Liu, X.; Levy, P.; Dore, A.; Fowler, D. Bulk deposition of organic and inorganic nitrogen in southwest China from 2008 to 2013. *Env. Pollut.* **2017**, *227*, 157–166. [CrossRef]
52. Dong, L.L.; Sun, T.; Berg, B.; Zhang, L.L.; Zhang, Q.Q.; Wang, Z.W. Effects of different forms of N deposition on leaf litter decomposition and extracellular enzyme activities in a temperate grassland. *Soil Biol. Biochem.* **2019**, *134*, 78–80. [CrossRef]
53. Du, E.Z.; Terrer, C.; Pellegrini, A.F.A.; Ahlstrom, A.; van Lissa, C.J.; Zhao, X.; Xia, N.; Wu, X.H.; Jackson, R.B. Global patterns of terrestrial nitrogen and phosphorus limitation. *Nat. Geosci.* **2020**, *13*, 221. [CrossRef]
54. Liu, G.C.; Yan, G.Y.; Huang, B.B.; Sun, X.Y.; Xing, Y.J.; Wang, Q.G. Long-term nitrogen addition alters nutrient foraging strategies of *Populus davidiana* and *Betula platyphylla* in a temperate natural secondary forest. *Eur. J. For. Res.* **2022**, *141*, 307–320. [CrossRef]
55. Vitousek, P.M.; Porder, S.; Houlton, B.Z.; Chadwick, O.A. Terrestrial phosphorus limitation: Mechanisms, implications, and nitrogen-phosphorus interactions. *Ecol. Appl.* **2010**, *20*, 5–15. [CrossRef] [PubMed]
56. Peñuelas, J.; Poulter, B.; Sardans, J.; Ciais, P.; van der Velde, M.; Bopp, L.; Boucher, O.; Godderis, Y.; Hinsinger, P.; Llusia, J.; et al. Human-induced nitrogen–phosphorus imbalances alter natural and managed ecosystems across the globe. *Nat. Commun.* **2013**, *4*, 2934. [CrossRef] [PubMed]
57. Huang, T.; Liu, W.; Long, X.E.; Jia, Y.; Wang, X.; Chen, Y. Different Responses of Soil Bacterial Communities to Nitrogen Addition in Moss Crust. *Front. Microbiol.* **2021**, *12*, 665975. [CrossRef] [PubMed]
58. Nie, Y.; Wang, M.; Zhang, W.; Ni, Z.; Hashidoko, Y.; Shen, W. Ammonium nitrogen content is a dominant predictor of bacterial community composition in an acidic forest soil with exogenous nitrogen enrichment. *Sci. Total Env.* **2018**, *624*, 407–415. [CrossRef]
59. Wang, J.; Shi, X.; Zheng, C.; Suter, H.; Huang, Z. Different responses of soil bacterial and fungal communities to nitrogen deposition in a subtropical forest. *Sci. Total Env.* **2021**, *755*, 142449. [CrossRef]
60. Wang, Z.H.; Wang, Z.R.; Li, T.P.; Wang, C.; Dang, N.; Wang, R.Z.; Jiang, Y.; Wang, H.Y.; Li, H. N and P fertilization enhanced carbon decomposition function by shifting microbes towards an r-selected community in meadow grassland soils. *Ecol. Indic.* **2021**, *132*, 108306. [CrossRef]
61. Turlapati, S.A.; Minocha, R.; Bhiravarasa, P.S.; Tisa, L.S.; Thomas, W.K.; Minocha, S.C. Chronic N-amended soils exhibit an altered bacterial community structure in Harvard Forest, MA, USA. *FEMS Microbiol. Ecol.* **2013**, *83*, 478–493. [CrossRef]
62. Schwede, D.B.; Simpson, D.; Tan, J.; Fu, J.S.; Dentener, F.; Du, E.; deVries, W. Spatial variation of modelled total, dry and wet nitrogen deposition to forests at global scale. *Env. Pollut.* **2018**, *243*, 1287–1301. [CrossRef]
63. Yang, Y.; Cheng, H.; Gao, H.; An, S.S. Response and driving factors of soil microbial diversity related to global nitrogen addition. *Land Degrad. Dev.* **2020**, *31*, 190–204. [CrossRef]
64. Delgado-Baquerizo, M.; Reich, P.B.; Khachane, A.N.; Campbell, C.D.; Thomas, N.; Freitag, T.E.; Abu Al-Soud, W.; Sorensen, S.; Bardgett, R.D.; Singh, B.K. It is elemental: Soil nutrient stoichiometry drives bacterial diversity. *Env. Microbiol.* **2017**, *19*, 1176–1188. [CrossRef] [PubMed]
65. Liu, Z.F.; Fu, B.J.; Zheng, X.X.; Liu, G.H. Plant biomass, soil water content and soil N:P ratio regulating soil microbial functional diversity in a temperate steppe: A regional scale study. *Soil Biol. Biochem.* **2010**, *42*, 445–450. [CrossRef]
66. Lammel, D.R.; Barth, G.; Ovaskainen, O.; Cruz, L.M.; Zanatta, J.A.; Ryo, M.; de Souza, E.M.; Pedrosa, F.O. Direct and indirect effects of a pH gradient bring insights into the mechanisms driving prokaryotic community structures. *Microbiome* **2018**, *6*, 106. [CrossRef] [PubMed]
67. Rousk, J.; Brookes, P.C.; Baath, E. Investigating the mechanisms for the opposing pH relationships of fungal and bacterial growth in soil. *Soil Biol. Biochem.* **2010**, *42*, 926–934. [CrossRef]
68. Leff, J.W.; Jones, S.E.; Prober, S.M.; Barberan, A.; Borer, E.T.; Firn, J.L.; Harpole, W.S.; Hobbie, S.E.; Hofmockel, K.S.; Knops, J.M.; et al. Consistent responses of soil microbial communities to elevated nutrient inputs in grasslands across the globe. *Proc. Natl. Acad. Sci. USA* **2015**, *112*, 10967–10972. [CrossRef]
69. Dai, Z.; Su, W.; Chen, H.; Barberan, A.; Zhao, H.; Yu, M.; Yu, L.; Brookes, P.C.; Schadt, C.W.; Chang, S.X.; et al. Long-term nitrogen fertilization decreases bacterial diversity and favors the growth of Actinobacteria and Proteobacteria in agro-ecosystems across the globe. *Glob. Chang. Biol.* **2018**, *24*, 3452–3461. [CrossRef]
70. Krumins, J.A.; Dighton, J.; Gray, D.; Franklin, R.B.; Morin, P.J.; Roberts, M.S. Soil microbial community response to nitrogen enrichment in two scrub oak forests. *For. Ecol. Manag.* **2009**, *258*, 1383–1390. [CrossRef]
71. Rousk, J.; Bååth, E.; Brookes, P.C.; Lauber, C.L.; Lozupone, C.; Caporaso, J.G.; Knight, R.; Fierer, N. Soil bacterial and fungal communities across a pH gradient in an arable soil. *ISME J.* **2010**, *4*, 1340–1351. [CrossRef]
72. Bahram, M.; Hildebrand, F.; Forslund, S.K.; Anderson, J.L.; Soudzilovskaia, N.A.; Bodegom, P.M.; Bengtsson-Palme, J.; Anslan, S.; Coelho, L.P.; Harend, H.; et al. Structure and function of the global topsoil microbiome. *Nature* **2018**, *560*, 233–237. [CrossRef]

73. Cui, J.; Zhu, R.; Wang, X.; Xu, X.; Ai, C.; He, P.; Liang, G.; Zhou, W.; Zhu, P. Effect of high soil C/N ratio and nitrogen limitation caused by the long-term combined organic-inorganic fertilization on the soil microbial community structure and its dominated SOC decomposition. *J. Environ. Manag.* **2022**, *303*, 114155. [CrossRef]
74. Dimitriu, P.A.; Grayston, S.J. Relationship between soil properties and patterns of bacterial beta-diversity across reclaimed and natural boreal forest soils. *Microb. Ecol.* **2010**, *59*, 563–573. [CrossRef] [PubMed]
75. Jones, R.T.; Robeson, M.S.; Lauber, C.L.; Hamady, M.; Knight, R.; Fierer, N. A comprehensive survey of soil acidobacterial diversity using pyrosequencing and clone library analyses. *ISME J.* **2009**, *3*, 442–453. [CrossRef] [PubMed]
76. Shen, C.C.; Xiong, J.B.; Zhang, H.Y.; Feng, Y.Z.; Lin, X.G.; Li, X.Y.; Liang, W.J.; Chu, H.Y. Soil pH drives the spatial distribution of bacterial communities along elevation on Changbai Mountain. *Soil Biol. Biochem.* **2013**, *57*, 204–211. [CrossRef]
77. Wang, C.; Liu, D.W.; Bai, E. Decreasing soil microbial diversity is associated with decreasing microbial biomass under nitrogen addition. *Soil Biol. Biochem.* **2018**, *120*, 126–133. [CrossRef]
78. Ramirez, K.S.; Lauber, C.L.; Knight, R.; Bradford, M.A.; Fierer, N. Consistent effects of nitrogen fertilization on soil bacterial communities in contrasting systems. *Ecology* **2010**, *91*, 3463–3470; discussion 3503–3414. [CrossRef] [PubMed]
79. Fierer, N.; Jackson, R.B. The diversity and biogeography of soil bacterial communities. *Proc. Natl. Acad. Sci. USA* **2006**, *103*, 626–631. [CrossRef]
80. Wang, C.; Lu, X.K.; Mori, T.; Mao, Q.G.; Zhou, K.J.; Zhou, G.Y.; Nie, Y.X.; Mo, J.M. Responses of soil microbial community to continuous experimental nitrogen additions for 13 years in a nitrogen-rich tropical forest. *Soil Biol. Biochem.* **2018**, *121*, 103–112. [CrossRef]
81. Zhou, J.; Jiang, X.; Zhou, B.K.; Zhao, B.S.; Ma, M.C.; Guan, D.W.; Li, J.; Chen, S.F.; Cao, F.M.; Shen, D.L.; et al. Thirty four years of nitrogen fertilization decreases fungal diversity and alters fungal community composition in black soil in northeast China. *Soil Biol. Biochem.* **2016**, *95*, 135–143. [CrossRef]

Article

Changes in Soil Ectomycorrhizal Fungi Community in Oak Forests along the Urban–Rural Gradient

Hongyan Shen [1], Baoshan Yang [1], Hui Wang [1,*], Wen Sun [1], Keqin Jiao [1] and Guanghua Qin [2]

[1] School of Water Conservancy and Environment, University of Jinan, Jinan 250022, China; ecogroup@126.com (H.S.); stu_yangbs@ujn.edu.cn (B.Y.); 15192764883@163.com (W.S.); 15244147901@163.com (K.J.)

[2] Office of Academic Research, Shandong Academy of Forestry Sciences, Jinan 250000, China; guanghuaqin@163.com

* Correspondence: hwang_118@163.com

Abstract: The ectomycorrhizal fungi communities of forests are closely correlated with forest health and ecosystem functions. To investigate the structure and composition of ectomycorrhizal fungi communities in oak forest soil and their driving factors along the urban–rural gradient, we set up a *Quercus acutissima* forest transect and collected samples from the center to the edge of Jinan city (urban, suburban, rural). The results showed that the ectomycorrhizal fungal community composition at the phyla level mainly included Basidiomycota and Ascomycota in three sites. At the genus level, the community compositions of ectomycorrhizal fungi, along the urban–rural gradient, exhibited significant differences. *Inocybe, Russula, Scleroderma, Tomentella, Amanita* and *Tuber* were the dominant genera in these *Quercus acutissima* forests. Additionally, the diversity of ectomycorrhizal fungi was the highest in rural *Quercus acutissima* forest, followed by urban and suburban areas. Key ectomycorrhizal fungi species, such as *Tuber, Russula* and Scrdariales, were identified among three forests. We also found that pH, soil organic matter and ammonium nitrogen were the main driving factors of the differences in ectomycorrhizal fungi community composition and diversity along the urban–rural gradient. Overall, the differences in composition and diversity in urban–rural gradient forest were driven by the differences in soil physicochemical properties resulting from the forest location.

Keywords: *Quercus acutissima* forest; soil physicochemical properties; community structure; LEfSe analysis; driving factors

Citation: Shen, H.; Yang, B.; Wang, H.; Sun, W.; Jiao, K.; Qin, G. Changes in Soil Ectomycorrhizal Fungi Community in Oak Forests along the Urban–Rural Gradient. *Forests* 2022, 13, 675. https://doi.org/10.3390/f13050675

Academic Editor: Christopher Gough

Received: 21 March 2022
Accepted: 25 April 2022
Published: 27 April 2022

Publisher's Note: MDPI stays neutral with regard to jurisdictional claims in published maps and institutional affiliations.

1. Introduction

Ectomycorrhizal (ECM) fungi are one of the taxa of fungi that are among the most abundant and widespread in forests [1]. There are about 20,000~25,000 ECM fungi in the world, and they can associate with 6000 species of trees and shrubs mainly distributed in tropical and temperate forests [2–4]. In particular, the tree species from Fagaceae, Betulaceae, Pinaceae and Legumes are the dominant symbiotic plants of ECM fungi [1]. These plants provide carbon synthesized by photosynthesis for the fungi, while ECM fungi supply nutrients to the host plants, such as nitrogen and phosphorus [5,6]. Moreover, ECM fungi play critical roles in global carbon cycles by secreting extracellular enzymes and organic acids to facilitate the decomposition of soil organic matter [7]. In addition, previous studies have confirmed that ECM fungi play a significant role in the enhancement of plant tolerance to adverse environments [8] and the sustainment of forest ecosystem diversity and functions [1,9]. Generally, ECM diversity and community structure are considered to be the important indicators to evaluate the health and stability of forest ecosystems [10]. In this sense, studies on ECM fungi diversity and its driving factors can provide us with a better understanding of ecological processes such as geochemical element cycling, plant nutrient acquisition, sustainable forest development and global environmental change [1,7].

The diversity and richness of ECM fungi are affected by different soil conditions [11,12]. It has been found that higher levels of organic matter and more phosphorus available in soil usually inhibit ECM fungi occurrence [13,14]. Lower pH has been found to be conducive to increases in ECM diversity and richness [15]. ECM communities are significantly different in the cold temperate zone, warm temperate zone and Mediterranean climate zone, indicating that climatic conditions are important driving factors for ECM composition and structure [11]. Furthermore, on the basis of their specificity to host plants, some ECM fungi are able to associate with specific tree species, which is another important factor affecting ECM fungi structure and diversity in different forest ecosystems [1]. Therefore, understanding the influence of environmental factors on the ECM community is critical for the exploitation and protection of ECM fungi function in forest ecosystems.

Oak (*Quercus* Spp.) is one of the dominant deciduous tree genera in boreal forests around the world, and it has great ecological and economic value for environmental protection, seedling establishment and wood yield [16]. Its roots rely on obligate symbiosis with ECM fungi. It cannot survive and establish itself without ECM fungi [17]. Oak roots can improve nutrient absorption and plant adaptability once they are associated with ECM fungi, especially in adverse conditions [16,17]. Thus, the composition and diversity of ECM fungi in oak forests have been paid much attention in previous studies. He et al. [18] observed that the compositions of ECM fungal communities varied with the stand ages of oak forests, but their diversities were broadly similar across the 20-, 30- and 40-year-old *Quercus Mongolica* forests. Furthermore, the ECM fungal community structure of oak forest soil was obviously different in different habitats. The soil fungal biodiversity had a great impact on hosts, especially at the nursery stage, where it could influence the quality of the plating material [19]. Jin et al. [20] found that the ECM fungi in oak nurseries were mainly composed of Boletales while Agaricales dominated in the Guizhou wild oak forest. Moreover, previous studies have shown that ECM fungal diversity and richness in oak forests significantly change with seasonal shifts and that they generally present the highest abundance in summer [20,21]. Although the changes in ECM fungal communities have been extensively studied, there is little information on the composition of ECM fungal communities and its driving factors in *Quercus acutissima* forests along the urban–rural gradient. As an effective ecological research method, the rural–urban gradient method has been widely used to study ecological problems in various regions of the world [22,23]. Here, we investigated the ECM community in *Quercus acutissima* forests along the urban–rural gradient in Jinan city, northern China. The aims were to: (1) explore the changes in ECM community composition and diversity along the urban–rural gradient and (2) reveal the effects of soil physicochemical properties on ECM community compositions and diversity. Our results provide practical guidance for urban and suburban forest ecosystem management.

2. Materials and Methods

2.1. Sample Site

Jinan city, the capital of Shandong Province, is located between 34°46′–37°32′ N and 116°13′–117°58′ E, in the middle and lower reaches of the Yellow River. The terrain in the south is higher than that in the north in this area. Jinan has a typical temperate monsoon climate characterized by distinct seasons, a mean annual temperature of 13.8 °C and mean precipitation of 650–700 mm. Jinan city has abundant forest resources, dominated by *Quercus acutissima*, *Platycladus orientalis*, *Robinia pseudoacacia* and *Pinus thunbergii Parl* [24]. The urbanization level in Jinan reached 73.46% as of 2021 [25]. The typical urban areas, ecologically sensitive areas and southern mountain control areas were designed in Jinan along with the development of the economy. Here, we focused on the *Quercus acutissima* forest, which is a typical type of vegetation in Jinan. The study areas were Quancheng Park (36°38′ N, 117°38′ E) in the urban area, Liubu Forest Farm (36°27′ N, 117°11′ E) in the suburbs and Yaoxiang Forest Park (36°12′ N, 117°4′ E) in the countryside. The geographic information for the study areas is shown in Figure 1.

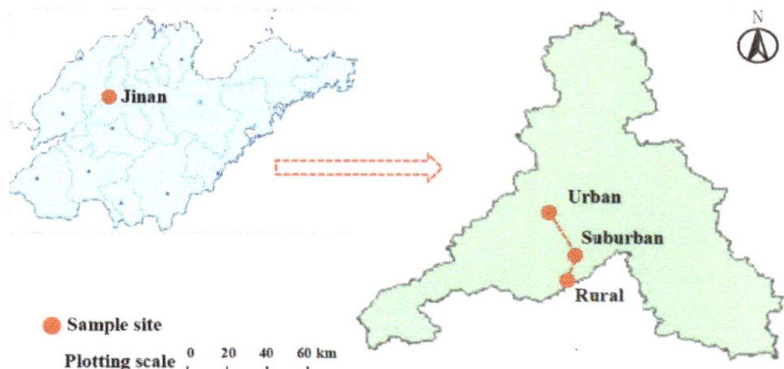

Figure 1. Maps of the sampling site and study area location.

2.2. Sample Collection and Treatment Process

In October 2020, three independent sample plots (20 m × 20 m each; the distances were more than 1000 m from each other) were set up in the urban, suburban and rural *Quercus acutissima* forests of Jinan city, respectively. Five healthy trees with similar tree heights and diameters at breast height were randomly selected in each plot. The distances between the trees were more than 10 m. Humus such as dead branches and leaves was removed from the soil surface, and then four soil cores were collected from the east, west, south and north directions around each tree with 3.5 cm diameter soil augers at 20 cm depths. All the soil samples were evenly combined into one in each plot and a total of nine composite soil samples were thus obtained from these urban, suburban and rural *Quercus acutissima* forests (three replicated samples per forest × three forests). Afterward, the collected samples were stored in cryogenic ice boxes and quickly transported back to the laboratory. The fresh soil samples were sieved through a 2 mm sieve after roots and residues were removed. After being mixed evenly, samples were divided into two parts: one was stored at 4 °C for the determination of soil physicochemical properties and the other was frozen at −80 °C until molecular testing of ECM fungi.

2.3. Soil Physicochemical Properties Determination

The soil pH (water: soil = 1: 2.5) was measured using an UltraBasic pH meter (Denver Instruments, UB-10, Bohemia, NY, USA) and conductivity (EC) (water: soil = 1:5) was measured using a portable conductivity meter. Microbial biomass carbon (MBC) and microbial biomass nitrogen (MBN) in the soil samples were determined by chloroform fumigation. [26]. Soil organic matter (SOM) was determined using the H_2SO_4-K_2CrO_7 oxidation method [27]. Total nitrogen (TN) was analyzed using an automatic Kjeldahl apparatus (KDY-9820, Tongrunyuan, Beijing, China) [27]. Soil ammonium nitrogen (NH_4^+-N) and nitrate nitrogen (NO_3^--N) in the KCl extraction were measured using a UV-visible spectrophotometer (UV-5200PC, Yuanxi, Shanghai, China) [28]. The ratio of C/N in the soil was calculated based on soil organic carbon and total nitrogen contents. Soil available phosphorus (AP) and available potassium (AK) were determined with atomic absorption spectrometers (SHIMADIV AA-7000, Shimadzu, Tokyo, Japan) after being extracted with 0.5 mol/L $NaHCO_3$ and 1 mol/L CH_3COONH_4 (pH = 7), respectively [29].

2.4. Soil DNA Extraction, PCR Amplification and Sequencing

DNA extraction, PCR amplification and high-throughput sequencing of the fungal ITS sequences of the soil samples were completed by Hangzhou LC-Bio Technology Co., Ltd. DNA extraction was performed using an OMEGA Soil DNA Kit (OMEGA, Norcross, GA, USA) according to the manufacturer's instructions. The ITS1 and ITS2 regions were used to identify the fungal species, and the analysis was carried out with ITS1FI2 (5′-GTGARTCATCGAATCTTTG-3′), ITS2 (5′-TCCTCCGCTTATTGATATGC-3′), F (5′-GAACCWGCGGARGGATCA-3′) and R (5′-GCTGCGTTCTTCATCGATGC-3′) primers.

PCR amplification was performed in a reaction mixture with a total volume of 25 µL containing 25 ng of template DNA, 12.5 µL PCR Premix, 2.5 µL of each primer and PCR-grade water to adjust the volume. The PCR conditions to amplify the ITS fragments consisted of an initial denaturation at 98 °C for 30 s; 32 cycles of denaturation at 98 °C for 10 s, annealing at 54 °C for 30 s and extension at 72 °C for 45 s; and then final extension at 72 °C for 10 min. The PCR products were confirmed with 2% agarose gel electrophoresis. The PCR products were purified with AMPure XT beads (Beckman Coulter Genomics, Danvers, MA, USA) and quantified with Qubit (Invitrogen, Carlsbad, CA, USA). The amplicon pools were prepared for sequencing and the size and quantity of the amplicon library were assessed with an Agilent 2100 Bioanalyzer (Agilent, Palo Alto, CA, USA) and the Library Quantification Kit for Illumina (Kapa Biosciences, Woburn, MA, USA), respectively. The libraries were sequenced on a NovaSeq PE250 platform. Samples were sequenced on an Illumina NovaSeq platform according to the manufacturer's recommendations, provided by LC-Bio. QIIME software was used to divide the OTU of Illumina Miseq sequencing data. The RDP database and UNITE database (https://unite.ut.ee/ (accessed on 1 March 2021)) were applied to classify and annotate ECM fungi, and Mothur software was used to calculate the Chao1 index, Shannon index and Simpson index.

2.5. Data Analysis

IBM SPSS Statistics 23.0 (SPSS Inc., Chicago, IL, USA) was used for one-way analysis of variance (ANOVA) and multiple comparison to evaluate the significant differences in the different forest soils ($p < 0.05$). Origin 2021 (OriginLab Inc., Northampton, MA, USA) was used to draw a histogram of community relative abundance at the order and genus levels (others < 0.01). Redundancy analysis (RDA) was performed with CANOCO 5.0 to examine the habitat differences in key ECM fungi species. We utilized linear discriminant analysis effect size (LEfSe) on the website of LC-Bio Technology Co., Ltd., to identify the ECM fungi that might explain the differences among the three forests along the urban–rural gradient (LDA > 4, $p < 0.05$). In addition, Pearson's correlation analysis was applied with Origin 2021 to examine the relative effects of ECM fungi characteristics (diversity and richness) and soil factors (pH, SOM, TN, NH_4^+-N, NO_3^--N, AP, AK, EC).

3. Results

3.1. Differences in Soil Physicochemical Properties

There were significant differences in the soil physicochemical properties among the three sampling areas along the urban–rural gradient in Jinan (Table 1). Specifically, the pH value and the contents of NH_4^+-N, AK and AP in the urban forest soil were significantly higher than those in the suburban and rural forests ($p < 0.05$), whereas TN and NO_3^--N in the urban forest were clearly lower than those in suburban and rural forests, and the difference between suburban and rural forests was not significant. In addition, the content of SOM in the suburban forest soil was 2.50 and 2.19 times higher than that in urban and rural forests, respectively. No significant differences in C/N among any of the sampling sites were observed.

Table 1. Soil physicochemical properties in *Quercus acutissima* forests along the urban–rural gradient.

Sampling Site	pH	SOM ($g \cdot kg^{-1}$)	TN ($g \cdot kg^{-1}$)	NH_4^+-N ($mg \cdot kg^{-1}$)	NO_3^--N ($mg \cdot kg^{-1}$)	AP ($g \cdot kg^{-1}$)	AK ($g \cdot kg^{-1}$)	C/N
Urban	7.87 ± 0.03 a	10.88 ± 4.13 b	2.67 ± 0.09 b	7.28 ± 0.03 b	11.65 ± 1.59 a	17.92 ± 1.11 a	297.61 ± 9.38 a	3.26 ± 1.28 a
Suburban	4.88 ± 0.11 b	27.16 ± 9.88 a	2.92 ± 0.03 a	8.59 ± 0.19 a	3.81 ± 0.68 b	5.22 ± 0.06 b	242.02 ± 15.24 b	5.44 ± 1.05 a
Rural	4.50 ± 0.31 b	12.40 ± 8.40 b	2.89 ± 0.01 a	8.81 ± 0.13 a	3.48 ± 0.20 b	4.42 ± 0.10 b	228.14 ± 9.80 b	2.50 ± 1.70 a

Different letters in each column indicate significant differences in soil samples ($p < 0.05$). The values in the table represent the mean ± standard error.

3.2. MBC and MBN Contents in the Quercus acutissima Forest Soil

Both the MBC and MBN in the forest soil exhibited significant differences in *Quercus acutissima* forests along the urban–rural gradient (Figure 2). Generally, the average content of MBC in urban, suburban and rural soil was 84.21, 53.68 and 205.96 mg·kg^{-1}, respectively, showing a descending order of rural > urban > suburban. With the same trend as the MBC, the average content of MBN in urban, suburban and rural forest soil was 18.18, 10.78 and 48.15 mg·kg^{-1}, respectively.

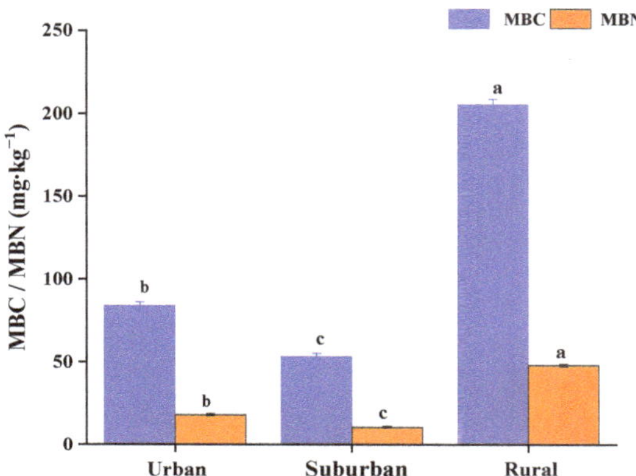

Figure 2. MBC and MBN contents in *Quercus acutissima* forest soil along the urban–rural gradient. Different letters in each column indicate significant differences in soil samples ($p < 0.05$).

3.3. The Change in ECM Fungi Diversity

All the Good's coverage estimators were greater than 99%, which suggested that there was an overall good sampling (Table 2). The ECM fungal community diversity in rural forest soil was the highest, with a Shannon index of 2.76 and a Simpson index of 0.92, followed by urban forest (2.54 and 0.89) and suburban forest (2.12 and 0.83). However, Chao1 values were broadly similar in the forest soil of urban, suburban and rural forests, which tat indicated habitat changes had little influence on the ECM fungi richness along the urban–rural gradient. This also showed that the effects of the urban–rural gradient on ECM fungi diversity and abundance were significantly different.

Table 2. Community diversity indices of ECM fungi in *Quercus acutissima* forest soil along the urban–rural gradient.

Sampling Site	Shannon Index	Simpson Index	Chao1 Index	Good's Coverage
Urban	2.54 ± 0.06 b	0.89 ± 0.01 b	58.00 ± 1.76 a	0.992± 0.002 a
Suburban	2.12 ± 0.04 c	0.83 ± 0.01 c	63.00 ± 10.50 a	0.993± 0.001 a
Rural	2.76 ± 0.02 a	0.92 ± 0.00 a	65.72 ± 2.97 a	0.993± 0.001 a

Different letters in each column indicate significant differences in soil samples ($p < 0.05$). The values in the table represent the mean ± standard error.

3.4. ECM Fungal Community Compositions Vary with Urban–Rural Gradient

A total of 282 ECM fungal OTUs were retrieved from the ITS sequence of soil samples collected from *Quercus acutissima* forests in Jinan along the urban–rural gradient. These ECM fungi were distributed in 3 phyla, 5 classes, 11 orders and 26 genera. At the phyla level, the most dominant were Basidiomycota (81.11%), followed by Ascomycota (13.83%) and Mucoromycota (1.06%). Agaricomycetes were the most dominant guild at the class level, accounting for up to 86.53%, 99.94% and 72.69% of the total community in urban, suburban and rural sampling sites, respectively. Sordariomycetes were the second class, accounting for 0.56%, 0.12% and 27.57% in urban, suburban and rural forests, respectively. Pezizomycetes, Eurotiomycetes and Endogonomycetes were occasionally detected, implying they were the minor classes.

Major differences were observed in ECM fungal community compositions along the urban–rural gradient (Figure 3). At the order level, Agaricales had the highest relative abundance in urban and rural *Quercus acutissima* forests (Figure 3a), which indicates that it can survive in broad habitats and plays crucial roles in sustaining ecosystem balance [30]. Russulales were extensively found in these *Quercus acutissima* forests along the urban–rural gradient, especially in suburban areas where the highest proportion was 56.02%, implying that Russulales may play an important role in the growth and health of oak trees in the sensitive areas disturbed by human activities. The top 13 genera of ECM fungi in *Quercus acutissima* forest soil are shown in Figure 3b. *Inocybe* was the dominant genus in the urban site and its relative abundance was 37.87%, followed by 5.49% for the rural site and 2.84% for the suburban site. However, the relative abundance of *Russula* was highest in the suburban area but lowest in the urban area. High proportions of *Amanita* were observed in the soil samples from the rural area, whereas *Amanita* was almost absent in urban and suburban areas. *Tomentella* was also one of the representative genera in the soil of *Quercus acutissima* forest. The order of its relative abundance in the study sites along the urban–rural gradient was suburban > urban > rural. Overall, the dominant ECM fungi in *Quercus acutissima* forest soil showed remarkable differences along the urban–rural gradient.

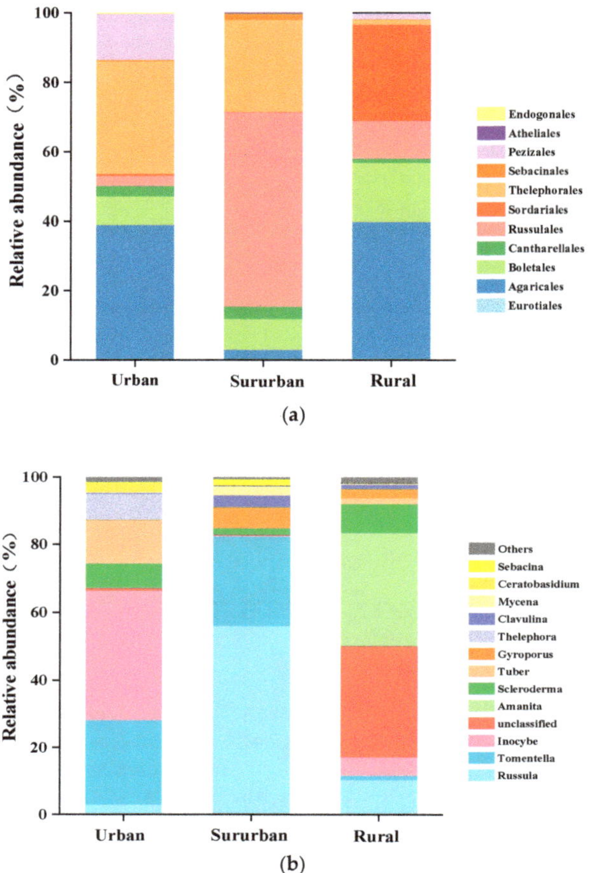

Figure 3. The relative abundance of ECM fungi at the order (**a**) and genus (**b**) levels in the soil of *Quercus acutissima* forests along the urban–rural gradient.

3.5. LEfSe Analysis of ECM Fungi Community

As shown in Figure 4, remarkable differences in key ECM fungi taxa were observed among the urban, suburban and rural *Quercus acutissima* forest soils. Specifically, *Tuber* and an unclassified genus from Thelephoraceae were two distinct ECM fungi members explaining habitat variations that showed significant differences in relative abundance in the urban soil. Meanwhile, *Russula* was the taxon of ECM fungi present in the suburban soil. In addition, there were two characteristic groups of ECM fungi with significant differences in relative abundance in the rural site compared to the urban and suburban sites. They were an unclassified genus from Sordariales and an unclassified genus from Thelephoraceae, respectively. The LDA value also further indicated that *Tuber*, *Russula* and Sordariales were the key groups in responses to more pronounced habitat shifts in the urban, suburban and rural *Quercus acutissima* forest soils, respectively.

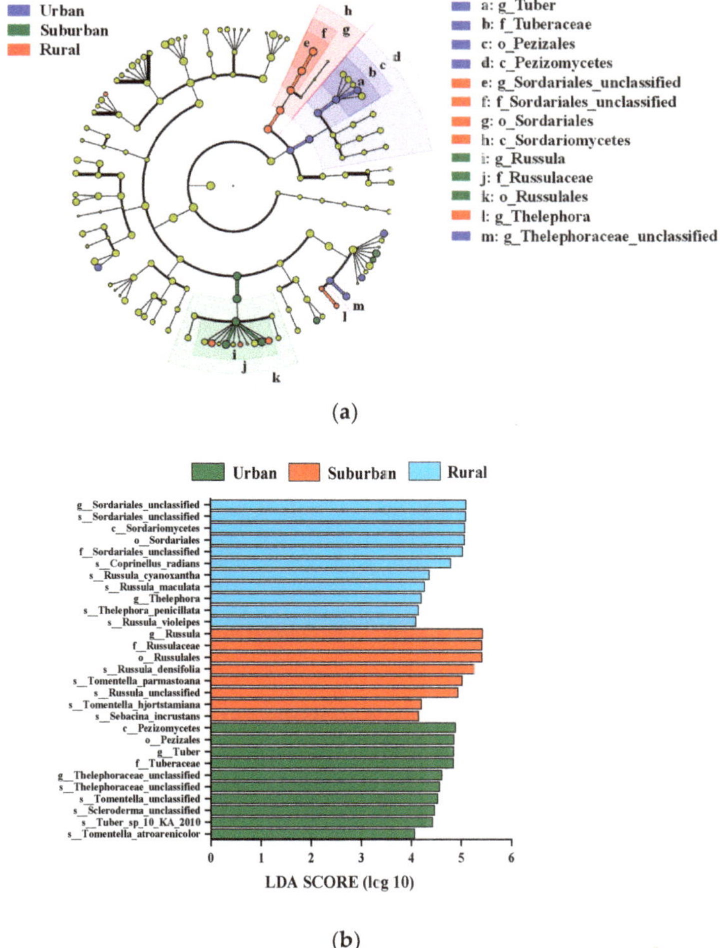

Figure 4. Cladogram (**a**) and LDA distribution histogram (**b**) for ECM fungi in the soil of *Quercus acutissima* forests along the urban–rural gradient.

3.6. Driving Factors of ECM Fungal Composition

Redundancy analysis confirmed that soil physicochemical properties had the greatest effect on dominant ECM fungi at the genus level (top six relative abundances) (Figure 5). The first axis and second axis explain the variations of 54.34% and 27.56%, respectively. The dominant *Russula* exhibited a strongly positive relationship with SOC and C/N, while *Scleroderma* was negatively correlated with SOC and C/N, indicating that soil carbon could drive much more variation in *Russula* and *Scleroderma*. *Inocybe* and *Tuber* showed a remarkably positive relationship with NO_3^--N, AK, AP and pH, but a strongly negative relationship with NH_4^+-N and TN. In addition, *Tomentella* weakly positively correlated with SOC, NO_3^--N, AK and AP. *Amanita* significantly positively correlated with MBC and MBN, but weakly related with NH_4^+-N and TN. The three sampling sites for urban, suburban and rural forests were distributed in different quadrants, and the three sampling plots in the same area showed obvious clustering phenomena, implying that the ECM fungal community composition differed significantly in the *Quercus acutissima* forests along the urban–rural gradient (Figure 3).

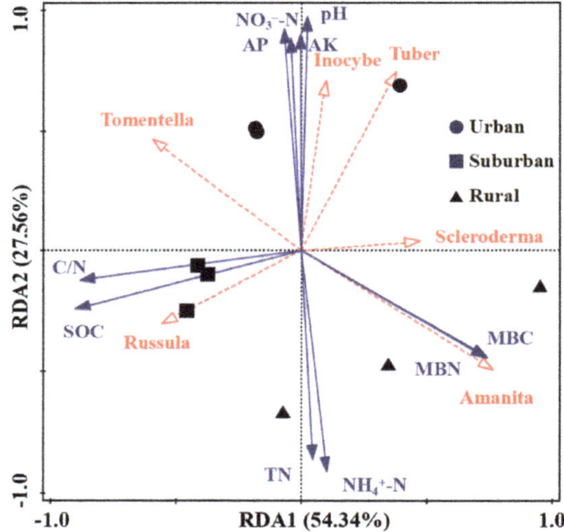

Figure 5. RDA of dominant ECM fungi and soil physicochemical properties in *Quercus acutissima* forests. The solid and dashed arrows represent ECM fungi and soil properties.

3.7. Relationships of ECM Fungal Diversity with Soil Factors

Among all the soil factors examined (Figure 6), the diversity indices (Shannon index and Simpson index) and richness index (Chao1 index) were strongly positively related to TN, NH_4^+-N, MBC and MBN, but negatively related to NO_3^--N, AP and AK, which was consistent with the research conclusions of Erlandson et al. [11] and He et al. [18]. Moreover, NH_4^+-N was negatively correlated with AP and AK, while NO_3^--N was positively correlated with AP and AK. It is worth noting that SOM and TN showed positive relationships with ECM fungal diversity, but negative relationships with ECM fungal richness, suggesting that the higher SOM and TN contents in soil were probably associated with the higher ECM fungi diversity.

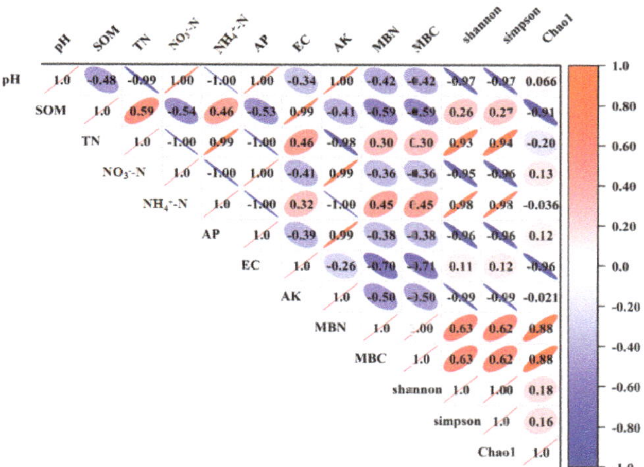

Figure 6. Pearson correlation heat map of soil physicochemical properties and ECM fungal diversity. Red is positive, blue is negative; the darker color means the correlation is stronger.

4. Discussion

4.1. Diversity and Composition of ECM Fungal Community in the Quercus acutissima Forest Soil along the Urban–Rural Gradient

ECM fungi are one of the most important functional phyla in the forest soil ecosystem [31–33]. *Quercus acutissima* is a typical tree species associated with ECM fungi and can form a huge mycelium network to participate in the nutrient cycle and energy metabolism of the host [21]. ECM fungal community composition and diversity are commonly used to evaluate the stability of forest ecosystems because their changes are strongly related to soil health [10,34]. In this study, ECM fungal diversity exhibited more pronounced shifts along the urban–rural gradient. It was obvious that there were minimum levels of MBC and MBN in the suburban forest soil, indicating that microorganism growth was significantly inhibited in this ecotone. The contents of TN, NH_4^+-N and SOM were higher in the soil of the suburban *Quercus acutissima* forest than in urban and rural forests, whereas the Shannon index, Simpson index and MBC and MBN contents of ECM fungi were the lowest, implying that the ECM fungi were not apt to occur in more fertile soil [35]. This finding could have been related to the higher contents of carbon and nitrogen in the suburban forest soil, where ECM fungi were not necessary for the host to obtain more nutrients. Moreover, the decrease in ECM fungi diversity also resulted in a weaker decomposition effect on the soil organic matter or litter [25,26]. Additionally, the ECM fungal diversity and soil microbial biomass in the urban *Quercus acutissima* forest were significantly lower than in rural forest, which implied that human activities in urban areas likely had a negative influence on ECM fungal diversity and even on other kinds of microbial diversity [36]. Previous studies have also shown that lower abundance and diversity of soil microorganisms can result in the limitation of community function. Thus, conservation of ECM diversity is of great significance for maintaining the function of urban forest ecosystems [1].

Key species are generally critical indicators of environment conditions and community functions [36]. This study of ECM fungal community composition in forests could provide a basis for the exploration of the ectomycorrhizal symbiotic mechanism. Earlier studies of ECM fungal communities were only based on morphological and anatomical observations [1,37]. In recent years, use of molecular biology methods has provided new insights into the below-ground ECM fungal community and a more precise approach to the study of ECM fungal diversity [1]. Here, the ECM fungal community compositions in *Quercus acutisana* forest soil were investigated in different habitats (urban, suburban and rural forests) in Jinan city. Some genera, such as *Inocybe*, *Russula* and *Tomentella*, were found to be dominant, which was similar to the results reported in other oak forests [18]. It was confirmed that these taxa may have important ecological significance in terms of improving the environmental adaptability of trees due to their high genetic diversity and probable functional diversity [38]. *Inocybe, Russula, Tomentella, Amanita* and *Scleroderma* belonging to Basidiomycetes and *Tuber* from Ascomycetes were identified as the key ECM fungi taxa. They were closely associated with the community function; thus, changes in these ECM fungi's relative abundance could potentially alter forest structure and function [20]. Interestingly, *Inocybe* frequently demonstrated a relatively high abundance in urban forest soil, but it was also occasionally observed in suburban and rural forests soil. Similarly, Jin [20] also found that the relative abundance of *Inocybe* in artificial forests was significantly higher in comparison to that growing in natural environment. Our result may be attributable to the presence of simple tree species in urban artificial forests, the conditions of which are favorable for *Inocybe* colonization. *Russula* is a widely distributed ECM fungi genus that can associate and be symbiotic with *Quercus, Pine, Abies* and *Spruce* throughout the world, especially in boreal temperate forests [39]. In our study, *Russula* showed a high relative abundance in the *Quercus acutissima* forest ecosystem along the urban–rural gradient, notably accounting for up to 56.01% in the suburbs. *Amanita* was the most abundant ECM fungi in the rural *Quercus acutissima* forest. Previous studies have confirmed that both *Russula* and *Amanita* are late successional species [18,39,40]. In the current study, the suburban and rural *Quercus acutisana* forests were older than the urban forest, which might have been the reason why the relative abundances of *Russula* and *Amanita* were higher in suburban and rural forest ecosystems. Yang et al. [41] found that *Scleroderma* and *Tomentella* had a strong Cd tolerance under the stress of Cd and benomyl. Both were also prevalent in this study, indicating that they play an essential role in resisting environmental stress, as well as maintaining plant growth and metabolism. In addition, some ECM fungi with lower abundances, such as *Elaphomyces, Helvella, Boletus, Amphinema* and *Hydnum*, were only observed in the rural *Quercus acutissima* forest, while *Pisolithus* was occasionally identified in the suburban forest. The functions of these rare or specific ECM fungi in different habitats need to be further studied.

4.2. Influencing Factors of ECM Fungal Community Composition in Quercus acutissima Forests along the Urban–Rural Gradient

The changes in ECM fungal community composition and diversity are often used to predict ecosystem health [1,20]. Here, there was an obvious correlation between soil physicochemical properties and ECM fungal community structure in the *Quercus acutissima* forests along the urban–rural gradient. Soil physicochemical properties can affect the growth and distribution of EMC fungi, thereby resulting in different EMC fungal communities [1,42]. In particular, pH exerts an important role in soil nutrient transformation and cycling by influencing the microbial community composition and activity [43]. It has been reported that most EMC fungi can survive in weakly acidic conditions, while their growth is inhibited in alkaline conditions [21]. Studies have found that the seedling roots associated with mycorrhizal fungi not only perform normal respiration to produce CO_2, but also secrete organic acids and H^+, leading to soil acidification in forests [44,45]. In this study, the soil pH ranged from 4.50 to 5.00 in the suburban and rural forests, while it was about 7.80 in the urban forest. The ECM fungal Shannon and Simpson indices in the rural forest were higher than those in the urban forest, suggesting that soil pH was significantly negatively correlated with diversity indices. This conclusion is consistent with the previous results obtained by Kyaschenko, who also stated that the presence of a large number of ECM fungi likely reduces soil pH values and that the diversity of a fungal community decreases as the pH values increase [46].

SOM is one of the main sources of metabolic substances and energy for the colonization of mycorrhizal fungal [47]. In turn, the interaction between fungal communities can also regulate the SOM accumulation and soil fertility [47,48]. Remarkably higher SOM content was detected in the suburban *Quercus acutissima* forest soil in Jinan. A possible reason was that the proportion of other mixed tree species was larger in the *Quercus acutissima* forests, which probably enhanced the input of organic matter from the aboveground litter layer into the soil and provided sufficient sources during SOM retention. In contrast, the *Quercus acutissima* forest in the urban area was strongly disturbed by human activities, which resulted in a low level of litter input and affected the formation and accumulation of soil SOM. Nitrogen is widely recognized as a key limiting nutrition in controlling species composition and diversity in forest ecosystems [49]. The soil nitrate nitrogen content in the *Quercus acutissima* forest was significantly higher than that of ammonium nitrogen. The NO_3^--N has a negative charge and the adsorption with soil is weak [50]. Thus, NO_3^--N is likely to be leached from soil, which is not conducive to fertilizer conservation [50]. However, the content of NH_4^+-N was more than twice that of NO_3^--N in the suburban and rural *Quercus acutissima* forest soils, which could have effectively prevented soil nitrogen loss and maintained soil fertility [50]. Hao et al. [51] found that most EMC fungi have a preference for NH_4^+-N in the growth process. Similar results were found in this study: ECM fungal diversity was significantly positively correlated with NH_4^+-N and negatively correlated with NO_3^--N, indicating that most ECM fungi also preferred NH_4^+-N in the *Quercus acutissima* forest soil. However, a number of studies have reported that excessive nitrogen deposition (mainly nitrate) also inhibits the growth of EMC fungi in temperate broad-leaved oak forests and leads to a decrease in their diversity [52]. Moreover, EMC fungi have different preferences and adaptability in different soils [34]. In the present study, there were different correlations between soil properties and key ECM fungi taxa. For example, *Russula* and *Scleroderma* were extremely sensitive to SOM content and *Russula* was suitable for growing in suburban forests with high SOM content. However, *Scleroderma* preferred to colonize in urban and rural forests with relatively low SOM content. It also had a wide adaptive capacity in the environment and its abundance was not affected by the contents of nitrogen, phosphorus and potassium in soil. *Scleroderma* was dominant in low-nutrient-level soil, which means that this genus may have a unique function in nutrient uptake. In comparison with *Scleroderma*, *Inocybe* is generally considered as a later successional ECM fungus [53], and it showed a very significant positive correlation with AP, AK and NO_3^--N contents in the soil, suggesting that the growth of *Inocybe* requires abundant nutrient supply and a stable ecosystem. These results indicate that the occurrence of ECM fungi in soil is affected by different soil physicochemical properties, which may be related to ECM fungal species and their ecological adaptability [31]. In turn, ECM fungi also greatly alter the chemical form and availability of soil nutrients by secreting functional substances, thus playing a regulatory role in the host rhizosphere environment [20]. Hence, ECM fungi in forests along the urban–rural gradient strongly change with the various soil properties. The effects of soil physicochemical properties on soil health and ECM fungi should be considered in the cultivation of *Quercus acutissima* forests. As far as ECM fungi protection is concerned, it is necessary to detect in a timely manner the changes in soil physicochemical properties in forests. In order to obtain detailed information on ECM fungi along the urban–rural gradient, further studies need to be implemented including focuses on sampling sites, frequency and the driving force of human activities.

5. Conclusions

Soil ECM fungi are fundamental components of forest ecosystems. The three *Quercus acutissima* forests demonstrated significant shifts in ECM fungal communities and diversity in different habitats, suggesting that different forest management strategies may be needed for microbial biodiversity conservation along the urban–rural gradient. Moreover, the distribution of ECM fungi in the oak forest soil implies that the dominance of ECM fungi differs depending on the spatial location. Our results provide basic information on forest protection and management along the urban–rural gradient in Jinan city. More abundant information about the location-specific relationship between environmental factors and microbial community is urgently needed for effective forest soil management.

Author Contributions: Conceptualization, H.S. and W.S.; methodology, H.S.; software, H.S. and K.J.; validation, B.Y., G.Q. and H.W.; formal analysis, H.S.; investigation, H.S. and B.Y.; resources, H.S.; data curation, H.S.; writing—original draft preparation, H.S.; writing—review and editing, H.S.; visualization, H.S.; supervision, H.W.; project administration, H.W.; funding acquisition, H.W. All authors have read and agreed to the published version of the manuscript.

Funding: This research was funded by the National Natural Science Foundation of China (41877424; 31870606), the Research Leader Studio Project (2021GXRC094), the Key R & D project of Shandong Province (2021LZGC005-02-02) and the Fundamental Research Funds for the Central Universities, CHD (300102351505).

Acknowledgments: The authors acknowledge Hangzhou LC-Bio Technology Co., Ltd. (Hangzhou, China), for providing the sequencing platform and technical support.

Conflicts of Interest: The authors declare no conflict of interest.

References

1. Kumar, J.; Atri, N.S. Studies on ectomycorrhiza: An appraisal. *Bot. Rev.* **2018**, *84*, 108–155. [CrossRef]
2. Roy-Bolduc, A.; Laliberté, E.; Hijri, M. High richness of ectomycorrhizal fungi and low host specificity in costal sand dune ecosystem revealed by network analysis. *Ecol. Evol.* **2016**, *6*, 349–362. [CrossRef] [PubMed]
3. Brundrett, M.C. Mycorrhizal associations and other means of nutrition of vascular plants: Understanding the global diversity of host plants by resolving conflicting information and developing reliable means of diagnosis. *Plant Soil* **2009**, *320*, 37–77. [CrossRef]
4. Rinaldi, A.C.; Comadini, O.; Kuyper, T.W. Ectomycorrhizal fungal diversity: Separating the wheat from the chaff. *Fungal Diversi.* **2008**, *33*, 1–45.
5. Simard, S.W.; Durall, D.M. Mycorrhizal networks: A review of their extent, function, and importance. *Can. J. Bot.* **2004**, *82*, 1140–1165. [CrossRef]
6. Tedersoo, L.; Smith, M.E.; May, T.W. Ectomycorrhizal lifestyle in fungi: Global diversity, distribution, and evolution of phylogenetic lineages. *Mycorrhiza* **2010**, *20*, 217–263. [CrossRef]
7. Finlay, R.D. Ecological aspects of mycorrhizal symbiosis: With special emphasis on the functional diversity of interactions involving the extraradical mycelium. *J. Exp. Bot.* **2008**, *59*, 1115–1128. [CrossRef]
8. Sebastiana, M.; Martins, J.; Figueiredo, A.; Monteiro, F.; Sardans, J. Oak protein profile alterations upon root colonization by an ectomycorrhizal fungus. *Mycorrhiza* **2017**, *27*, 109–128. [CrossRef]
9. Rudawska, M.; Pietras, M.; Smutek, I.; Strzeliński, P.; Leski, T. Ectomycorrhizal fungal assemblages of *Abies alba* Mill. outside its native range in Poland. *Mycorrhiza* **2016**, *26*, 57–65. [CrossRef]
10. Delgado-Baquerizo, M.; Reich, P.B.; Trivedi, C.; Eldridge, D.J. Multiple elements of soil biodiversity drive ecosystem functions across biomes. *Nat. Ecol. Evol.* **2020**, *4*, 210–220. [CrossRef]
11. Erlandson, R.S.; Savage, J.A.; Cavender-Bares, J.M.; Peay, K.G. Soil moisture and chemistry influence diversity of ectomycorrhizal fungal communities associating with willow along a hydrologic gradient. *FEMS Microbiol. Ecol.* **2016**, *92*. [CrossRef]
12. Garcia, M.; Jane, O.; Smith, E.; Daniel, E.; Luoma, L.; Melanie, L.; Jones, D. Ectomycorrhizal communities of *ponderosa pine* and *lodgepole pine* in the south-central Oregon pumice zone. *Mycorrhiza* **2016**, *26*, 275–286. [CrossRef]
13. Corrales, A.; Arnold, A.E.; Ferrer, A.; Turner, B.L.; Dalling, J.W. Variationin ectomycorrhizal fungal communities associated with *Oreomunnea mexicana* (Juglandaceae) in a Neotropical montane forest. *Mycorrhiza* **2016**, *26*, 1–17. [CrossRef]
14. Cox, F.; Barsoum, N.; Lilleskov, E.A.; Bidartondo, M.I. Nitrogen availability is a primary determinant of conifer mycorrhizas across complex environmental gradients. *Ecol. Lett.* **2010**, *13*, 1103–1113. [CrossRef]
15. Benucci, G.M.N.; Bonito, G.M. The Truffle Microbiome: Species and geography effects on bacteria associated with fruiting bodies of hypogeous Pezizales. *Microb. Ecol.* **2016**, *72*, 4–8. [CrossRef]
16. Szuba, A. Ectomycorrhiza of *Populus*. *Forest Ecol. Manag.* **2015**, *347*, 156–169. [CrossRef]
17. Walker, J.; Miller, O.; Horton, J. Hyperdiversity of ectomycorrhizal fungus assemblages on oak seedlings in mixed forests in the southern Appalachian Mountains. *Mol. Ecol.* **2010**, *14*, 829–838. [CrossRef]
18. He, F.; Yang, B.S.; Wang, H.; Yan, Q.L.; Cao, Y.N.; He, X.H. Changes in composition and diversity of fungal communities along *Quercus mongolica* forests developments in Northeast China. *Appl. Soil Ecol.* **2016**, *100*, 162–171. [CrossRef]
19. Behnke-Borowczyk, J.; Kowalkowski, W.; Kartawik, N.; Baranowska, M.; Barzdajn, W. Soil fungal communities in nurseries producing *Abies alba*. *Balt For.* **2020**, *26*, 426. [CrossRef]
20. Jin, W.; Yang, Y.Z.; Sun, H.J.; Yuan, Z.L. Diversity of Ectomycorrhizal Fungi a Seed Collecting Forest of *Quercus virginiana*. *Sci. Silvae Sin.* **2020**, *56*, 1001–7488.
21. Voříšková, J.; Brabcová, V.; Cajthaml, T.; Baldrian, P. Seasonal dynamics of fungal communities in a temperate Oak forest soil. *New Phytol.* **2014**, *201*, 269–278. [CrossRef]
22. Salvati, L.; Ranalli, F.; Gitas, I. Landscape fragmentation and the agro-forest ecosystem along a rural-to-urban gradient: An exploratory study. *Int. J. Sustain. Dev. World Ecol.* **2014**, *21*, 160–167. [CrossRef]

23. Cardou, F.; Aubin, I.; Bergeron, A.; Shipley, B. Functional markers to predict forest ecosystem properties along a rural-to-urban gradient. *J. Veg. Sci.* **2020**, *31*, 416–428. [CrossRef]

24. Bu, F.Q.; Yan, H.; Fan, M.J. Suburban communities in ecologically sensitive areas of vegetation Optimization—Case study of jinan city, the southern mountains. *Shandong For. Sci. Technol.* **2014**, *44*, 16–20.

25. The 7th National Census of Jinan. Available online: http://m.iqilu.com/pcarticle/4882711?ivk_sa=1024320u (accessed on 16 June 2021).

26. Vance, E.D.; Brookes, P.C.; Jenkinson, D.S. An extraction method for measuring soil microbial biomass C. *Soil Biol. Biochem.* **1987**, *19*, 703–707. [CrossRef]

27. Zhang, J.; Zhang, M.; Huang, S.Y.; Zha, X. Assessing spatial variability of soil organic carbon and total nitrogen in eroded hilly region of subtropical China. *PLoS ONE* **2020**, *15*, e0244322. [CrossRef]

28. Gao, G.F.; Li, P.F.; Zhong, J.X.; Shen, Z.J.; Chen, J.; Li, Y.T.; Isabwe, A.; Zhu, X.Y.; Ding, Q.S.; Zhang, S.; et al. Spartina alterniflflora invasion alters soil bacterial communities and enhances soil N_2O emissions by stimulating soil denitrififlcation in mangrove wetland. *Sci. Total Environ.* **2019**, *653*, 231–240. [CrossRef]

29. Foster, J.C. Soil sampling, handling, storage and analysis. In *Methods in Applied Soil Microbiology and Biochemistry*; Academic Press: New York, NY, USA, 1995; pp. 49–121.

30. Richard, F.; Roy, M.; Shahin, O. Ectomycorrhizal communities in a Mediterranean forest ecosystem dominated by *Quercus ilex*: Seasonal dynamics and response to drought in the surface organic horizon. *Ann. Forest Sci.* **2011**, *68*, 57–68. [CrossRef]

31. Garcia, K.; Doidy, J.; Zimmermann, S.D.; Wipf, D.; Courty, P.E. Take a trip through the plant and fungal transportome of mycorrhiza. *Trends. Plant Sci.* **2016**, *21*, 937–950. [CrossRef] [PubMed]

32. Van Dam, N.M.; Bouwmeester, H.J. Metabolomics in the rhizosphere: Tapping into belowground chemical communication. *Trends. Plant Sci.* **2016**, *21*, 256–265. [CrossRef] [PubMed]

33. Li, M.; Gao, M.H. Community structure and driving factors for rhizosphere ectomycorrhizal fungi of Betula platyphylla in Daqing Mountain. *Chin. J. Ecol.* **2021**, *40*, 1244–1252.

34. Treseder, K.K. A meta-analysis of mycorrhizal responses to nitrogen, phosphorus, and atmospheric CO_2 in field studies. *New Phytol.* **2004**, *164*, 347–355. [CrossRef]

35. Schmidt, D.J.E.; Pouyat, R.; Szlavecz, K.; Setälä, H.; Kotze, D.J.; Yesilonis, I.; Sarel, C. Urbanization erodes ectomycorrhizal fungal diversity and may cause microbial communities to converge. *Nat. Ecol. Evol.* **2017**, *1*, 439–447. [CrossRef]

36. Ashley, S.; Hannes, P.; Steven, A.D.; Baho, D.L.; Mercè, B. Fundamentals of microbial community resistance and resilience. *Front. Microbiol.* **2012**, *3*, 417. [CrossRef]

37. Tedersoo, L.; Suvi, T.; Larsson, E.; Kõljalg, U. Diversity and community structure of ectomycorrhizal fungi in a wooded meadow. *Mycol. Res.* **2006**, *110*, 734–748. [CrossRef]

38. Yang, R.H.; Li, Y.; Wu, Y.Y.; Tang, L.H.; Shang, J.J.; Bao, D.P. Genome-based analysis of lignocellulose-degrading enzyme systems in different Lentinus edodes strains. *Acta Edulis Fungi* **2018**, *25*, 15–22.

39. Wieg, B.D.; Durall, D.M.; Simard, S.W. Ectomycorrhizal fungal succession in mixed temperate forests. *New Phytol.* **2007**, *176*, 437–447.

40. Redecker, D.; Szaro, T.M.; Bowman, R.J.; Bruns, T.D. Small genets of *Lactarius xanthogalactus*, *Russula cremoricolor* and *Amanita francheti* in late-stage ectomycorrhizal successions. *Mol. Ecol.* **2001**, *10*, 1025–1034. [CrossRef]

41. Yang, B.S.; He, F.; Zhao, X.H.; Wang, H.; Xu, X.H.; He, X.H.; Zhu, Y.D. Composition and function of soil fungal community during the establishment of *Quercus acutissima* seedlings in a Cd-contaminated soil. *J. Environ. Manag.* **2019**, *246*, 150–156. [CrossRef]

42. Jiang, Z.Y.; Zou, W.Q.; Yang, L.; Li, W.Z.; Zhang, H.D.; Chen, X.W.; Wang, X.W. Root exudation rate and rhizosphere effect of different mycorrhizal associations of tree species in typical black soil area. *Chin. J. Ecol.* **2021**, *33*, 2810–2816. [CrossRef]

43. Liu, X.L.; Dou, L.; Ding, X.H.; Sun, T.; Zhang, H.J. Influences of different afforestation systems on the soil properties of limestone mountains in the mid-eastern region of China. *Catena* **2021**, *201*, 105198. [CrossRef]

44. Fang, F.; Wu, C.Z.; Hong, W.; Fang, H.L.; Song, P. Study on relationship between plant rhizosphere and non-rhizosphere soil enzymes and microorganisms. *Subtrop. Agric. Res.* **2016**, *3*, 209–215.

45. Wang, X.; Liu, J.; Long, D.; Han, Q.; Huang, J. The ectomycorrhizal fungal communities associated with *Quercus liaotungensis* in different habitats across northern China. *Mycorrhiza* **2017**, *27*, 441–449. [CrossRef] [PubMed]

46. Kyaschenko, J.; Clemmensen, K.E.; Karltun, E.; Lindahl, B.D. Below-ground organic matter accumulation along a boreal forest fertility gradient relates to guild interaction within fungal communities. *Ecol. Lett.* **2017**, *20*, 1546–1555. [CrossRef]

47. Zhou, Y.; Hartemink, A.E.; Shi, Z.; Liang, Z.; Lu, Y. Land use and climate change effects on soil organic carbon in North and Northeast China. *Sci. Total Environ.* **2019**, *647*, 1230–1238. [CrossRef]

48. Dong, W.Y.; Zhang, X.Y.; Liu, X.Y.; Fu, X.L.; Chen, F.S.; Wang, H.M.; Sun, X.M.; Wen, X.M. Responses of soil microbial communities and enzyme activities to nitrogen and phosphorus additions in Chinese fir plantations of subtropical China. *Biogeosciences* **2016**, *12*, 5537–5546. [CrossRef]

49. Dong, S.K.; Wen, L.; Li, Y.Y.; Wang, X.X.; Zhu, L.; Li, X.Y. 2012. Soil-quality effects of grassland degradation and restoration on the Qinghai-Tibetan Plateau. *Soil Sci. Soc. Am. J.* **2012**, *76*, 2256–2264. [CrossRef]

50. Wu, Z.Z.; Cheng, H.G.; Wang, J.T.; Cheng, Q.D. Effects of biochar addition ratio and freezing-thawing on nitrogen leaching from ditched soil. *J. Agro-Environ. Sci.* **2019**, *39*, 1295–1302.

51. Hao, L.F.; Hao, W.Y.; Wang, X.F.; Liu, H.; Liu, T.Y. Response of four ectomycorrhizal fungi to nitrogen sources. *South. For. Sci.* **2020**, *48*, 20–24.
52. Simkin, S.M.; Allen, E.B.; Bowman, W.D.; Clark, C.M. Conditional vulnerability of plant diversity to atmospheric nitrogen deposition across the United States. *Proc. Natl. Acad. Sci. USA* **2016**, *113*, 4086–4091. [CrossRef]
53. Olchowik, J.; Hilszczańska, D.; Bzdyk, R.M.; Studnicki, M.; Malewski, T.; Borowski, Z. Effect of deadwood on ectomycorrhizal colonisation of old-growth oak forests. *Forests* **2019**, *10*, 480. [CrossRef]

Article

Soil Chemical Properties Strongly Influence Distributions of Six Kalidium Species in Northwest China

Decheng Liu [1,†], Zongqiang Chang [2,†] , Xiaohui Liang [1] and Yuxia Wu [1,*]

1 State Key Laboratory of Herbage Improvement and Grassland Agro-Ecosystems, College of Ecology, Lanzhou University, Lanzhou 730000, China
2 Northwest Institute of Eco-Environment and Resources, Chinese Academy of Sciences, Lanzhou 730000, China
* Correspondence: wuyx@lzu.edu.cn
† These authors contributed equally to this work.

Abstract: The degrees of adaptive responses of different halophytes to saline–alkali soil vary substantially. *Kalidium* (Amaranthaceae), a genus comprised of six species of succulent euhalophytes with significantly differing distributions in China, provides ideal material for exploring the ecophysiological relationships involved in these variations. Thus, in a large-scale field survey in 2014–2018, samples of soil (at 20 cm depth intervals spanning 0 to 100 cm) and seeds were collected from areas where these six species are naturally distributed. Chemical properties of soils in the areas and germinability of the species' seeds in media with 0–500 mM NaCl and 0–250 mM Na_2SO_4 were then analyzed to test effects of salinity-related factors on the species' distributions. The pH of the soil samples mainly ranged between 8.5 and 10.5 and positively correlated with their mean total salt contents. Germination rates of all six species' seeds were negatively correlated with concentrations of NaCl and Na_2SO_4 in the media, and their recovery germination rates in distilled water were high (>74%). The results show that the species' distributions and chemical properties of their saline soils are strongly correlated, notably the dominant cation at all sites is Na^+, but the dominant anions at *K. cuspidatum* and *K. caspicum* sites are Cl^- and SO_4^{2-}, respectively. Species-associated variations in concentrations of Ca^{2+} were also detected. Thus, our results provide clear indications of major pedological determinants of the species' geographic ranges and strong genotype-environment interactions among *Kalidium* species.

Keywords: germination percentage; *Kalidium*; halophytes; pH; ion content; total salt contents; saline soil

Citation: Liu, D.; Chang, Z.; Liang, X.; Wu, Y. Soil Chemical Properties Strongly Influence Distributions of Six Kalidium Species in Northwest China. *Forests* **2022**, *13*, 2178. https://doi.org/10.3390/f13122178

Academic Editors: Xiao-Dong Yang and Nai-Cheng Wu

Received: 21 November 2022
Accepted: 16 December 2022
Published: 19 December 2022

Publisher's Note: MDPI stays neutral with regard to jurisdictional claims in published maps and institutional affiliations.

1. Introduction

One of the diverse environmental factors that strongly affect terrestrial plants' natural distributions is soil salinity, which has growth-impairing and lethal effects on all plants when it exceeds species-dependent thresholds [1–3]. Further, salinity reportedly reduces crop yields on about a fifth of all irrigated land and, in combination with increasing global scarcity of water resources, salinization of soil and water is seriously threatening crop yields and future food production [4,5]. However, halophytes, comprising about 1% of the world's flora, can grow in saline environments with relatively high concentrations of electrolytes [6]. For example, the growth and development of glycophytes is severely inhibited by exposure to 100–200 mM of NaCl, while halophytes can tolerate and complete their life cycles at substantially higher concentrations [2,4,7]. This is due to anatomical adaptations in halophytes, such as salt bladders, salt hairs, and/or salt glands in the leaves [7,8] and various physiological tolerance mechanisms. For example, excess salt may be excreted through trichomes of halophytic grasses [8], or diluted by increases in the water content and thickness of succulent halophytes' leaves [9,10]. Moreover, different halophytes growing together on the same saline–alkali soil often have substantially differing elemental concentrations, indicating that their physiological selectivity varies [11]. Generally, the dominant ions in salty habitats are Na^+ and Cl^-, but other ions (including Ca^{2+}, Mg^{2+}, K+, SO_4^{2-}, and CO_3^{2-}) may also be abundant [9,12]. Moreover, both their absolute and relative concentrations may vary, and influence the composition of the associated plant communities [13–15].

Kalidium (Amaranthaceae) is a genus of succulent halophytes with five species (*Kalidium caspicum* (L.) Ung.-Sternb., *Kalidium cuspidatum* (Ung.-Sternb.) Grubov., *Kalidium foliatum* (Pall.) Moq., *Kalidium gracile* Fenzl, and *Kalidium schrenkianum* Bunge ex Ung.-Sternb.), mainly distributed as shrubs in Northwest Asia and Southeast Europe. Some authorities have also recognized *K. wagenitzii* as an endemic species in Turkey, but others include it in *K. foliatum* [16]. In addition, two varieties of *K. cuspidatum* (var. *cuspidatum* and *sinicum* A. J. Li) have been recognized as separate species [17]. Here these varieties are called species, mainly due to genetic differences in DNA barcodes [17], but partly because our results indicate that they have significant adaptive differences. All six *Kalidium* species (including the two varieties of *K. cuspidatum* as species) naturally grow in deserts in northwest China. These succulent halophytes provide important fodder for livestock in winter, and have high ecological value for soil and water conservation in their semi-arid and arid areas. An extensive field survey showed that their distributions in China significantly differ (Figure 1). *K. foliatum* is the most widespread. Most *K. cuspidatum* sites are in Ningxia, *K. gracile* is mostly located in Gansu, Qinghai and Inner Mongolia. *K. sinicum* is naturally distributed in East Gansu and West Ningxia, while *K. caspicum* and *K. schrenkianum* are mainly restricted to regions north and south of the Tianshan Mountains in Xinjiang, respectively.

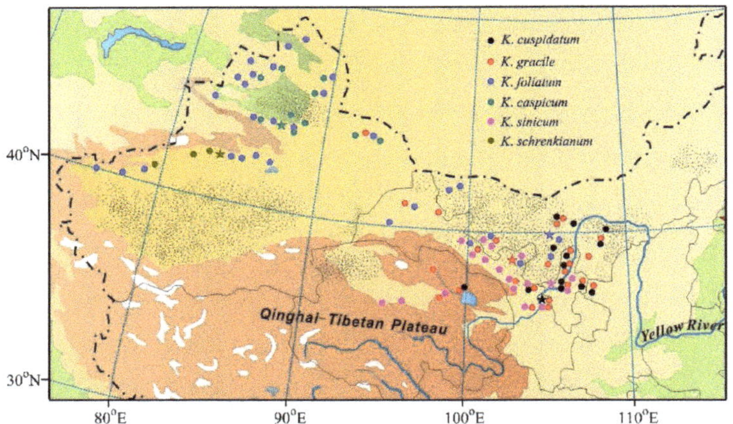

Figure 1. Map showing soil sampling sites in areas occupied by the six *Kalidium* species. Stars indicate seed sampling sites.

The adaptive evolution and origin of key halotolerance mechanisms have been intensively studied, as reviewed for example [2,18]. However, the NaCl concentration is not the sole stressor in saline environments. Thermal and water stresses are also often important [19,20]. Furthermore, salinity may be associated with extreme pH and/or variations in relative proportions of both cations and anions [21]. Effects of these variations have been less extensively studied, so this study focused on their impacts on distributions of the six *Kalidium* species in China. For this purpose, soil samples were collected from sites of the six species, across their ranges in China, then the pH, total salt contents, and ion contents of the soil were assayed at 20 cm depth intervals spanning 0 to 100 cm. In addition, the germinability of seeds of the six species was determined under different concentrations of NaCl and Na_2SO_4. Correlations between these abiotic factors and distributions of the

Kalidium species were then examined, to explore mechanisms affecting the relationship between biodiversity and ecosystem functions.

2. Materials and Methods

2.1. Field Investigation and Sampling

In a comprehensive field investigation of the areas where *Kalidium* Moq is distributed in northwest China during 2014–2018, we identified 103 representative sites, in total, of the six *Kalidium* species. These sites are mainly located in gravel deserts and/or gravel dunes, according to the FAO soil classification system (FAO 2016). Vegetation at the sites is dominated by *Kalidium* Moq and *Halocnemum strobilaceum* of the Amaranthaceae, *Halogeton arachnoideus* (Amaranthaceae), *Stipa glarecsa* (Gramineae) and various other halophytes. Distances between neighboring sites mainly ranged from 150 to 200 km. Soil samples were collected from centers of these areas using a soil auger, with three replications (three-point sampling), at 20 cm depth intervals from 0 to 100 cm. Furthermore, mature seeds were collected from the six *Kalidium* species and stored in a refrigerator at $-20\ °C$ before the start of experiments. The altitude and geographic coordinates of each site were measured using an Etrex GIS unit (Garmin, Taiwan). Locations of the collection sites are shown in Figure 1.

2.2. Determination of Soil Chemical Properties

The soil samples were air-dried, passed through a 1 mm sieve, then their pH was measured at a 1:5 soil: water ratio (w/v) using a PHS-25 pH meter (Shanghai Biocotek Inc., Shanghai, China), and their electrical conductivity (mS/cm) using a DDSJ-318 conductivity meter (Shanghai Biocotek Inc., Shanghai, China). Their contents of eight ions were also analyzed: Na^+, K^+, Ca^{2+}, and Mg^{2+} using a 180-80 Polarized Zeeman atomic absorption spectrophotometer (Hitachi Inc., Tokyo, Japan); CO_3^{2-} and HCO_3^- by the double indicator titration method; Cl^- by silver nitrate titration; and SO_4^{2-} by the turbidimetric method [22].

2.3. Determination of Seeds' Germinability

Mean seed mass was calculated by weighing 1000 seeds of each *Kalidium* species with three replications (Table S1). Then germination and recovery experiments were conducted in an LRH 550-G programmed controlled-environment chamber (Shaoguan taihong, China) providing 16 h light/8 h dark cycles with cool white fluorescent lamps 100 $\mu mol\ m^{-2}\ s^{-1}$ (Philips), and 25/19 °C day/night temperatures.

Seeds of the six *Kalidium* species were subjected to treatment with NaCl at seven concentrations by placing them on filter paper in Petri dishes (9 cm diameter) moistened with 10 mL of 0, 50, 100, 200, 300, 400, and 500 mM NaCl solution. Other batches were exposed to Na_2SO_4 with corresponding Na concentrations (0, 25, 50, 100, 150, 200 and 250 mM). Germination (regarded as emergence of the radicle from the seed by about 1 mm) was scored every day for 7 days. The germination rate at that point was calculated for each species, then ungerminated seeds were transferred to distilled water, and incubated under otherwise identical conditions. Germination of these seeds was scored for a further 7 days, after which the recovery germination percentage was calculated for each species.

2.4. Data Analysis

The soils' total salt contents were estimated from electrical conductivity measurements, using the following empirically determined linear relationship between NaCl concentration (y) and conductivity (x): y = 0.0159x (R^2 = 0.9811) [23]. Relationships between total salt contents and pH in 20 cm layers of the top 100 cm of soil at sites of the six species were examined by linear regression. The significance of differences in germination rates and recovery germination rates of seeds at different salt concentrations was tested by one-way ANOVA. Values of 18 bioclimatic factors covering most of the distributions of the six species were downloaded from the Global Climate Database (http://www.worldclim.org/bioclim, accessed on 9 November 2021). After excluding redundant bioclimatic factors, by applying a cumulative contribution ratio threshold of 80% [24], eight remained (Table S2). These were used in combination with altitude and two soil factors (pH and total salt concentration of the soil) in the Principal Component Analysis (PCA) of abiotic factors affecting distributions of each of the six studied *Kalidium* species. For this, SPSS (Version 19.0., Chicago, IL, USA) was used (with the significance threshold set at $p < 0.05$).

3. Results

3.1. pH and Total Salt Contents

The pH of soils in the areas occupied by the *Kalidium* species mainly ranged from 8.5 to 10.5, although the *K. foliatum* sites had a wider range (7.3–11). The mean pH of soil samples from areas occupied by *K. cuspidatum*, *K. gracile*, *K. sinicum*, *K. foliatum*, *K. caspicum*, and *K. schrenkianum* was 9.41, 9.05, 9.13, 9.03, 8.94, and 8.90, respectively (Figure 2). Mean total salt contents in areas occupied by *K. cuspidatum*, *K. gracile*, *K. foliatum*, *K. caspicum* and *K. sinicum* were 22.4, 19.5, 17.6, 16.5 and 14.3 g/kg, respectively (Figure 2), while the highest mean value for any layer in *K. schrenkianum* areas was just 13.0 g/kg (in the 0–20 cm layer) and the overall mean was just 7.08 g/kg (Figures 2 and 3a). Generally, there was a positive correlation between pH values and mean salt contents in the topsoil (0–20 cm), but at >40 cm depths the relationship turned negative in areas occupied by all six species (Figure 3a). However, salt contents of samples from *K. cuspidatum* areas (with the highest mean pH and salt contents) significantly varied with depth while they were much more constant in *K. caspicum* areas (Figure 3). In addition, pH and total salt contents varied much more widely in soil samples from *K. foliatum*, *K. gracile*, and *K. sinicum* areas than in samples from *K. schrenkianum* areas (Figures 2 and 3).

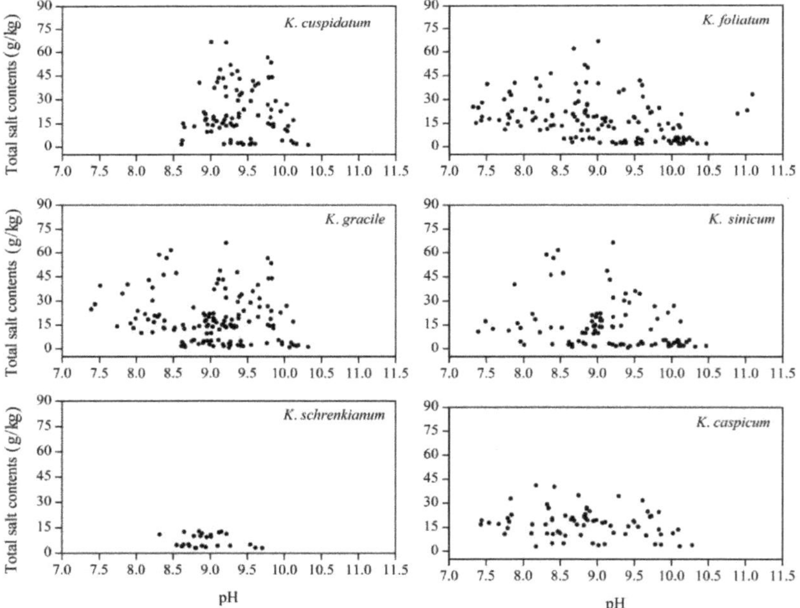

Figure 2. Scatter plots of total salt contents and pH of soil samples from areas occupied by indicated *Kalidium* species.

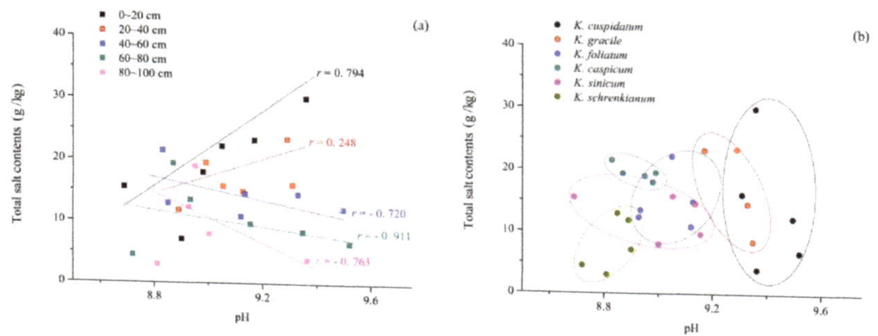

Figure 3. Linear regression plots showing the relationship between total salt contents and pH at indicated depths, and overall, in soil from areas occupied by the six *Kalidium* species. (**a**) total salt contents at different soil depth; (**b**) total salt contents at different species.

3.2. Na^+, Ca^{2+}, K^+, and Mg^{2+} Concentrations

Na^+ concentrations in areas occupied by the six species were all high (Figure 4), especially in *K. cuspidatum* areas, where they declined with increases in depth but even at 80–100 cm exceeded 2.5 g/kg, the upper limit for sensitive crops according to the FAO. In *K. gracile* and *K. schrenkianum* areas, Na^+ concentrations peaked at 20–40 cm depth. Mean Na^+ concentrations in samples from all soil layers in areas occupied by the six species ranged from 0.90 to 4.22 g/kg. As shown in Table 1, Na^+ contents were also significantly positively correlated ($r > 0.89$) with total salt contents in soil from the six species' areas, except for 20–40 cm samples ($r = 0.632$), strongly suggesting that Na^+ made the largest contributions to the total salt contents.

Table 1. Correlation coefficient (r) between total salt contents and concentrations of indicated ions in soil from indicated depths in areas occupied by the six *Kalidium* species.

Soil Depth	Ca^{2+}	Mg^{2+}	K^+	Na^+	HCO_3^-	Cl^-	SO_4^{2-}
0~20 cm	−0.841 *	0.725	0.018	0.896 *	0.992 **	0.856 *	0.793
20~40cm	−0.093	−0.156	−0.680	0.632	−0.969 **	0.689	0.031
40~60 cm	0.677	0.980 **	−0.140	0.986 **	−0.809	0.880 *	0.784
60~80 cm	0.934 *	0.991 **	0.567	0.972 **	−0.873	0.933 *	0.966 **
80~100 cm	0.965 **	−0.958 *	0.660	0.973 **	−0.502	0.167	0.344

*, $p < 0.05$; **, $p < 0.01$.

Ca^{2+} was the second most abundant cation in all tested soil samples (Figure 4). In *K. foliatum* and *K. caspicum* areas, mean concentrations increased as soil depth increased, reaching 0.74 and 0.98 g/kg at 80–100 cm, respectively. In contrast, in *K. cuspidatum* and *K. sinicum* areas, mean Ca^{2+} concentrations declined as depth increased and then increased (Figure 4), to <86 mg/kg at 40–60 cm in *K. cuspidatum* areas (Figure 4). Overall, as shown in Table 1, there was a negative correlation between mean Ca^{2+} concentrations and total salt contents in the topsoil ($r = −0.841$, $p < 0.05$), but a positive correlation between them at 60–80 cm ($r = 0.934$, $p < 0.05$) and 80–100 cm ($r = 0.965$, $p < 0.01$).

K^+ concentrations did not exceed 260 mg/kg in any tested soil samples. In *K. gracile* and *K. sinicum* areas, as depth increased they first declined and then increased, and were almost twice as high in the former as the latter at each soil depth (0–60 cm) (Figure 4). K^+ concentrations significantly declined with depth (from 216, and 257 mg/kg, respectively, in topsoil) at all sites occupied by *K. schrenkianum* and *K. cuspidatum* (Figure 4).

Mg^{2+} concentrations were also substantially lower than Na^+ and Ca^{2+} concentrations (consistently < 300 mg/kg). In areas occupied by *K. caspicum*, *K. gracile*, and *K. sinicum* they first increased and then declined with increases in soil depth (Figure 4), peaking at 268 mg/kg at 40–60 cm in samples from *K. caspicum* areas (Figure 4). Overall, as shown in

Table 1, at sites of all six species, Mg^{2+} concentrations were positively correlated with total salt contents at 40–60 cm ($r = 0.980$, $p < 0.01$) and 80–100 cm ($r = 0.991$, $p < 0.01$) depths.

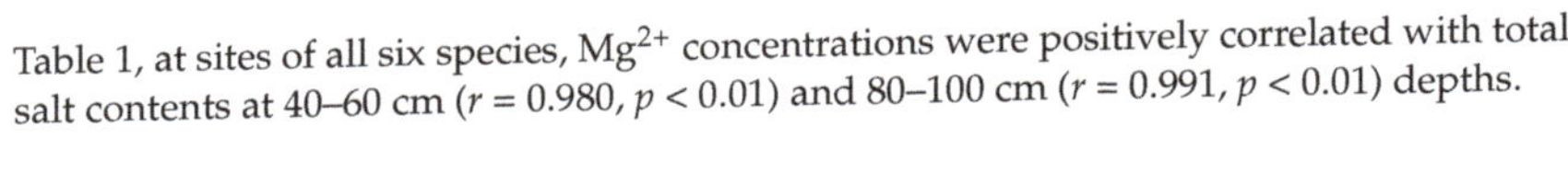

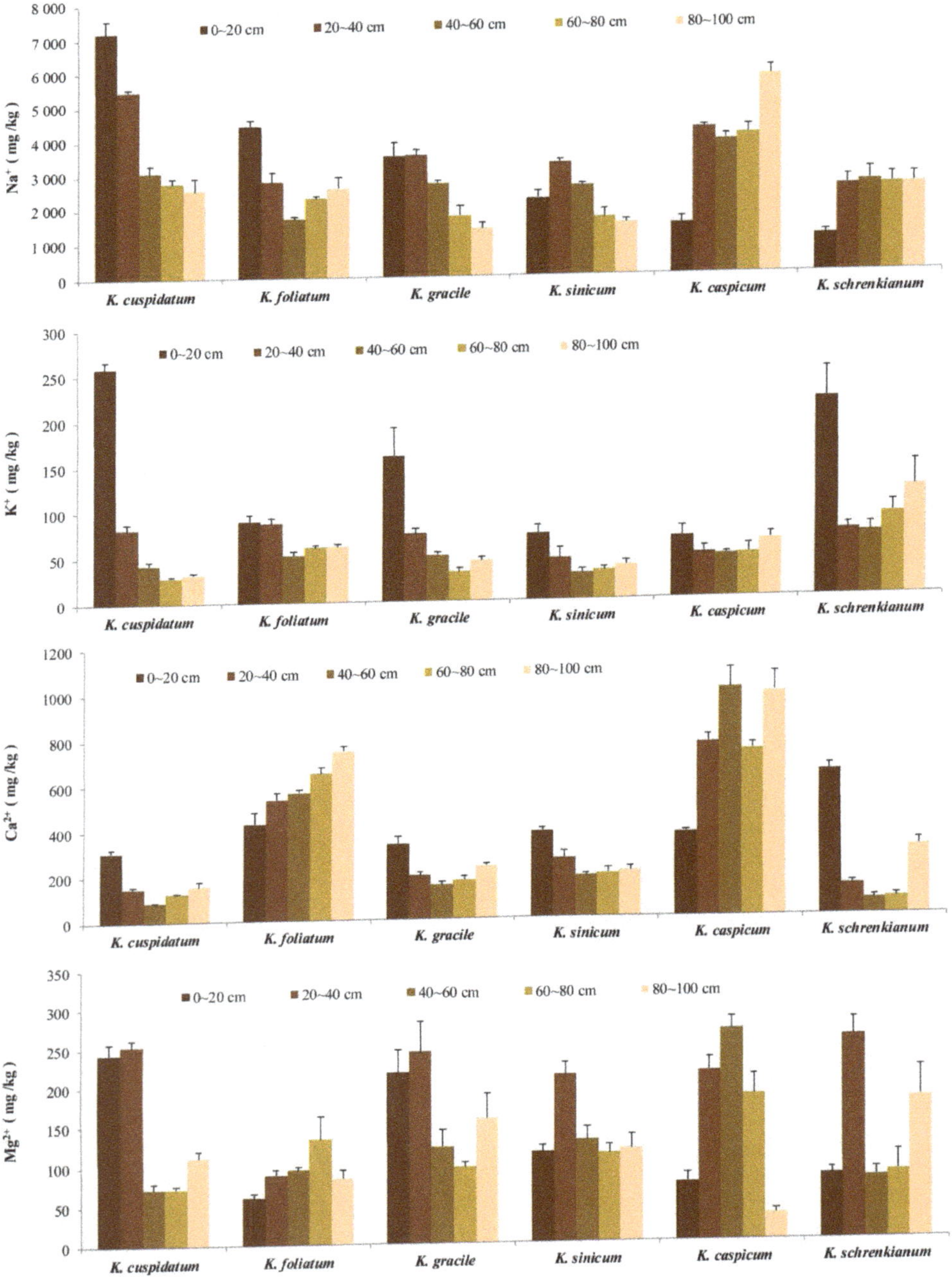

Figure 4. Concentrations of Na^+, Ca^{2+}, K^+, and Mg^{2+} in soil from indicated depths in areas occupied by the six *Kalidium* species. The values shown are means with SE ($n = 3$).

3.3. Cl^-, SO_4^{2-} and HCO_3^- Concentrations

Mean Cl^- concentrations were highest in *K. cuspidatum* areas, where they declined from 10.62 g/kg in topsoil but still exceeded 2.5 g/kg at 80–100 cm depth, and lowest in *K. caspicum* areas, where the mean topsoil concentration was only 1.53 g/kg (Figure 5). It was also a major anion in *K. sinicum*, *K. foliatum* and *K. gracile* areas, where mean concentrations were 2.6, 2.7 and 3.2 g/kg, respectively (Figure 5). Overall, as shown in Table 1, Cl^- concentrations were positively correlated with total salt contents at 0–20 cm ($r = 0.856$, $p < 0.05$), 40–60 cm ($r = 0.880$, $p < 0.05$) and 80–100 cm ($r = 0.933$, $p < 0.05$) depths.

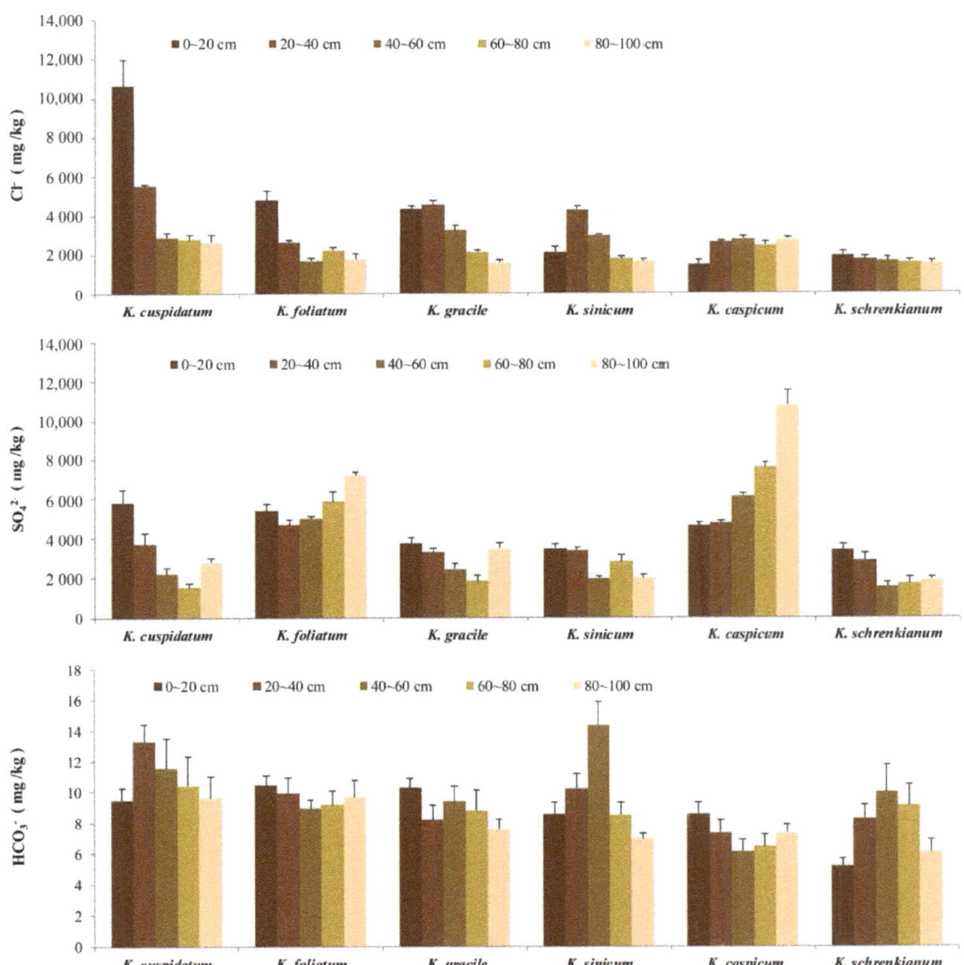

Figure 5. Concentrations of Cl^-, SO_4^{2-}, and HCO_3^- in soil from indicated depths in areas occupied by the six *Kalidium* species. The values shown are means with SE ($n = 3$).

In *K. caspicum* areas, SO_4^{2-} was the main anion, and its mean concentration increased as soil depth increased, reaching 11.78 g/kg at 80–100 cm, while concentrations were much lower in *K. cuspidatum* areas (just 1.52 g/kg at 60–80 cm depth) (Figure 5). In areas occupied by the other four species—*K. gracile*, *K. foliatum*, *K. sinicum*, and *K. schrenkianum*—the mean concentration first declined (from 3.7, 5.4, 3.5 and 3.43 g/kg, respectively, in topsoil) and then increased as depth increased. Overall, it was only correlated with total salt contents at 60–80 cm depth (r = 0.966, *p* < 0.01) in areas occupied by the six *Kalidium* species (Figure 5, Table 1).

HCO_3^- concentrations in soil from areas occupied by all six species were very low. Its mean concentrations first increased then declined as depth increased in *K. schrenkianum*, *K. cuspidatum* and *K. sinicum* areas (Figure 5), and were highest in the 40–60 cm layer of soil in *K. sinicum* areas, at just 14 mg/kg. Moreover, HCO_3^- concentrations were negatively correlated with total salt contents at depths below 20 cm in areas occupied by all species (Figure 5, Table 1). CO_3^{2-} was undetectable with the applied equipment in most samples.

3.4. Germination Rates

Germination rates of seeds of the six *Kalidium* species were negatively correlated with concentrations of both NaCl and Na_2SO_4 in the media (*p* < 0.05, Figures 6 and 7). In distilled water their germination rates ranged from 90.8% for *K. sinicum* to 100% for *K. caspicum*. However, less than half of all species' seeds germinated when the concentration exceeded 200 mM NaCl, and no *K. sinicum* seeds germinated at higher concentrations (400 or 500 mM NaCl) (Figure 6). Germination rates also declined with increases in Na_2SO_4 concentrations in the medium, and at 150 mM exceeded 50% (56.3%) for seeds of only one species (*K. caspicum*) (Figure 7).

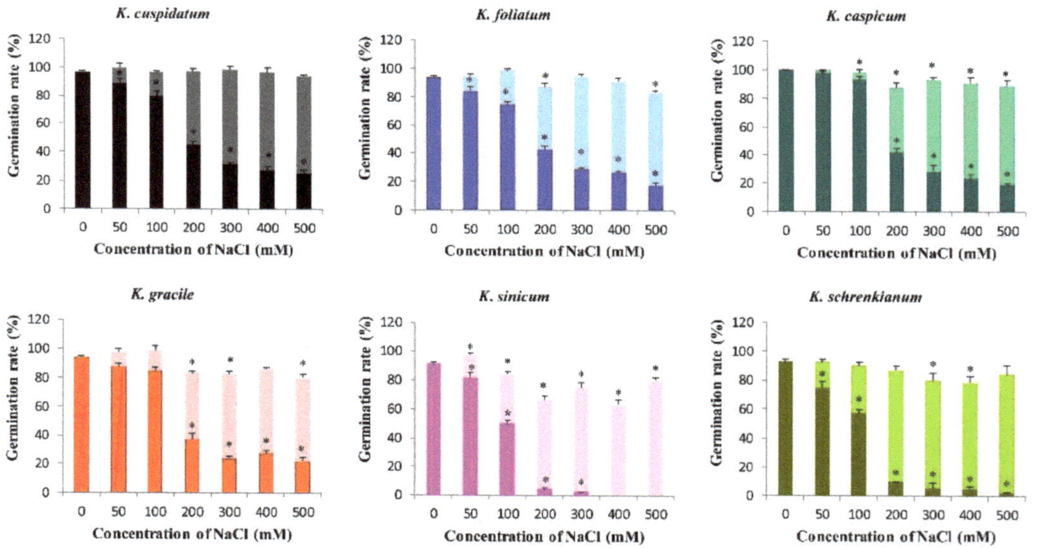

Figure 6. Total germination percentages after 14 days at the indicated NaCl concentrations in the six studied *Kalidium* species. Germination rates (%, means and standard deviations) of the six *Kalidium* species after treatments with indicated concentrations of NaCl (0, 50, 100, 200, 300, 400, and 500 mM NaCl) after 7 days. The dark bars indicate germination rates after the treatments and light bars the total percentages that germinated during the recovery treatment in distilled water with additional 7 days. Asterisks indicate significant differences in each condition respect to the corresponding control (according to Dunnet test, *p* < 0.05).

The recovery germination rates of seeds of all six *Kalidium* species after treatment with both salts were high. The recovery germination rates of *K. gracile* seeds following the NaCl treatment were negatively correlated with the NaCl concentration during the treatment, declining from 92.2 to 74.2% following exposure to 500 and 200 mM NaCl, respectively (Figure 6). However, the recovery germination rate of *K. sinicum* seeds exposed to 500 mM NaCl was high (>80%). Moreover, there were no significant differences in recovery germination rates of *K. cuspidatum* seeds exposed to different NaCl concentrations (p >0.05) and those of the other three species were all above 78% at 500 mM NaCl (Figure 6). Following treatment with Na_2SO_4, the recovery germination percentages of *K. sinicum* and *K. schrenkianum* seeds were lower than those of the other four species at the same concentrations, but were still high (76.6 and 83.2%, respectively), following exposure to the highest Na_2SO_4 concentration, 250 mM (Figure 7).

All six species produce small seeds (<0.5 g/1000 seeds, Table S1), and there was no linear relationship between the mean mass and germination rate of their seeds.

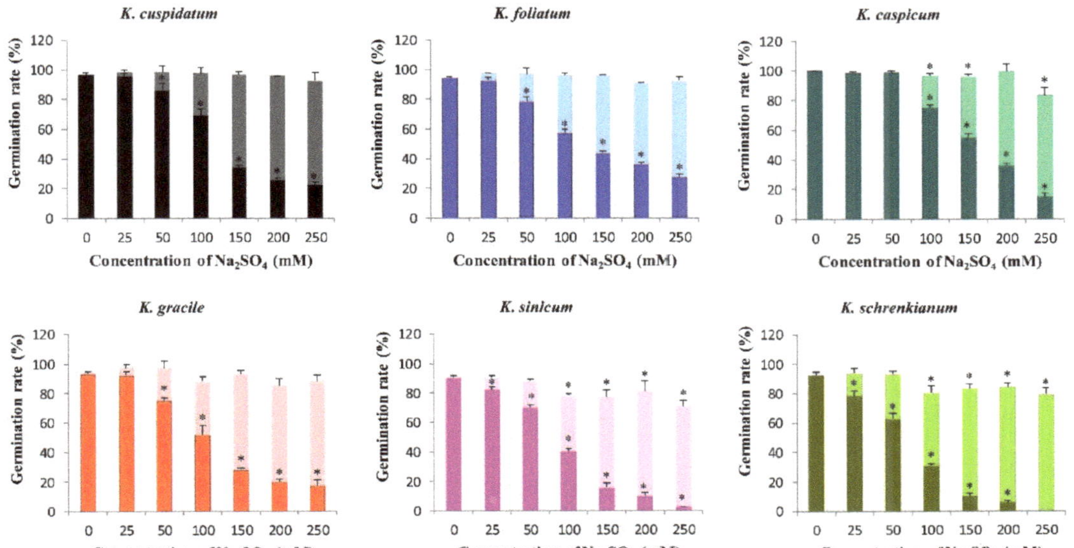

Figure 7. Total germination percentages after 14 days at the indicated Na_2SO_4 concentrations in the six studied *Kalidium* species. Germination rates (%, means and standard deviations) of the six *Kalidium* species after treatments with indicated concentrations of Na_2SO_4 (0, 25, 50, 100, 150, 200, and 250 mM) after 7 days. The dark bars indicate germination rates after the treatments and light bars the total percentages that germinated during the recovery treatment in distilled water with additional 7 days. Asterisks indicate significant differences in each condition respect to the corresponding control (according to Dunnet test, $p < 0.05$).

3.5. Principal Component Analyses

Two soil factors (pH and total salt contents), altitude, four temperature factors and two precipitation factors were used in PCA to explore relationships between abiotic factors and spatial distributions of each of the six *Kalidium* species (Figure 8 and Table S2). Principal Component (PC1) explained 40.4–50.5% of the variance and was most strongly influenced by temperature factors (maximum temperature of the warmest month, annual mean temperature, and mean temperature of the warmest quarter; Tables 2 and S2). PC2 explained 20.8–28.4% of the variance and largely reflected effects of precipitation factors on four of the six species. In addition, the two soil factors (total salt contents and pH) strongly contributed to PC3, explaining 9.96–18.4% of the variation in distribution of the six species. In total, PCs 1–3 explained >85% of the variation in the six species' distributions (Figure 8, Tables 2 and S2). The results strongly indicate that the most important ecological variables for adaptation to the species' saline environments are temperature during the driest month and precipitation. They also indicate that the selected ecological variables are strongly associated with the six *Kalidium* species' spatial distributions through their effects on the soil environment (Figure 1).

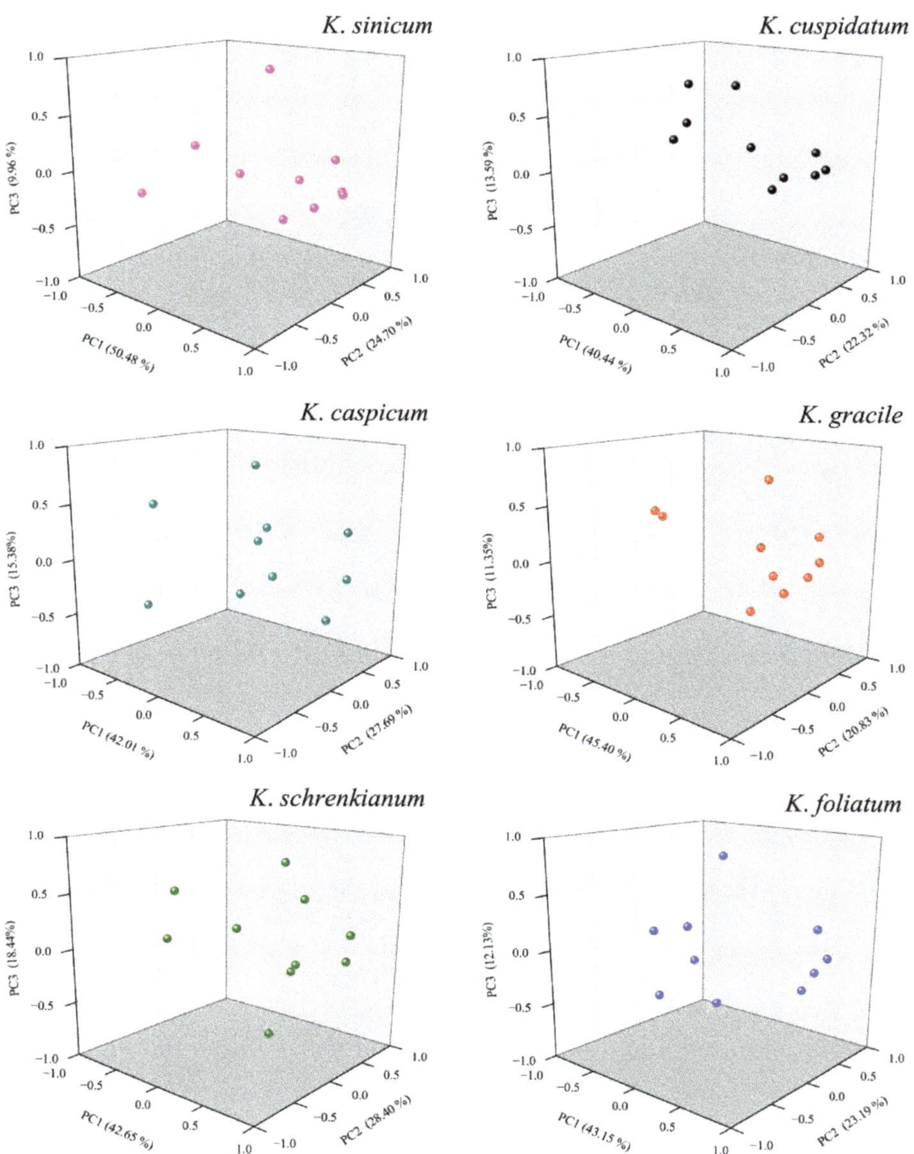

Figure 8. Score plots obtained from Principal Component Analysis (PCA) of effects of abiotic factors on distributions of the six *Kalidium* species. *x* axis, *y* axis, and *z* axis indicate the first, the second and the third Principle Component, respectively.

Table 2. Loadings of the main factors influencing the first three Principle Components obtained from Principle Component Analysis of the relationships between abiotic factors and distributions of the six *Kalidium* species.

Species	Main Factors and Loadings (Correlation Coefficients)		
	PC1	PC2	PC3
K. schrenkianum	bio5 (0.971)	pH (0.797)	TS (0.719)
K. sinicum	bio9 (0.975)	bio14 (0.812)	TS (0.906)
K.cuspidatum	bio5 (0.892)	bio14 (0.707)	TS (0.766)
K. capsicum	bio1 (0.949)	bio14 (0.870)	pH (0.833)
K. gracile	bio1 (0.993)	bio12 (0.832)	bio14 (0.658)
K. foliatum	bio9 (0.924)	bio5 (0.713)	TS (0.840)

bio1, annual mean temperature; bio5, max temperature of the warmest month; bio9, mean temperature of the driest quarter; bio12, annual average precipitation; bio14, precipitation of the driest month; TS, total salt contents.

4. Discussion

The mean pH was high (8.9–9.4) in soil samples from areas occupied by all six of the *Kalidium* species. Mean total salt contents were also high, but covered a substantial range (22.4, 19.5, 17.6, 16.5, 14.3, and 7.08 g/kg in soils from *K. cuspidatum*, *K. gracile*, *K. foliatum*, *K. caspicum*, *K. sinicum*, and *K. schrenkianum* areas, respectively). The indication that *K. schrenkianum* has relatively low salt tolerance, according to the low mean salt content of samples from areas it occupies, may at least partly explain why this species is restricted to a narrow range to the south of the Tianshan Mountains [18]. In areas occupied by all six species, total salt concentrations were positively related to pH at 0–20 cm and 20–40 cm soil depths ($r = 0.794$ and $r = 0.248$, respectively), indicating that salt contents are particularly strongly correlated to pH in topsoil in the study region.

In areas occupied by all the species, Na^+ was the main cation, and its concentrations were significantly correlated with total salt contents at all soil depths except 20–40 cm ($r = 0.632$), showing that they either require high Na^+ concentrations for optimal growth and development or at least tolerate them [6]. Ca^{2+} was the next most abundant cation at 0–40 cm soil depths, indicating that high concentrations of Ca^{2+} and Na^+ likely accumulate in all the *Kalidium* species, as previously found in roots and leaves of *K. foliatum*, *K. cuspidatum*, and various other halophytes [25,26]. As a kind of antagonist, absorption of large amounts of Ca^{2+} by roots could potentially alleviate damage to plants by other ions [27]. Concentrations of the major nutrient K^+, which is required by all living cells and often deficient in barren soil [28,29], ranged from 28.2 to 256.9 mg/kg in soil from *K. sinicum* and *K. cuspidatum* areas, respectively. Generally, in soils from areas of all six species the K^+ concentration was much lower than the Na^+ concentration and (hence) the Na^+/K^+ ratio was high (>30:1). In addition, there was no significant correlation between the K^+ concentration and total salt contents, indicating that the species' requirements for Na^+ and K^+ ions significantly differ. Mean Mg^{2+} concentrations were highest and lowest in soil from *K. gracile* and *K. foliatum* areas (ca. 217 mg/kg and 2.3-fold lower, respectively), and substantially lower than concentrations of the other measured cations in all surveyed areas.

The most abundant anion was Cl^- in *K. cuspidatum* areas (where concentrations of both Na^+ and Cl^- were highest) and SO_4^{2-} in *K. caspicum* areas (where Cl^- concentrations were lowest). Thus, anions in soils in these areas were strongly dominated by Cl^- and SO_4^{2-}, respectively (mainly balanced in both cases by Na^+). These were also the major ions in habitats of the other species, but there was less dominance by Cl^- or SO_4^{2-}, e.g., mean SO_4^{2-}, Cl^- and Na^+ concentrations in *K. foliatum* areas were 5.5, 2.6, and 2.8 g/kg, respectively. There were also wide variations in pH and total salt contents in soil samples from *K. foliatum* areas, indicating that the species has strong adaptive ability and, thus, can thrive in relatively diverse habits. Concentrations of CO_3^{2-} were nondetectable and HCO_3^- concentrations were very low (with no significant differences) in areas of the *Kalidium* species.

Plants must be sufficiently adapted to the salinity of their environments to germinate [19,30] and establish [31–34]. In our assays, the germination rates of seeds of the six *Kalidium* species were all negatively correlated with NaCl and Na$_2$SO$_4$ contents of the medium. Soil salinity fluctuates with precipitation, and can be alleviated in periods with high precipitation, so high proportions of seeds of many halophytes stored in highly saline soil may germinate during such periods [30,35,36]. Moreover, their recovery germination parameters may be major determinants of their distributions. The recovery germination rates of all six *Kalidium* species exceeded 74% after the NaCl and Na$_2$SO$_4$ treatments, corroborating the conclusions that the quality of topsoil is the first selective barrier affecting plants' distributions [14,35]. In addition, PCA showed that maximum temperature, summer rainfall and total salt contents of the soil strongly affect geographic distributions of the six *Kalidium* species. Similarly, distributions and yields of wild barley are clearly related to climatic factors, especially precipitation [37], and distributions of *Arabidopsis halleri* and *A. lyrate* are apparently linked to differences in their tolerance of the heavy metals Zn and Cd [38]. Clearly, therefore, ecological factors (and genetically-based adaptations to them) are key determinants of plants' distributions [18,37,38].

5. Conclusions

Adaptation to topsoil salinity in early stages is a major determinant of the six *Kalidium* species' geographic distributions in the study region. The dominant cation at all sites is Na$^+$, but the dominant anions at *K. caspicum* and *K. cuspidatum* sites are SO4^{2-} and Cl$^-$, respectively. Both salinity and their distributions are affected by numerous interacting factors. Inter alia, temperatures during the driest month and precipitation directly and/or indirectly affect soils' salt contents and pH, which also strongly influence the six *Kalidium* species' distributions. Clearly, high tolerance of salinity stress is a key adaptive trait of halophytes, which has multiple evolutionary origins. Moreover, major changes have occurred in plants' distributions and population sizes during desertification, following which halophytes may occupy extensive semi-arid and arid regions.

Supplementary Materials: The following supporting information can be downloaded at: https://www.mdpi.com/article/10.3390/f13122178/s1:, Table S1: Masses of 1000 seeds of the six *Kalidium* species; Table S2: Loading matrix of the Principle Components (PCs).

Author Contributions: Y.W. and Z.C. conceived and managed the project; Y.W., D.L. and X.L. performed the experiments; Y.W., Z.C., D.L. and X.L. analyzed the data; Z.C. and D.L. performed statistical analyses; Y.W. and D.L. wrote the original manuscript; and Y.W. and Z.C. reviewed and edited the final manuscript. All authors have read and agreed to the published version of the manuscript.

Funding: This work was supported by the Central Government Guides Local Scientific and Technological Development Programs of Gansu Province (Grant Number 22ZY2QG001), National Key Research and Development Program of China (Grant Number 2022YFF1303301), the National Natural Science Foundation of China (NSFC, Grant Number 41871092), and Science and Technology Project of Forestry and Grassland Bureau of Gansu Province (Grant Number 2022kj063).

Data Availability Statement: The data presented in this study are available on request from the corresponding author.

Acknowledgments: We are grateful to Kuibing Meng and Fengzhu Zhang for collecting samples in the field.

Conflicts of Interest: The authors declare that they have no conflict of interest.

References

1. Munns, R.; Tester, M. Mechanisms of salinity tolerance. *Annu. Rev. Plant Biol.* **2008**, *59*, 651–681. [CrossRef] [PubMed]
2. Cheeseman, J.M. The integration of activity in saline environments: Problems and perspectives. *Unct. Plant Biol.* **2013**, *40*, 759–774. [CrossRef] [PubMed]
3. Hanin, M.; Ebel, C.; Ngom, M.; Laplaze, L.; Masmoudi, K. New insights on plant salt tolerance mechanisms and their potential use for breeding. *Front. Plant Sci.* **2016**, *7*, 1787. [CrossRef] [PubMed]
4. Horie, T.; Karahara, I.; Katsuhara, M. Salinity tolerance mechanisms in glycophytes: An overview with the central focus on rice plants. *Rice* **2012**, *5*, 11. [CrossRef]
5. FAO. FAO Soils Portal. 2016. Available online: http://www.fao.org/soils-portal/soil-management/management-of-some-problem-soils/salt-affected-soils/more-information-on-salt-affected-soils/en (accessed on 9 November 2021).
6. Flowers, T.J.; Colmer, T.D. Plant salt tolerance: Adaptations in halophytes. *Ann. Bot.* **2015**, *115*, 327–331. [CrossRef]
7. Kachout, S.S.; Mansoura, A.B.; Jaffel, K.; Leclerc, J.C.; Rejeb, M.N. The effect of salinity on the growth of the halophyte *Atriplex hortensis* (Chenopodiaceae). *Appl. Ecol. Env. Res.* **2009**, *7*, 319–332. [CrossRef]
8. Shabala, S. Learning from halophytes: Physiological basis and strategies to improve stress tolerance in crops. *Ann. Bot.* **2013**, *112*, 1209–1221. [CrossRef]
9. Waisel, Y. *Biology of Halophytes*; Academic Press: New York, NY, USA, 1972.
10. Parida, A.K.; Jha, B. Salt tolerance mechanisms in mangroves: A review. *Trees* **2010**, *24*, 199–217. [CrossRef]
11. Zhao, K.F.; Fan, H.; Jiang, X.Y.; Song, J. Improvement and utilization of saline soil by planting halophytes. *Chin. J. Appl. Env. Biol.* **2002**, *8*, 31–35.
12. Cram, W.J. Negative feedback regulation of transport in cells. The maintenance of turgor, volume and nutrient supply. In *Encyclopaedia of Plant Physiology*; Luttge, U., Pitman, M.G., Eds.; Springer: Berlin, Germany, 1976; pp. 284–316.
13. Richau, K.H.; Schat, H. Intraspecific variation of nickel and zinc accumulation and tolerance in the hyperaccumulator *Thlaspi caerulescens*. *Plant and Soil* **2009**, *314*, 253–262. [CrossRef]
14. Turner, T.L.; Bourne, E.C.; Von Wettberg, E.J.; Hu, T.T.; Nuzhdin, S.V. Population resequencing reveals local adaptation of *Arabidopsis lyrata* to serpentine soils. *Nat. Genet* **2010**, *42*, 260–263. [CrossRef] [PubMed]
15. Lazarus, B.E.; Richards, J.H.; Claassen, V.P.; O'Dell, R.E.; Ferrell, M.A. Species-specific plant-soil interactions influence plant distribution on serpentine soils. *Plant and Soil* **2011**, *342*, 327–344. [CrossRef]
16. Piirainen, M. *Kalidium* (Chenopodiaceae) the information resource for Euro-Mediterranean plant diversity (Uotila, P., Eds.). *Eur. Med. Plantbase* **2009**, *13*, 146.
17. Liang, X.H.; Wu, Y.X. Identification of *Kalidium* species (Chenopodiaceae) by DNA barcoding. *Sci. Cold. Ari. Regi.* **2017**, *9*, 89–96.
18. Flowers, T.J.; Galal, H.K.; Bromham, L. Evolution of halophytes: Multiple origins of salt tolerance in land plants. *Funct. Plant Biol.* **2010**, *37*, 604–612. [CrossRef]
19. Tobe, K.; Li, X.M.; Omasa, K. Seed germination and radicle growth of a halophyte, *Kalidium caspicum* (Chenopodiaceae). *Ann. Bot.* **2000**, *85*, 391–396. [CrossRef]
20. Song, J.; Feng, G.; Zhang, F.S. Salinity and temperature effects on germination for three salt resistant euhalophytes, *Halostachys caspica*, *Kalidium foliatum* and *Halocnemum Strobilaceu*. *Plant and Soil* **2006**, *279*, 201–207. [CrossRef]
21. Haris, S.; Xia, H.; Elisabeth, B.; Camile, M.; Lindell, B. Predicting species' tolerance to salinity and alkalinity using distribution data and geochemical modelling: A case study using Australian grasses. *Ann. Bot.* **2015**, *115*, 343–351.
22. Lu, R.K. *Analytical Methods for Soil and Agro-Chemistry*; China Agricultural Science and Technology Press: Beijing, China, 1999.
23. Li, R.A.; Wang, F.; Qin, F.J.; Lou, F.; Wu, D.Y. Studies on the best curve equation between the total salts and the electrical conductivity of the coastal saline soil. *J. Agric.* **2015**, *5*, 59–62.
24. Wang, W.X.; Li, Z.Z.; Chang, Z.Q. Principal component analysis of geological factors related to landforms-hydrological system of Qilian mountain region. *Acta Bot. Boreali-Occident. Sin.* **2004**, *24*, 533–537. (In Chinese)
25. Tobe, K.; Li, X.M.; Omasa, K. Effects of sodium, magnesium and calcium salts on seed germination and radicle survival of a halophyte, *Kalidium caspicum* (Chenopodiaceae). *Aust. J. Bot.* **2002**, *50*, 163–169. [CrossRef]
26. Gao, R.R.; Zhao, R.H.; Du, X.M.; Huang, Z.Y.; Yang, X.J.; Wei, X.Z.; Huang, P.Y. Characteristics of root systems of two halophytes for adaptability to salinity. *Sci. Silvae Sin.* **2010**, *46*, 176–182.
27. Gul, B.; Khan, M.A. Role of calcium in alleviating salinity effects in coastal halophytes. In *Ecophysiology of High Salinity Tolerant Plants*; Khan, M.A., Weber, D.J., Eds.; Springer: Netherlands, Switzerland, 2006; pp. 107–114.
28. Very, A.A.; Sentenac, H. Molecular mechanisms and regulation of K^+ transport in higher plants. *Annu. Rev. Plant Biol.* **2003**, *54*, 575–603. [CrossRef]
29. Shabala, S.; Cuin, T.A. Potassium transport and plant salt tolerance. *Physiol. Plant.* **2007**, *133*, 651–669. [CrossRef] [PubMed]
30. El-Keblawy, A.; Elnaggar, A.; Tammam, A.; Mosa, K.A. Seed provenance affects salt tolerance and germination response of the habitat-indifferent *Salsola drummondii* halophyte in the arid Arabian deserts. *Flora* **2020**, *266*, 151592. [CrossRef]
31. Attia, H.; Al-Yasi, H.; Alamer, K.; Esmat, F.; Hassan, F.; Elshazly, S.; Hessini, K. Induced anti-oxidation efficiency and others by salt stress in *Rosa damascena* Miller. *Sci. Hortic.* **2020**, *274*, 109681. [CrossRef]
32. Yasir, T.A.; Khan, A.; Skalicky, M.; Wasaya, A.; Rehmani, M.; Sarwar, N.; Mubeen, K.; Aziz, M.; Hassan, M.M.; Hassan, F.A.S. Exogenous sodium nitroprusside mitigates salt stress in lentil (*Lens culinaris* medik.) by affecting the growth, yield, and biochemical properties. *Molecules* **2021**, *26*, 2576. [CrossRef]

33. Hassan, F.; Al-Yasi, H.; Ali, E.F.; Alamer, K.; Hessini, K.; Attia, H.; El-Shazly, S. Mitigation of salt stress effects by moringa leaf extract or salicylic acid through motivating antioxidant machinery in damask rose. *Can. J. Plant Sci.* **2021**, *101*, 157–165. [CrossRef]
34. Hassan, F.; Ali, E.; Gaber, A.; Fetouh, M.I.; Mazrou, R. Chitosan nanoparticles effectively combat salinity stress by enhancing antioxidant activity and alkaloid biosynthesis in *Catharanthus roseus* (L.) G. Don. *Plant Physiol. Biochem.* **2021**, *162*, 291–300. [CrossRef]
35. Bhatt, A.; Gallacher, D.J.; Jarma-Orozco, A.; Fernandes, D.; Pompelli, M.F. Seed provenance selection of wild halophyte seeds improves coastal rehabilitation efficiency. *Estuar. Coast. Shelf Sci.* **2021**, *265*, 107657. [CrossRef]
36. Vicente, O.; Boscaiu, M.; Naranjo, M.O.; Estrelles, E.; Bellés, J.M.; Soriano, P. Responses to salt stress in the halophyte *Plantago crassifolia* (Plantaginaceae). *J. Arid Environ.* **2004**, *58*, 463–481. [CrossRef]
37. Cronin, J.K.; Bundock, P.C.; Henry, R.J.; Nevo, E. Adaptive climatic molecular evolution in wild barley at the Isa defense locus. *PNAS* **2007**, *104*, 2773–2778. [CrossRef] [PubMed]
38. Alcazar, R.; Pecinka, A.; Aarts, M.G.; Fransz, P.F.; Koornneef, M. Signals of speciation within *Arabidopsis thaliana* in comparison with its relatives. *Curr. Opin. Plant Biol.* **2012**, *15*, 205–211. [CrossRef] [PubMed]

Article

Potential Geographical Distribution of Medicinal Plant *Ephedra sinica* Stapf under Climate Change

Kai Zhang [1,*] , Zhongyue Liu [2], Nurbiya Abdukeyum [1] and Yibo Ling [3]

1 College of Life and Geographic Sciences, Kashi University, Kashi 844006, China
2 Senior High School Department, Kashi Special Zone Experimental School, Kashi 844006, China
3 Department of Agriculture and Rural Affairs of Xinjiang, Urumqi 830000, China
* Correspondence: bzxychxzk@163.com

Abstract: *Ephedra sinica* Stapf is an important traditional medicinal plant. However, in recent years, due to climate change and human activities, its habitat area and distribution area have been decreasing sharply. In order to provide better protection for *E. sinica*, it is necessary to study the historical and future potential zoning of *E. sinica*. The maximum entropy model (MaxEnt) was used to simulate the potential geographical distribution patterns of *E. sinica* under historical and future climatic conditions simulated using two Shared Socio-economic Pathways. The main results were also analyzed using the jackknife method and ArcGIS. The results showed that: (1) the potential suitable distribution area of *E. sinica* in China is about 29.18×10^5 km^2—high-suitable areas, medium-suitable areas, and low-suitable areas cover 6.38×10^5 km^2, 8.62×10^5 km^2, 14.18×10^5 km^2, respectively—and *E. sinica* is mainly distributed in Inner Mongolia; (2) precipitation and temperature contribute more to the distribution of *E. sinica*; (3) under two kinds of SSPs, the total suitable area of *E. sinica* increased significantly, but the differences between 2021–2040, 2041–2060, 2061–2080, and 2081–2100 are not obvious; (4) the barycentre of *E. sinica* moves from the historical position to its southwest. The results show that *E. sinica* can easily adapt to future climates well, and its ecological value will become more important. This study provides scientific guidance for the protection, management, renewal and maintenance of *E. sinica*.

Keywords: dryland plant; shared socio-economic pathways; MaxEnt; spatial distribution pattern; barycentre migration

Citation: Zhang, K.; Liu, Z.; Abdukeyum, N.; Ling, Y. Potential Geographical Distribution of Medicinal Plant *Ephedra sinica* Stapf under Climate Change. *Forests* **2022**, *13*, 2149. https://doi.org/10.3390/f13122149

Academic Editors: Mark Vanderwel and Stefan Arndt

Received: 25 October 2022
Accepted: 13 December 2022
Published: 15 December 2022

Publisher's Note: MDPI stays neutral with regard to jurisdictional claims in published maps and institutional affiliations.

1. Introduction

Ephedra sinica Stapf is a herbaceous shrub species of the *Ephedra* genus in the *Ephedraceae* family. It grows on hillsides, plains, dry wastelands, riverbeds, and grasslands. It has no strict requirements on soil and is strongly resistant to drought and cold. *E. sinica* is mainly distributed in Liaoning, Jilin, Inner Mongolia, Hebei, Shanxi, Gansu, Henan, and Shaanxi provinces of China. It often forms large simple communities in the distribution areas, with important ecological value in desert community maintenance and stability [1]. *Ephedra* is a traditional herb [2] that is rich in ephedrine [3]. According to the *Pharmacopoeia of the People's Republic of China* [4], the original medical plants of the *Ephedra* genus include *E. sinica*, *Ephedra intermediate* Schrenk et C.A.Mey., *Ephedra equisetina* Bge. *Ephedra* is a traditional drug proven to treat wind-cold syndrome and is anti-inflammatory and anti-arthritic [5]. Experiments have shown that *Ephedra* can effectively reduce body temperature and improve metabolism and the immune level, and *E. sinica* has the best effect among the three kinds of *Ephedra* medicinal plants [6–8]. However, due to climate change and human activities, the total suitable habitat of wild *Ephedra* is decreasing sharply [9]. Therefore, it is vital to study the impact of climate change on the distribution prediction of wild medicinal plant resources and the impact of future climate change on species distribution for the protection of wild medicinal plant resources.

Many studies have shown that global climate change has an increasing impact on plant growth and distribution and also affects the pattern and function of ecosystems and biota by changing water and heat distribution [10,11]. Some scholars, e.g., Yan et al., found that the *Hydrangea macrophylla* distribution range is mainly affected by precipitation and temperature [10]. Under a future scenario of increased greenhouse gas emissions, the area of suitable habitats would increase, and the barycentre would have the longest migration distance [10]. For two genetic lineages of *Mikania micrantha* Kunth, the range will expand for one and decrease for another because of future climate change [12]. The extinction risk of species is related to their limited plasticity and ability to adapt to rapid changes in environmental factors (e.g., temperature or precipitation [13]. However, when involving more species, scholars have found varied changes in distribution in response to climate change or human activities [14–17]. Therefore, under global climate change, it is a topic worth discussing whether the distribution range and area of species will become larger or smaller.

According to Kim et al.'s [18] research, *E. sinica* was used to treat symptoms caused by external stressors, and its extract significantly reduces body temperature rise and improves weight loss. Park et al. [19] and Lv et al. [20] studied *E. sinica* extract, proving that the extract had a preventive effect on ulcerative colitis and had a role in the treatment of adipocyte browning and obesity, respectively. At present, the main focus is on pharmacology, although the potential distribution and distribution pattern changes under historical and future climate scenarios are rare.

For some species, detailed presence/absence occurrence data are available, allowing the use of a variety of standard statistical techniques. However, absence data are not available for most species [21]. The maximum entropy (MaxEnt) model has been widely used to predict the potential geographical distribution based on limited species distribution and bioclimate data under the climate change scenario; the bioclimate data are from the World Climate Database (WorldClim), which contains different time periods and SSPs; as the website is updated, the new database can be better used for climate change analysis [22–34]. He et al. [9] analyzed the important environmental variables affecting the distribution of three *Ephedra* species using the MaxEnt model, establishing a linear relationship between environmental variables and chemical components and determining which habitats can be used as priority conservation areas, providing a theoretical basis for the restoration, protection and cultivation of *Ephedra*. However, these papers did not analyze how *Ephedra* would respond to future climate change. Therefore, under different SSPs, the maximum and minimum potential distribution areas and regional changes of *Ephedra* need to be clarified.

In recent years, due to climate change and human activities, its habitat area and distribution area have been decreasing sharply. Therefore, it is of great significance to study the suitable habitat of *E. sinica* under historical and future climate change for its protection, development, and utilization. The MaxEnt model and bioclimatic data in WorldClim were used to analyze the potential suitable distribution and spatio-temporal evolution of *E. sinica* under two extreme climate scenarios (SSP126, sustainable path and SSP585, unsustainable path) in four different time periods (2021–2040, 2041–2060, 2061–2080, and 2081–2100), indicate the limit range of the distribution of *E. sinica* under the different climate scenarios, and analyze its main influencing factors. This study provides a scientific basis for the protection, development and utilization of wild medicinal plant resources.

2. Materials and Methods

2.1. Acquisition of Species Distribution Data

This paper selected China as the study area to study the distribution of *E. sinica* in China. The geographic distribution data (longitude and latitude) of *E. sinica* were gathered from the Chinese Virtual Herbarium (CVH, https://www.cvh.ac.cn/, (accessed on 10 September 2022)) and the Global Biodiversity Information Facility (GBIF, https://doi.org/10.15468/dl.rc84eb, (accessed on 16 September 2022)). Then, removing duplicate sample points and sample points without accurate longitude and latitude information. Data autocorrelation was disabled in ArcGIS 10.3 (Esri, Redlands, CA, USA), and the resolution was set at 10 km; finally, 56 reliable distribution points were obtained to execute the MaxEnt model program (Figure 1).

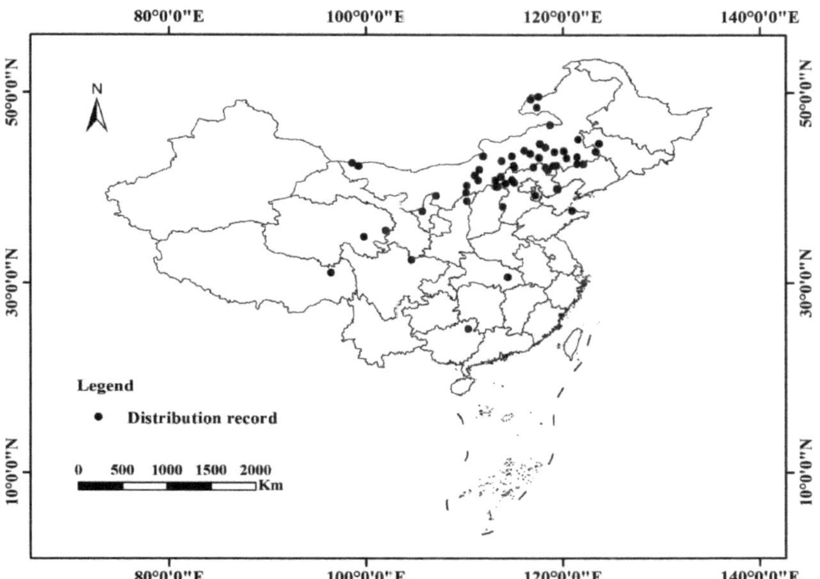

Figure 1. Distribution of *E. sinica* data points. This map was made based on the standard map No. GS (2019) 1822 downloaded from the National Administration of Surveying, Mapping and Geoinformation (NASG) of China. The base map is unchanged, and the geographical coordinate is WGS84. The same is below.

2.2. Acquisition of Environment Variable Data

The climate data in this study were all from the global climate database (https://www.worldclim.org/, (accessed on 6 September 2022)), with a spatial resolution of 30″, about 1 km × 1 km. The historical environmental data were 19 bioclimatic factors from 1970 to 2000. The MIROC6 (Model for Interdisciplinary Research On Climate version 6) model was developed by East Asian scholars and can be used to simulate the climate scenarios in East Asia.

This study downloaded the data of SSP126 (maximum) and SSP585 (minimum) climate scenarios, corresponding to the MIROC6 model in CMIP6 (Coupled Model Intercomparison Project Phase 6) in 4 different time periods of 2021–2040, 2041–2060, 2061–2080, and 2081–2100.

2.3. Build MaxEnt Model

MaxEnt mainly constructs models based on the longitude and latitude data of species' existence points and the data of species' living environment factors and expresses the degree of habitat suitability of species in the form of probability [30]. Under the condition that the sample size of species points is small and the correlation between various climatic and environmental factors is not clear, the prediction results of the MaxEnt model are better than those of other models, and the MaxEnt model could produce robust and accurate distribution maps [35,36]. During the operation of the MaxEnt model, the importance of variables can be measured by jackknife to avoid the influence of correlation among factors, which truly reflects the importance of each factor [37].

MaxEnt (version 3.4.1, Steven Phillips et al., New York, NY, USA) was used for this study (https://biodiversityinformatics.amnh.org/open_source/maxent/, (accessed on 5 July 2022)). When building the model, the geographic location information data of the sample were divided into 2 parts, of which 75% was randomly selected for model simulation, and the remaining 25% was used for model testing. The 10 simulation results of the MaxEnt model were averaged to determine the contribution of various environmental factors [38,39].

The receiver operating characteristic (ROC) curve analysis method was used to test the accuracy of the model. The area under the curve (AUC) can easily explain the accuracy of the model simulation [40]. The AUC value range is [0, 1]. If the AUC value is between 0.5 and 0.7, the prediction accuracy is poor. If the AUC value is between 0.7 and 0.9, the prediction accuracy is medium. If the AUC value is >0.9, the prediction accuracy of the model is very high [41,42].

ArcGIS 10.3 (Esri, Redlands, CA, USA) was used to visualize the simulation results and divide the habitat suitability into 4 levels (Jenks' natural breaks) [27] and reclassify the simulation results into 4 categories: unsuitable areas, low-suitable areas, medium-suitable areas, and high-suitable areas.

3. Results

Firstly, the distribution of *E. sinica* in China was analyzed. Then, the MaxEnt model was used to analyze the importance of 19 bioclimatic factors on *E. sinica*. Then, ArcGIS was used to analyze the past potential geographical distribution, adaptability degree, future distribution area, and migration of the distribution center of *E. sinica*.

3.1. Geographical Distribution

E. sinica is mainly distributed in the Inner Mongolia Autonomous Region, Hebei Province, Shanxi Province, Shaanxi Province, Beijing City, Gansu Province, and other provinces, with a total of 56 record points; and its distribution scope is concentrated in the semi-humid and semi-arid areas near the Hu Huanyong Line (Figure 1). *E. sinica* was usually distributed on hillsides, plains, arid wastelands, river beds and grasslands. It has no strict requirements on soil and has strong drought resistance and cold resistance. The provinces where the species distribution points are located meet the main conditions.

3.2. MaxEnt Model Accuracy Test

In this study, the ROC curve analysis method was used to test the accuracy of the MaxEnt model. The average AUC value of the model training set was 0.926 ± 0.022, and the AUC value of 10 simulation training sets was high (Figure 2, Table 1), indicating that the predictive accuracy of the model was good.

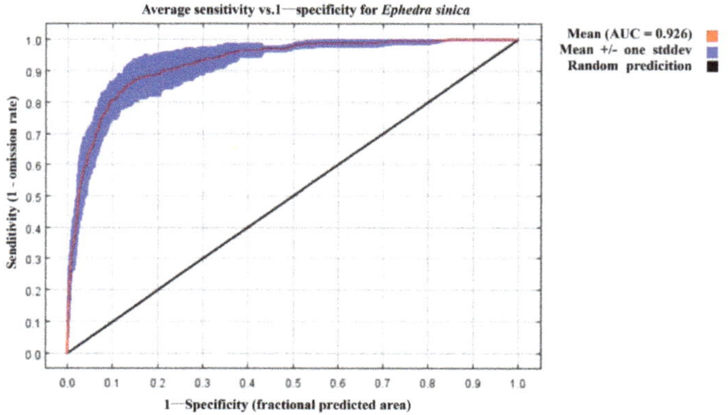

Figure 2. AUC training values for *Ephedra sinica* Stapf.

Table 1. AUC value of ten simulations.

| | AUC Value of Ten Simulations | |
No.	Training	Test
1	0.94	0.94
2	0.95	0.88
3	0.89	0.91
4	0.95	0.91
5	0.89	0.91
6	0.90	0.94
7	0.94	0.91
8	0.93	0.88
9	0.94	0.82
10	0.94	0.94

3.3. Contribution Percentage of Environment Variables

During the operation of the MaxEnt model, the importance of variables can be measured by jackknife to avoid the influence of correlation among various factors, which truly reflects their importance. The average value according to 10 simulation results of the MaxEnt model was taken to determine the contribution degree and ranking importance of each environmental factor (Table 2).

Table 2. Relative contributions of environmental variables to the Maxent model for *E. sinica*.

	Bioclimate Variable	Contribution (%)	Permutation Importance
BIO15	Precipitation Seasonality (Coefficient of Variation)	23.30	14.50
BIO4	Temperature Seasonality (standard deviation $\times$ 100)	21.00	0.70
BIO13	Precipitation of Wettest Month	19.30	26.70
BIO6	Min Temperature of Coldest Month	8.60	1.30
BIO5	Max Temperature of Warmest Month	5.30	0.40
BIO19	Precipitation of Coldest Quarter	4.80	0.50
BIO12	Annual Precipitation	3.00	4.40
BIO16	Precipitation of Wettest Quarter	2.70	17.60
BIO3	Isothermality (BIO2/BIO7) ($\times$100)	2.50	1.00
BIO2	Mean Diurnal Range (Mean of monthly (max temp—min temp))	2.00	2.60
BIO18	Precipitation of Warmest Quarter	1.80	12.70
BIO8	Mean Temperature of Wettest Quarter	1.70	9.10
BIO14	Precipitation of Driest Month	1.30	0.60
BIO17	Precipitation of Driest Quarter	1.10	4.00
BIO10	Mean Temperature of Warmest Quarter	0.70	0.70
BIO7	Temperature Annual Range (BIO5–BIO6)	0.40	1.10
BIO9	Mean Temperature of Driest Quarter	0.40	1.70
BIO1	Annual Mean Temperature	0.00	0.10
BIO11	Mean Temperature of Coldest Quarter	0.00	0.20

It can be seen from Table 2 that, among the contribution rates of various climate factors obtained by running the MaxEnt model, the top factors are precipitation seasonality (BIO15, 23.3%), temperature seasonality (BIO4, 21.0), the precipitation of the wettest quarter (BIO13, 19.3%), the minimum temperature of the coldest quarter (BIO6, 8.6%), the maximum temperature of the hottest quarter (BIO5, 5.3%), and the precipitation of the coldest quarter (BIO19, 4.8%), with a cumulative contribution rate of 82.3%; thus, the prediction of *E. sinica* can yield rich information.

In terms of the importance of the arrangement, the precipitation of the wettest month (BIO13, 26.7%), the precipitation of the wettest quarter (BIO16, 17.6%), the precipitation of the hottest quarter (BIO18, 12.7%), the precipitation seasonality (BIO15, 14.5%), and the average temperature of the wettest quarter (BIO8, 9.1%) add up to 80.6%, which also reflects the importance of precipitation and temperature on the distribution of *E. sinica*.

Furthermore, based on the training gains of different climate factors analyzed by the jackknife method (Figure 3), the average value of the 10 simulation results shows that the environment variable with the highest gain when simulated alone is BIO12; thus, it itself has the most useful information. BIO15 will reduce the benefit to the greatest extent when omitted; therefore, it has the most information that does not exist in other variables.

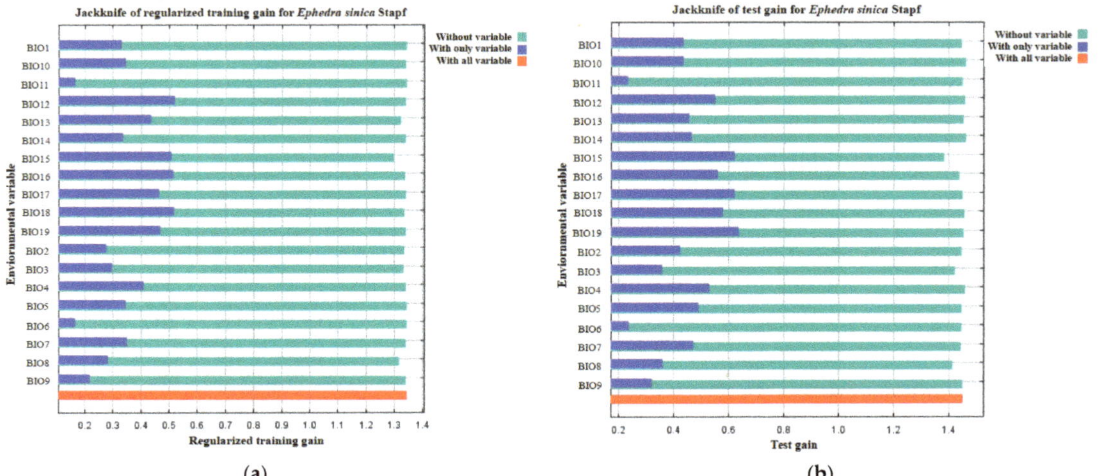

(a) (b)

Figure 3. The Jackknife test with (**a**) training gain and (**b**) test gain for variable importance. The blue bar (left part of each variable row) indicates model gain with a single variable, and the cyan bar (right part of each variable row) indicates model gain with the other remaining variables. The red bar (base, all variables) represents the total gain with all variables in the final model.

3.4. Threshold Analysis of Important Environmental Variables

It can be seen from the response curve analysis that the climate characteristics of *E. sinica* in China's distribution area are precipitation seasonality (BIO15; 25.4–150.6), temperature seasonality (BIO4; 308.1%–1773.9%), the precipitation of the wettest month (BIO13; 3.0–768.0 mm), the minimum temperature of the coldest month (BIO6; 37.2–16.6 °C), the maximum temperature of the hottest month (BIO5; 1.8–42.1 °C), and the precipitation of the coldest month (BIO19; 0–461.0 mm). Six bioclimatic factors mainly determine the living range of *E. sinica* and also show the species' basic niches formed in the process of adapting to the environment. This shows that the growth of *E. sinica* is mainly affected by "extreme" precipitation and "extreme" temperature. It has a wide range of temperature tolerance, can withstand a certain degree of low temperature, and is suitable for growing in arid and semi-arid deserts, grassland, sandy land, river beaches, etc. (Figure 4).

3.5. Evaluation of Potential Geographical Distribution and Suitable Areas

The total distribution area of *E. sinica* in China is about 29.18×10^5 km^2, accounting for 30.33% of the land area. High-suitable areas are mainly located in the middle of the Inner Mongolia Autonomous Region, northern Shaanxi Province, northern and central Shanxi Province, central and southern Hebei Province, western Liaoning Province, western Jilin Province, southwestern Heilongjiang Province, northwestern Beijing, Gansu Province, and the Ningxia Hui Autonomous Region of Qinghai Province (Figure 5). The optimal area based on model simulation is about 6.38×10^5 km^2.

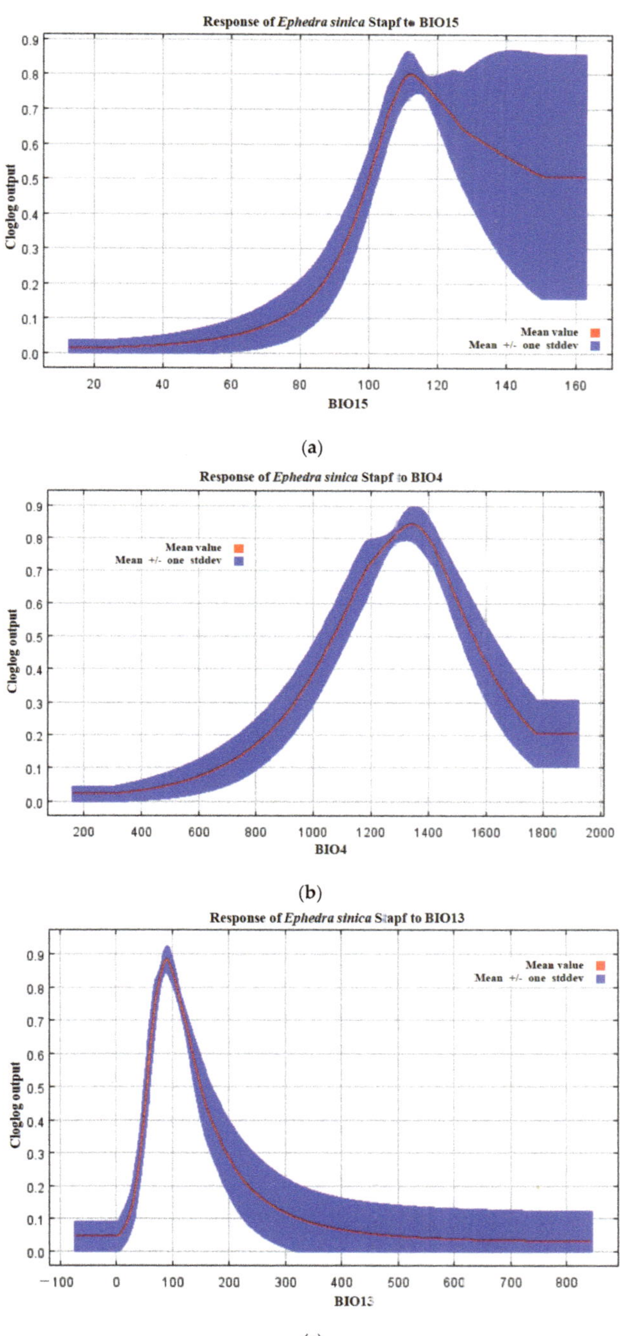

Figure 4. *Cont.*

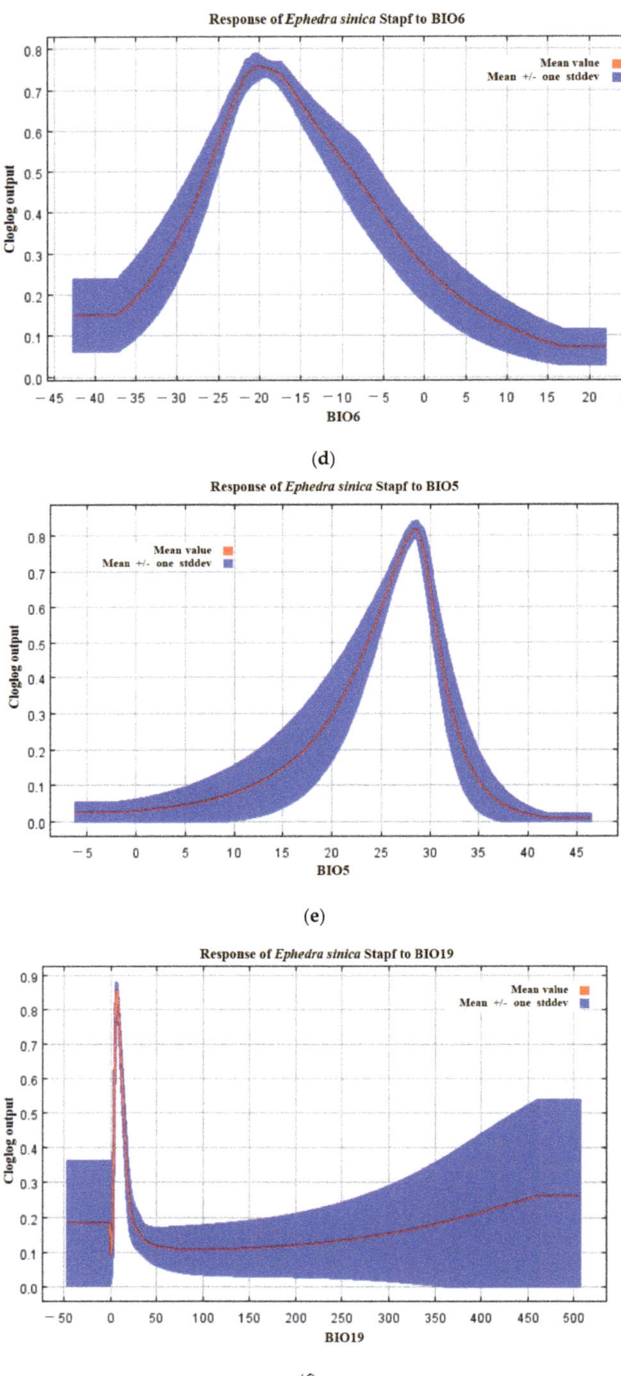

(d)

(e)

(f)

Figure 4. The response curve of dominant environment variables. (**a–f**) represent the response curves of ephedra under the influence of BIO15, BIO4, BIO13, BIO6, BIO5 and BIO19 respectively.

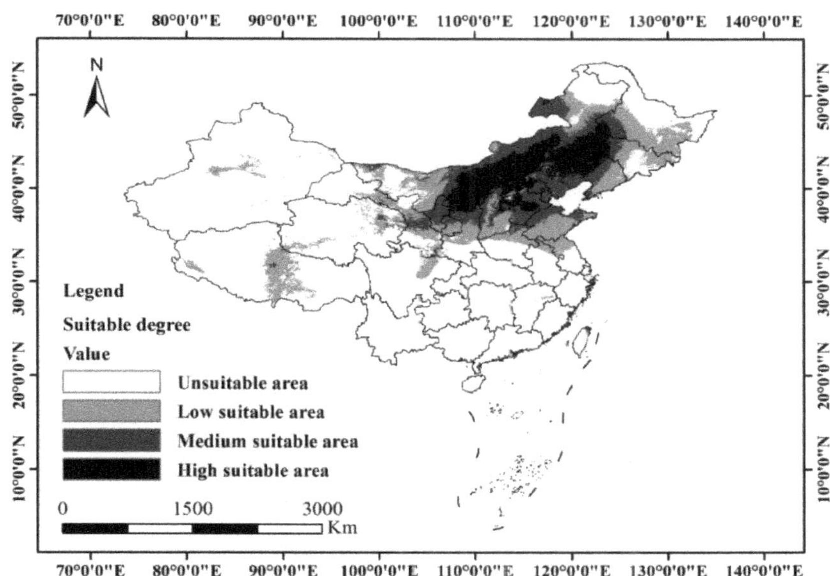

Figure 5. Distribution of suitable areas for *E. sinica*. The map source is the same as in Figure 1.

The medium-suitable areas are located around the most suitable regions. There are also some suitable regions in the center of the Tibet Autonomous Region, the center of the Xinjiang Uygur Autonomous Region, Gansu Province, Taiwan and Qinghai Province. The distribution of the medium-suitable area covers about 8.62×10^5 km^2 (Figure 5).

The distribution of low-suitability area covers about 14.18×10^5 km^2, mainly concentrated in the east and west of the Inner Mongolia Autonomous Region, the east and center of Xinjiang, the west and center of Tibet, the east and west of Qinghai Province, the center and east of Gansu Province, the south of the Ningxia Hui Autonomous Region, the center of Shaanxi Province, the south and center of Shanxi Province, the north and center of Henan Province, the north of Anhui Province, the north of Jiangsu Province, most of Shandong Province, the east of Liaoning Province, and northeast and center of Jilin Province. There are also a few in Hebei Province, Jiangxi Province and Taiwan Province in the east-central part of Heilongjiang Province (Figure 5).

3.6. Future Trends in the Barycentre of Suitable Habitat

Through simulation, it was found that, under future climate scenarios, the distribution area of *E. sinica* will become larger, and the distribution range will continue to increase (Table 3). The results infer that the ecological value of *E. sinica* will become more and more important. Using ArcGIS to analyze the center of gravity transfer trajectories of *E. sinica* at different suitability levels, it was found that the center of gravity coordinates of high-suitable, medium-suitable, and low-suitable areas were 115.7090° E, 42.0735° N; 115.0260° E, 41.2850° N; and 110.7560° E, and 38.8274° N, respectively, under historical climate conditions. Under SSP126, the distribution barycentre of the three types of suitable areas will shift from northeast to southwest, compared with historical climate conditions. Under SSP585, the gravity center of different suitable areas will also move to the southwest. It shows that climate change will cause the barycentre of the *E. sinica* distribution area to migrate (Figure 6).

Table 3. Potential distribution area changes in different periods.

Shared Socio-Economic Pathways	Time Periods	Area ($\times 10^4$ km^2)
Historical	1970–2000	291.76
SSP126	2021–2040	611.40
	2041–2060	617.67
	2061–2080	624.19
	2081–2100	644.53
SSP585	2021–2040	630.14
	2041–2060	648.23
	2061–2080	621.06
	2081–2100	644.25

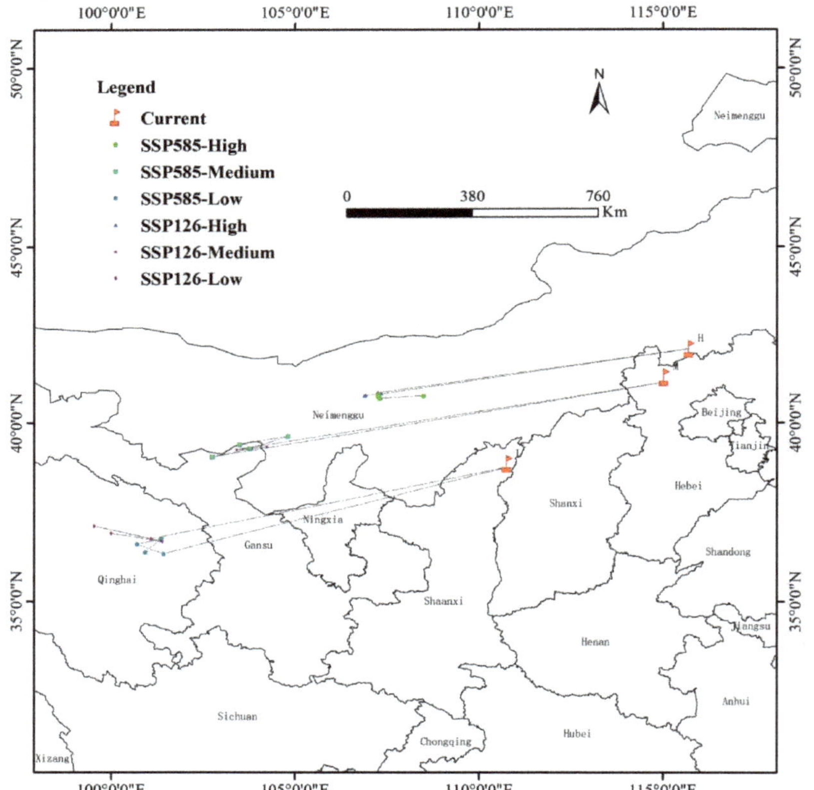

Figure 6. The change trends in the gravity points of the suitable areas for *E. sinica* under different climatic conditions. The map source is the same as in Figure 1.

4. Discussion

In this study, the MaxEnt model was selected to simulate the historical potential distribution and future potential distribution of *E. sinica*; it was found that the simulation accuracy was very good. Some scholars also found that the prediction accuracy indicators of the MaxEnt model are all greater than 0.90 [41,43], which is consistent with these results. Therefore, this model can be used as a powerful tool to study the potential distribution of species and plant distribution under future climate change scenarios [44–46]. Of course, some scholars (e.g., Cotto et al.) believed that the high fitting degree of MaxEnt fitting results does not mean that the simulation results can accurately reflect the actual distribution and potential distribution of species [47]. In this regard, we should be more careful in interpreting the results.

The main bioclimatic factors affecting the distribution of *E. sinica* are precipitation seasonality, temperature seasonality, the precipitation of the wettest month, the minimum temperature of the coldest month, the maximum temperature of the hottest month, and the precipitation of the coldest quarter. The order of importance was as follows, the precipitation of the wettest month, the precipitation of the wettest quarter, the precipitation of the hottest quarter, precipitation seasonality, and the average temperature of the wettest quarter are more important. It can be seen from this that the distribution of *E. sinica* in China is limited by water and temperature, indicating that *E. sinica* has a wide range of temperature ecology, can withstand a certain degree of drought, and can grow in arid and semi-arid deserts, grasslands, sandy lands, river beaches, etc.

The distribution center of *E. sinica* is Inner Mongolia Autonomous Region-Ningxia Hui Autonomous Region–Qinghai Province. *Flora of China* [1] records that the flowering period of *E. sinica* is concentrated from May to June, and the seed maturity is concentrated from August to September; there, it can be inferred that *E. sinica* completes its life cycle in a relatively short time (2–3 months), which is consistent with the hydrothermal conditions in the growing season of *E. sinica*. The environment in these areas is mainly semi-humid and semi-arid desert and sandy land, with four distinct seasons and harsh conditions. In the assessment area, the very serious and very serious interference areas are mainly distributed in the south and east of Inner Mongolia, and the middle interference areas are concentrated in Gansu and Inner Mongolia [10]. Due to the dry climate, cold waves and frequent wind and sand dust disasters, plants in some areas are difficult to survive, which can be proved by these results and the results of the MaxEnt model.

Withstanding extinction while facing rapid climate change depends on a species' ability to track its ecological niche or to evolve a new one [47]. The adaptation of plants to arid environments can be reflected by their functional traits [48]. However, when it is difficult to obtain functional traits, the distribution area and distribution area of plants can reflect the adaptation of plants to the environment. This paper has chosen two extreme pathways of social and economic sharing. Under SSP126 and SSP585, the distribution range and area of *E. sinica* increased significantly, but the two different SSP pathways had no significant impact on the distribution of *E. sinica*. In the future, due to the impact of human socio-economic activities, the potential distribution of *E. sinica* will change significantly, *E. sinica* can adapt well both in the sustainable development pathways and the unsustainable development way, and its distribution area could rapidly expand and be maintained, indicating that the survival prospects of *E. sinica* are improving under climate change. This may be related to future climate warming and the redistribution of hydrothermal conditions in the temperate zone where *E. sinica* grows. This is consistent with Wang et al.'s research on the increase in the suitable habitat of *Tricholoma matsutake* in the western Sichuan Plateau under the future climate scenario [49].

The barycenters of different suitability levels of *E. sinica* in the future climate were extracted using ArcGIS, and their transfer tracks were analyzed. It was found that the range of *E. sinica* will move to the southwest in the future. When Xu et al. [26] studied the future distribution of the traditional medicinal plant *Rheum nanum* Siev. ex Pall., they also found that its distribution center will shift to the south. Similarly, the southwest movement of *E. sinica* may be due to the redistribution of temperature and precipitation caused by global warming; thus, it moves to the southwest to expand new distribution areas.

The rapid economic growth, overexploitation of natural resources, habitat degradation, pollution and pressure related to global climate change are all serious challenges to plant protection in the new millennium. At the same time, management problems (such as lack of protection awareness of government officials and local people, imperfect legal system) and insufficient basic research on endangered species are also obstacles to success [50]. The protection and sustainable use of biodiversity is an important task of nature conservation.

Mankind's predatory development and utilization of *E. sinica* may reduce its numbers to a certain extent [9]; for the protection, development, and utilization of *E. sinica*, therefore, we should not only pay attention to climate change but also consider the impact of human beings.

This paper analyzes the potential distribution of *E. sinica*, the bioclimatic factors affecting its distribution, and the future distribution changes, obtains some good results, and puts forward some useful suggestions. Whether the simulation results of the MaxEnt model and other niche models are consistent? If not, what is the reason and how to improve? How about the impact of human beings on the distribution of *E. sinica*, and how to measure the impact of human beings? Are more species consistent with climate change? How can biodiversity be maintained? How does the higher classification unit "genus" adapt to climate change? These all need to be further studied.

5. Conclusions

The MaxEnt model can be used to simulate the potential distribution of species, but attention should be paid to the interpretation of the model results, and excessive inference is not allowed. According to the data collection points and simulated niches of *E. sinica*, the conditions of temperature and water, namely the combination of water and heat, limit the changes in the actual distribution and future distribution of *E. sinica*. Therefore, regional moisture and heat conditions should be considered when protecting *E. sinica*. The existing distribution area of *E. sinica* is greatly disturbed by human activities, so it is necessary to protect *E. sinica* pertinently, such as setting up natural reserves, carrying out publicity and education, raising people's awareness of *E. sinica* and enhancing the awareness of ecological environment protection.

This study provides a research model for the evolution of the distribution pattern of *E. sinica*, an important traditional medicinal plant in ethnic minority areas, and also provides a theoretical basis for the protection and management of *E. sinica*, a single community in desert areas, which is conducive to maintaining regional ecological balance and sustainable development. Of course, in view of the small sample size of research species, it is not enough to study only one kind of plant as an indicator of environmental change and a guide for government decision-making.

In the future, authors will continue to pay attention to plants in arid areas and deduce the response of plants to climate change through experiments and simulation calculations so as to provide better guidance for regional ecological balance and government decision-making. For example, these topics should continue to be studied in depth; the response of more plant species' geographical distribution patterns to climate change, the analysis of biodiversity maintenance mechanisms, and the adaptive changes in plant evolution should be analyzed from the higher taxon "genus", etc.

E. sinica, as an important wild medicinal plant resource, should take on-site protection measures for large populations as soon as possible, such as delimiting natural population protection areas and ecological protection areas of *E. sinica* in Inner Mongolia, which can not only protect the local ecological environment, but also protect the wild resources of *E. sinica*, and reserve natural germplasm resources for domestication, cultivation and resource development. In addition, *E. sinica* cultivation should be actively promoted in other potentially suitable areas as the main repair service for ecological restoration of sandy land, and the local people should be guided to cultivate so as to ensure species protection, desert control and economic development into a coordinated and sustainable state.

Author Contributions: Conceptualization, K.Z.; methodology, K.Z.; software, Z.L.; validation, K.Z. and Z.L.; formal analysis, K.Z. and Z.L.; investigation, K.Z. and Z.L.; resources, K.Z.; data curation, K.Z. and Z.L.; writing—original draft preparation, K.Z.; writing—review and editing, Z.L. and N.A.; visualization, Z.L. and N.A.; supervision, Y.L.; project administration, K.Z.; funding acquisition, K.Z. All authors have read and agreed to the published version of the manuscript.

Funding: This research was funded by the Kashi University High-level Talents Research Initiation Funding Project, grant number GCC2020ZK-002.

Data Availability Statement: Not applicable.

Conflicts of Interest: The authors declare no conflict of interest.

References

1. Flora of China. Available online: http://www.iplant.cn/foc (accessed on 2 September 2022).
2. Ibragic, S.; Sofic, E. Chemical composition of various ephedra species. *Bosn. J. Basic Med. Sci.* **2015**, *15*, 21–27. [CrossRef] [PubMed]
3. Lu, Q.; Wang, J.H.; Zhu, J.M. *Atlas of Desert Plants in China*, 1st ed.; China Forestry Publishing House: Beijing, China, 2012; p. 16.
4. Pharmacopoeia of the People's Republic of China. Available online: https://db.ouryao.com/yd2020/ (accessed on 2 September 2022).
5. Yeom, M.J.; Lee, H.C.; Kim, G.H.; Lee, H.J.; Shim, I.; Oh, S.K.; Kang, S.K.; Hahm, D.H. Anti-arthritic effects of *Ephedra sinica* STAPF herb-acupuncture: Inhibition of lipopolysaccharide-induced inflammation and adjuvant-induced polyarthritis. *J. Pharmacol. Sci.* **2006**, *100*, 41–50. [CrossRef] [PubMed]
6. Li, H.Y.; Ding, X.Y.; Zhang, D.; An, Q.; Jin, Y.; Zhan, Z.L.; Zheng, Y.G. Herbal textual research on ephedrae herba in famous classical formulas. *Chin. J. Exp. Tradit. Med. Formulae* **2022**, *28*, 102–110.
7. Zhang, B.B.; Zeng, M.N.; Zhang, Q.Q.; Jia, J.F.; Wang, R.; Guo, P.L.; Liu, M.; Jian, L.H.; Feng, W.S.; Zheng, X.K. Intervention effects of *Ephedra sinica* Stapf, *Ephedra intermedia* Schrenk et C. A. Mey., and *Ephedra equisetina* Bge. on rat model of fenghanbiaozheng. *Pharmacol. Clin. Chin. Mater. Med.* **2022**, *38*, 121–127.
8. Jia, J.F.; Zeng, M.N.; Zhang, B.B.; Guo, P.L.; Liu, M.; Zhang, Q.Q.; Wang, R.; Zheng, X.K.; Feng, W.S. Study on the differences of immune effects of water extracts of *Ephedra sinica* Stapf, *Ephedra intermedia* Schrenk et C. A. Mey., *Ephedra equisetina* Bge. on wind-cold superficies syndrome in rats. *Chem. Life* **2021**, *41*, 2265–2273.
9. He, P.; Li, J.Y.; Li, Y.F.; Xu, N.; Gao, Y.; Guo, L.F.; Huo, T.T.; Peng, C.; Meng, F.Y. Habitat protection and planning for three *Ephedra* using the MaxEnt and Marxan models. *Ecol. Indic.* **2021**, *133*, 108399. [CrossRef]
10. Yan, X.Y.; Wang, S.C.; Duan, Y.; Han, J.; Huang, D.H.; Zhou, J. Current and future distribution of the deciduous shrub *Hydrangea macrophylla* in China estimated by MaxEnt. *Ecol. Evol.* **2021**, *11*, 16099–16112. [CrossRef]
11. Zhang, Y.; An, M.T.; Wu, J.Y.; Liu, F.; Wang, W. Geographical distribution pattern and dominant climatic factors of the *Paphiopedilum Subgen. Brachypetalum* in China. *Chin. J. Plant Ecol.* **2022**, *46*, 40–50.
12. Banerjee, A.K.; Mukherjee, A.; Guo, W.X.; Ng, W.L.; Huang, Y.L. Combining ecological niche modeling with genetic lineage information to predict potential distribution of *Mikania micrantha* Kunth in South and Southeast Asia under predicted climate change. *Glob. Ecol. Cons.* **2019**, *20*, e00800. [CrossRef]
13. Antúnez, P. Main environmental variables influencing the abundance of plant species under risk category. *J. For. Res.* **2021**, *33*, 1209–1217. [CrossRef]
14. Alkhalifah, D.H.M.; Damra, E.; Khalaf, S.M.H.; Hozzein, W.N. Biogeography of black mold *Aspergillus niger*: Global situation and future perspective under several climate change scenarios using MaxEnt modeling. *Diversity* **2022**, *14*, 845. [CrossRef]
15. Nzei, J.M.; Ngarega, B.K.; Mwanzia, V.M.; Kurauka, J.K.; Wang, Q.F.; Chen, J.M.; Li, Z.Z.; Pan, C. Assessment of climate change and land use effects on Water Lily (*Nymphaea* L.) habitat suitability in South America. *Diversity* **2022**, *14*, 830. [CrossRef]
16. Cunningham, M.A. Climate change, Agriculture, and Biodiversity: How does shifting agriculture affect habitat availability? *Land* **2022**, *11*, 1257. [CrossRef]
17. Buendía-Espinoza, J.C.; Martínez-Ochoa, E.D.C.; Díaz-Aguilar, I.; Cahuich-Damián, J.E.; Zamora-Elizalde, M.C. Identifying potential planting sites for three non-native plants to be used for soil rehabilitation in the Tula Watershed. *Forests* **2022**, *13*, 270. [CrossRef]
18. Kim, W.; Lee, W.; Huh, E.; Choi, E.; Jang, Y.P.; Kim, Y.-K.; Lee, T.-H.; Oh, M.S. *Ephedra sinica* Stapf and gypsum attenuates heat-induced hypothalamic inflammation in mice. *Toxins* **2020**, *12*, 16. [CrossRef]
19. Park, S.-J.; Shon, D.-H.; Ryu, Y.-H.; Ko, Y. Extract of *Ephedra sinica* Stapf induces browning of mouse and human white adipocytes. *Foods* **2022**, *11*, 1028. [CrossRef]
20. Lv, M.; Wang, Y.; Wan, X.; Han, B.; Yu, W.; Liang, Q.; Xiang, J.; Wang, Z.; Liu, Y.; Qian, Y.; et al. Rapid screening of proanthocyanidins from the roots of *Ephedra sinica* Stapf and its preventative effects on dextran-sulfate-sodium-induced ulcerative colitis. *Metabolites* **2022**, *12*, 957. [CrossRef]
21. Phillips, S.J.; Anderson, R.P.; Schapire, R.E. Maximum entropy modeling of species geographic distributions. *Eco. Model.* **2006**, *190*, 231–259. [CrossRef]
22. Du, Q.; Wei, C.H.; Liang, C.T.; Yu, J.H.; Wang, H.M.; Wang, W.J. Future climatic adaption of 12 dominant tree species in Northeast China under 3 climatic scenarios by using MaxEnt modeling. *Acta Ecol. Sin.* **2022**, *42*, 1–14.
23. Ma, S.M.; Zhang, M.L.; Chen, X. Potential geographical distribution of Genus *Ammopitanthus* (Leguminosae) in the Eastern Central Asian Desert and its determinant environmental factors. *J. Desert. Res.* **2012**, *32*, 1301–1307.
24. Fungjanthuek, J.; Huang, M.J.; Hughes, A.C.; Huang, J.F.; Chen, H.H.; Gao, J.; Peng, Y.Q. Ecological niche overlap and prediction of the potential distribution of two sympatric ficus (Moraceae) species in the Indo-Burma Region. *Forests* **2022**, *13*, 1420. [CrossRef]

25. Liu, M.L.; Sun, H.Y.; Jiang, X.; Zhou, T.; Zhang, Q.J.; Su, Z.D.; Zhang, Y.N.; Liu, J.N.; Li, Z.H. Simulation and prediction of the potential geographical distribution of Acer cordatum Pax in different climate scenarios. *Forests* **2022**, *13*, 1380. [CrossRef]
26. Xu, W.; Zhu, S.; Yang, T.; Cheng, J.; Jin, J. Maximum entropy niche-based modeling for predicting the potential suitable habitats of a traditional medicinal plant (*Rheum nanum*) in Asia under climate change conditions. *Agriculture* **2022**, *12*, 610. [CrossRef]
27. Meng, Y.; Ma, J.M.; Wang, Y.Q.; Mo, Y.H. Prediction of distribution area of *Loropetalum chinense* based on Maxent model. *Acta Ecol. Sin.* **2020**, *40*, 8287–8296.
28. Qi, S.; Luo, W.; Chen, K.-L.; Li, X.; Luo, H.L.; Yang, Z.Q.; Yin, D.M. The prediction of the potentially suitable distribution area of *Cinnamomum mairei* H. Lév in China based on the MaxEnt model. *Sustainability* **2022**, *14*, 7682. [CrossRef]
29. Wan, J.N.; Mbari, M.J.; Wang, S.W.; Liu, B.; Mwangi, B.N.; Rasoarahona, J.R.E.; Xin, H.P.; Zhou, Y.D.; Wang, Q.F. Modeling impacts of climate change on the potential distribution of six endemic baobab species in Madagascar. *Plant Divers.* **2021**, *43*, 117–124. [CrossRef]
30. Zhang, L.; Wei, Y.Q.; Wang, J.N.; Zhou, Q.; Liu, F.G.; Chen, Q.; Liu, F. The potential geographical distribution of *Lycium ruthenicum* Murr under differernt climate change scenarios. *Chin. J. Appl. Envion. Biol.* **2020**, *26*, 969–978.
31. Zhang, L.X.; Chen, X.L.; Xin, X.G. Short commentary on CMIP6 Scenario Model Intercomparison Project (ScenarioMIP). *Clim. Change Res.* **2019**, *15*, 519–525.
32. Wang, Y.; Li, H.X.; Wang, H.J.; Sun, B.; Chen, H.P. Evaluation of CMIP6 model simulations of extreme precipitation in China and comparison with CMIP5. *Acta Meteorol. Sin.* **2021**, *79*, 369–386.
33. Xia, S.; Liu, P.; Jiang, Z.H.; Cheng, J. Simulation evaluation of AMO and PDO with CMIP5 and CMIP6 models in historical experiment. *Adv. Earth Sci.* **2021**, *36*, 58–68.
34. Hu, T.; Lv, S.H.; Chang, Y.; Yang, M.X.; Luo, J.X.; Cheng, X.Q. Analysis and prediction of permafrost changes in Qinghai-Xizang Plateau by CMIP6 climate models. *Plateau Meteorol.* **2022**, *41*, 363–375.
35. Antúnez, P. Influence of physiography, soil and climate on taxus globosa. *Nord. J. Bot.* **2021**, *39*, 03058. [CrossRef]
36. Kaky, E.; Nolan, V.; Alatawi, A.; Gilbert, F. A comparison between Ensemble and MaxEnt species distribution modelling approaches for conservation: A case study with Egyptian medicinal plants. *Ecol. Inform.* **2020**, *60*, 101150. [CrossRef]
37. Merow, C.; Smith, M.J.; Silander, J.A., Jr. A partical guide to MaxEnt for modeling species' distribution: What it does, and why inputs and settings matter. *Ecography* **2013**, *36*, 1–12. [CrossRef]
38. Zhang, H.; Zhao, H.X.; Wang, H. Potential geographical distribution of *Populus euphratica* in China under future climate change scenarios based on MaxEnt model. *Acta Ecol. Sin.* **2020**, *40*, 6552–6563.
39. Yan, H.; Ma, S.M.; Wei, B.; Zhang, H.X.; Zhang, D. Historical distribution patterns and environmental drivers of relict shrub *Amygdalus pedunculata. Chin. J. Plant Ecol.* **2022**, *46*, 766. [CrossRef]
40. Ying, L.X.; Liu, Y.; Chen, S.T.; Shen, Z.H. Simulation of the potential range of *Pistacia weinmannifolia* in Southwest China with climate change based on the maximum-entropy model. *Biodivers. Sci.* **2016**, *24*, 453–461. [CrossRef]
41. Chu, J.M.; Li, Y.F.; Zhang, L.; Li, B.; Gao, M.Y.; Tang, X.Q.; Ni, J.W.; Xu, X.Q. Potential distribution range and conservation strategies for the endangered species *Amygdalus pedunculata. Biodivers. Sci.* **2017**, *25*, 799–806. [CrossRef]
42. Kim, J.Y.; Kim, G.-Y.; Do, Y.; Park, H.-S.; Joo, G.-J. Relative importance of hydrological variables in predicting the habitat suitability of *Euryale ferox* Salisb. *J. Plant Ecol.* **2018**, *11*, 169–179.
43. Xiong, Q.L.; He, Y.L.; Deng, F.Y.; Li, T.Y.; Yu, L. Assessment of alpine mean response to climate change in Southwest China based on MaxEnt model. *Acta Ecol. Sin.* **2019**, *39*, 9033–9043.
44. Venne, S.; Currie, D.J. Can habitat suitability estimated from MaxEnt predict colonizations and extinctions? *Divers. Distrib.* **2021**, *27*, 873–886. [CrossRef]
45. Zhang, W.P.; Hu, Y.Y.; Li, Z.H.; Feng, X.P.; Li, D.W. Predicting suitable distribution areas of *Juniperus przewalskii* in Qinghai Province under climate change scenarios. *Chin. J. Appl. Ecol.* **2021**, *32*, 2514–2524.
46. Liu, Z.L.; Hu, L.L. Prediction of potential distribution and climate change of rare species *Cephalotaxus oliveri. For. Resour. Wanagement* **2022**, *1*, 35–42.
47. Cotto, O.; Wessely, J.; Georges, D.; Klonner, G.; Schmid, M.; Dullinger, S.; Thuiller, W.; Guillaume, F. A dynamic eco-evolutionary model predicts slow response of alpine plants to climate warming. *Nat. Commun.* **2017**, *8*, 15399. [CrossRef] [PubMed]
48. Yang, X.-D.; Anwar, E.; Zhou, J.; He, D.; Gao, Y.-C.; Lv, G.-H.; Cao, Y.-E. Higher association and integration among functional traits in small tree than shrub in resisting drought stress in an arid desert. *Environ. Exp. Bot.* **2022**, *201*, 104993. [CrossRef]
49. Wang, Q.L.; Wang, R.L.; Zhang, L.P.; Han, Y.J.; Wang, M.T.; Chen, H.; Chen, J.; Guo, W. Climatic ecological suitability and potential distribution of *Trichlolma matsutake* in western Sichuan Plateau, China Based on MaxEnt model. *Chin. J. Appl. Ecol.* **2021**, *32*, 2525–2533.
50. Ma, Y.-P.; Chen, G.; Grumbine, R.E.; Dao, Z.; Sun, W.; Guo, H. Conserving plant species with extremely small populations (psesp) in china. *Biodivers. Conserv.* **2013**, *22*, 803–809. [CrossRef]

Article

Leaf Stoichiometry of Halophyte Shrubs and Its Relationship with Soil Factors in the Xinjiang Desert

Yan Luo [1,2,3,4,*,†] , Cuimeng Lian [1,†], Lu Gong [1,2,3] and Chunnan Mo [1]

1 College of Ecology and Environment, Xinjiang University, Urumqi 830017, China
2 Key Laboratory of Oasis Ecology of Education Ministry, Urumqi 830017, China
3 Xinjiang Jinghe Observation and Research Station of Temperate Desert Ecosystem, Ministry of Education, Xinjiang University, Urumqi 830017, China
4 Ecological Postdoctoral Research Station, Xinjiang University, Urumqi 830017, China
* Correspondence: luoyan505@xju.edu.cn
† These authors contributed equally to this work.

Abstract: Desert halophytes are a special plant group widely distributed in desert ecosystems. Studying their ecological stoichiometric characteristics is helpful for understanding their nutrient utilization characteristics and survival strategies. In this study, three functional groups of halophyte shrubs (euhalophytes, pseudohalophytes, and secretohalophytes) were studied in the Xinjiang desert, and the ecological stoichiometric characteristics of their leaves and their relationships with soil factors were evaluated. The results showed that the C content in secretohalophytes (442.27 ± 3.08 mg g^{-1}) was significantly higher than that in the other functional groups ($p < 0.05$). The N and P contents in euhalophytes (22.17 ± 0.49 mg g^{-1} and 1.35 ± 0.04 mg g^{-1}, respectively) were significantly higher than those in halophytes ($p < 0.05$). The N/P results showed that the growth rates of euhalophytes and pseudohalophytes were more susceptible to P limitation, whereas that of secretohalophytes was more susceptible to both N and P limitations, indicating that there were differences in nutrient characteristics among different functional groups. The results of the redundancy analysis showed that the leaf C, N, and P contents of euhalophytes were most affected by electrical conductivity (EC), whereas those of pseudohalophytes and secretohalophytes were most affected by the soil C content, indicating that different functional groups of halophyte shrubs had different responses to soil factors. The results of this study revealed the nutrient utilization characteristics of different functional groups of halophyte shrubs in the Xinjiang desert and their response and adaptation mechanisms to soil factors, thereby providing a basis for desert ecosystem management.

Keywords: desert ecosystem; desert shrubs; halophytes; nutrient stoichiometry; edaphic factors

Citation: Luo, Y.; Lian, C.; Gong, L.; Mo, C. Leaf Stoichiometry of Halophyte Shrubs and Its Relationship with Soil Factors in the Xinjiang Desert. *Forests* **2022**, *13*, 2121. https://doi.org/10.3390/f13122121

Academic Editors: Timothy A. Martin and Choonsig Kim

Received: 13 October 2022
Accepted: 9 December 2022
Published: 11 December 2022

Publisher's Note: MDPI stays neutral with regard to jurisdictional claims in published maps and institutional affiliations.

1. Introduction

Ecological stoichiometry studies the balance between C, N, P, and other nutrients among plants during their adaptation to and interactions with their environment [1]. Studying the ecological stoichiometric characteristics of desert plants can effectively evaluate and analyze the nutrient supply potential of desert soils and the productivity of desert ecosystems [2]. As a special desert plant group widely distributed in desert ecosystems, desert halophytes play an important role in maintaining nutrients and biodiversity and improving the regional climate in desert ecosystems [3]. Studying the ecological stoichiometric characteristics of desert halophytes is helpful for understanding their nutrient utilization characteristics and survival strategies [4,5]. With the aggravation of global soil salinization and N and P nutrient imbalances, the nutrient characteristics of desert halophytes have also changed [6]. Therefore, it is of great significance for the management and restoration of desert ecosystems to study the stoichiometric characteristics of desert halophyte leaves and their relationship with soil factors and to elucidate the nutrient limitations of desert halophytes and their response mechanisms to soil factors.

In recent years, many scholars have conducted extensive research on desert plants, mainly focusing on the relationship between desert plant nutrient characteristics, the climate, and the physical and chemical factors of soil [7–10]. Some studies believed that most soil factors directly affect the N and P stoichiometry, while climate factors indirectly affect the N and P stoichiometry of desert plants. Therefore, soil factors are still key factors affecting the nutrient characteristics of desert plants [11]. Relevant studies have shown that EC has direct effects on the leaf N and P contents of desert plants, and that pH and STN (soil total nitrogen) have direct effects on leaf N:P ratios [11]. Soil water content, total carbon (TC), total nitrogen (TN), and available phosphorus (AP) content are positively correlated with leaf C and N content [12], and soil water content is positively correlated with N and P in plant leaves and negatively correlated with other stoichiometric ratios. Soil salinity is positively correlated with leaf C/P and negatively correlated with leaf P [13]. However, according to some studies, the leaf stoichiometric characteristics of desert plants are not directly determined by soil nutrient content but are more affected by the genetic characteristics of plants [14,15]. The C, N, and P stoichiometries of plant leaves in different functional groups vary to different degrees; for example, succulent woody plants have higher N contents and lower C contents than herbaceous plants and non-succulent woody plants, reflecting the unique adaptability of desert plants to extreme habitats [16].

Xinjiang is an extremely arid area with scarce precipitation, strong evaporation, and a fragile ecological environment [17]. It is an area with the largest distribution area of saline alkali land in China; under such natural environmental conditions, the widespread distribution of desert halophytes has been promoted [18]. Under the long-term adaptation to soil salt, the halophytes in Xinjiang mainly evolved into three categories (euhalophytes, species that accumulate and isolate salt within succulent leaf or stem tissues; secretohalophytes, species with salt-secreting glands; pseudohalophytes, species that restrict the entry of saline ions into the transpiration stream) [19]. However, up to now, the stoichiometric characteristics of nutrients in different halophyte species have still been unclear, and there have been few studies on the relationship between the ecological stoichiometries of desert halophytes and environmental factors.

With the intensification of drought and salinization, the ecological stoichiometric characteristics of desert halophytes have changed [20,21]. Desert halophyte shrubs play an irreplaceable role in maintaining the stability of biodiversity in desert ecosystems [19]. The stoichiometry of C, N, and P in the leaves of three different functional groups of halophyte shrubs and their relationship with soil factors were studied in this paper; we clarify their response characteristics to soil factors in extremely arid areas. We assume that (1) under arid and poor nutrient conditions, halophytes have higher C content and lower N and P content; (2) owing to the different physiological functions of desert halophytes, the stoichiometric ratios of C, N, and P of different functional groups may have different response characteristics to soil factors. This study reveals the adaptation and feedback mechanisms of different functional groups of desert halophytes shrubs, enriches the research on stoichiometry in the desert ecosystem, and provides a scientific basis for the management of desert plants.

2. Materials and Methods

2.1. Study Area

In this study, 25 sample plots were selected in the desert areas of the Junggar Basin and Tarim Basin in Xinjiang (Figure 1), spanning 80°39′–91°19′ E, 40°14′–46°14′ N, and 270–1025 m above sea level (Table S1). The study area has a typical temperate continental arid climate, with little rainfall and strong evaporation, and a high soil salt content. The mean annual precipitation (MAP) ranges from 45 mm to 159 mm, and the mean annual temperature (MAT) ranges from 5.85 to 11.87 °C. Our database consisted of 10 species from 6 families, including 225 samples (euhalophytes, 100 samples; pseudohalophytes, 40 samples; secretohalophytes, 85 samples). The dominant euhalophyte species include *Haloxylon ammodendron* and *Halostachys caspica*, the dominant pseudohalophyte species include *Cal-*

ligonum leucocladum and *Halimodendron halodendron*, and the dominant secretohalophyte species include *Tamarix ramosissima* and *Tamarix arceuthoides* (Table S2).

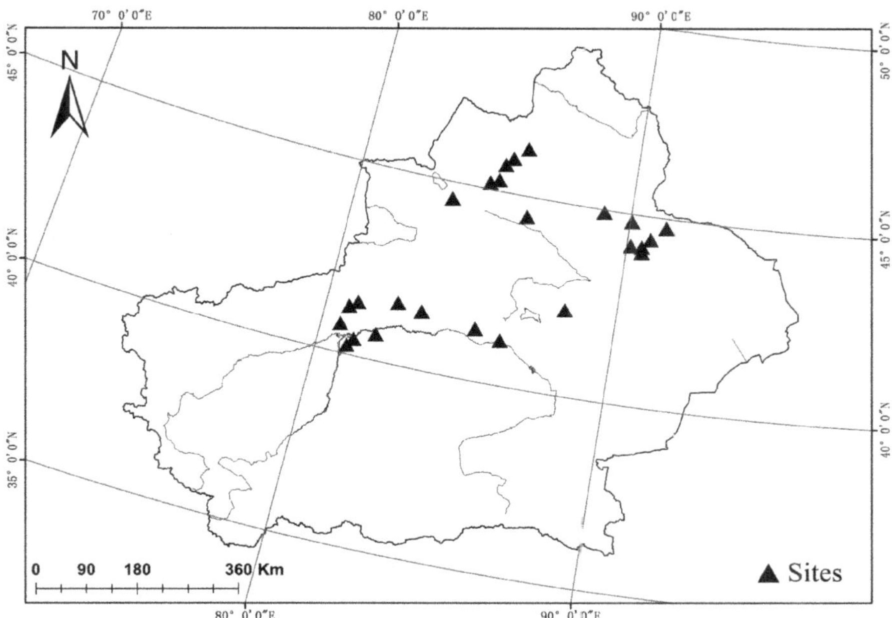

Figure 1. Locations of the 29 sampling sites in Xinjiang, China.

2.2. Field Sampling

Based on the data from the previous field survey and literature such as Xinjiang Saline Flora, typical sampling sites in Xinjiang desert were identified. In accordance with the dominance, the constructive species or common plant species within the sample area of the sampling site were selected so as to adequately represent the composition and structure of the plant community there. At the same sampling site (plant community), individuals of the same species were selected at a uniform distance for each species, so that the selected individuals could adequately represent the growth of the species in the community.

In July 2018, plant and soil surveys and sample collections were conducted at sites far from human disturbance. Three 10×10 m quadrats were randomly set up at each site and species composition was investigated, and the dominant species and soil samples were collected. At the species level, five individual plants of the same species were selected according to their growth status, and 50 g of fresh, healthy, and mature (non-destructive, sunny, petioles removed) plant leaves were collected from each individual plant. The collected plant leaves were loaded into an envelope with an appropriate amount of desiccant and labeled. In order to analyze soil nutrients, we randomly selected two sampling points from the four corners of each sample, took five 0–50 cm soil samples at each sampling point (remove impurities such as plants and rocks), and then fully mixed them to create a composite sample for each sample. Collected 100 g of soil samples from the composite samples, and put them into the sealed bag to take back to the laboratory for determination of soil nutrient content. In addition, we used GPS to record the altitude and geographical coordinates of each plot.

2.3. Chemical Analysis

The leaf samples were rinsed with deionized water to remove salt and dust from the leaf surface, then stored in a drying oven at 105 °C and dried for 30 min to reduce nutrient loss due to the respiration and decomposition of the leaves, and then dried to constant weight in an oven at 65 °C. The leaves were ground using a grinder and passed through an 80-mesh sieve, weighed, bagged, and sealed for the determination of C, N, and P contents. The collected soil samples were naturally dried, any remaining plant tissues were removed, and the soil samples were ground through a 60-mesh sieve. The sieved soil samples were divided into five equal parts and stored in numbered sample bags for the determination of soil nutrients and other indicators. An elemental analyzer (Elementar Inc., Hanau, Hessen, Germany) was used to determine the plant and soil carbon and nitrogen content (soil C and soil TN). The samples were digested with H_2SO_4-H_2O_2-HF, and the plant and soil phosphorus contents (soil TP) were analyzed by colorimetry. The ratio of soil to water in the test solution of soil pH value and electrical conductivity (EC) was 1:2.5 and 1:5, respectively, and then measured with a pH meter and electrical conductivity meter (Mettler-Toledo International Inc., Zurich, Greifensee, Switzerland).

2.4. Data Analysis

SPSS (Version 26.0, Armonk, NY, USA) was used for test data processing and statistical analysis. Levene's test was used to test the homogeneity of variance. Duncan's method was used when there was homogeneity of variance, and Tamhane's T2 method was used when their variance was not homogeneous (α = 0.05). Canono (Version 5.0, Ithaca, NY, USA) was used for redundancy analysis of the leaf stoichiometric characteristics and soil factors. Blue solid arrows represent the ecological stoichiometric characteristics of halophyte shrub leaves, and red hollow arrows represent soil factors. The arrow length of each environmental factor represents the length of its eigenvector; this can also be seen as the effect of soil factors on leaf ecological stoichiometry. The longer the line, the greater the absolute cosine value, indicating a greater impact, and vice versa. The angle between the environmental factor arrow and sorting axis indicates the size of the correlation. The smaller the angle, the higher the correlation; that is, the sorting axis reflects the gradient of the environmental factor. The data in the chart are presented as the mean $\pm$ standard error ($p < 0.05$). MAP and MAT data were obtained from WorldClim Version 2.0 (http://worldclim.org/version2 (accessed on 7 March 2020)). In this study, the data were interpolated with a precision of 30 s, and the longitude, latitude, and elevation data were measured with GPS during the sampling process.

3. Results

3.1. Ecological Stoichiometric Characteristics of C, N, and P in Desert Saline Shrub Leaves

The mean C, N, and P contents in the leaves of Xinjiang desert halophyte shrubs were 409.31, 19.92, and 1.25 mg g^{-1}, respectively (Table 1). The mean C/N, C/P, and N/P ratios were 22.25, 352.20, and 16.53, respectively. The coefficients of variation (CV) of C, N, and P content were 11.96, 29.42, and 28.18, respectively; the order from large to small was N > P > C. The coefficients of variation of C/N, C/P, and N/P ratios were 30.11, 29.01, and 28.67, respectively; the order from large to small was C/N > C/P > N/P.

Table 1. The overall descriptive statistical characteristics of C, N, P stoichiometric characteristics of desert saline shrub leaves.

Variable	n	Mean	SE	Minimum	Maximum	CV (%)
C (mg g^{-1})	225	409.31	3.26	303.64	512.29	11.96
N (mg g^{-1})	225	19.92	0.39	10.21	43.97	29.42
P (mg g^{-1})	225	1.25	0.24	0.59	2.39	28.18
C/N	225	22.25	0.45	9.40	44.38	30.11
C/P	225	352.20	6.81	145.12	654.44	29.01
N/P	225	16.53	0.32	7.02	37.39	28.67

Note: n, sample size.

There were significant differences in the C, N, and P contents and stoichiometric characteristics of desert halophyte shrubs in different functional groups (Figure 2). The leaf C content of secretohalophytes (442.27 ± 3.08 mg g^{-1}) was significantly higher than that of both pseudohalophytes (405.12 ± 7.49 mg g^{-1}) and euhalophytes (382.96 ± 4.70 mg g^{-1}) ($p < 0.05$), the N content of euhalophytes (22.17 ± 0.49 mg g^{-1}) was significantly higher than that of pseudohalophytes (18.64 ± 1.41 mg g^{-1}) and secretohalophytes (17.88 ± 0.43 mg g^{-1}) ($p < 0.05$), and the P content of euhalophytes (1.35 ± 0.04 mg g^{-1}) was significantly higher than that of both of pseudohalophytes (1.10 ± 0.06 mg g^{-1}) and secretohalophytes (1.21 ± 0.03 mg g^{-1}) ($p < 0.05$).

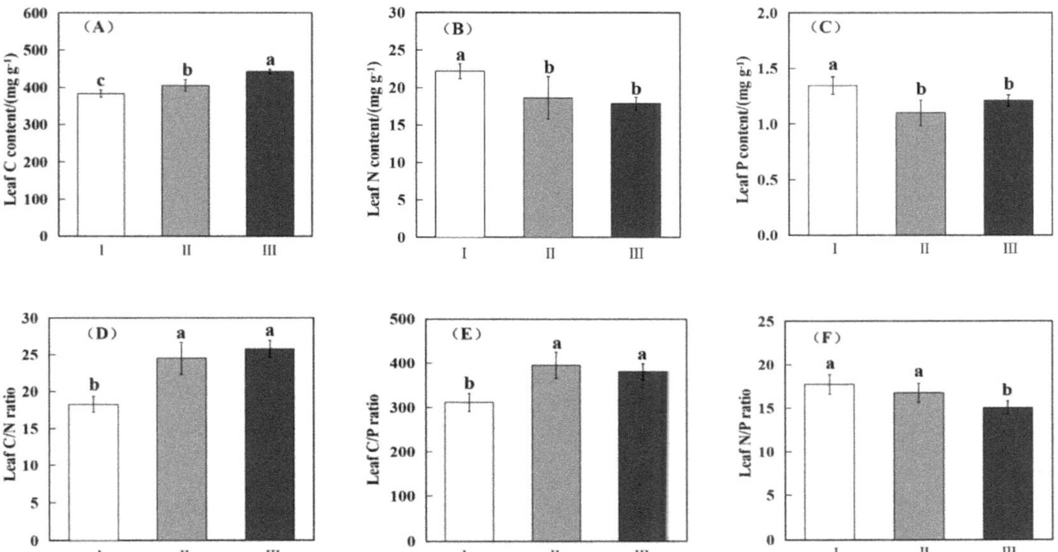

Figure 2. Comparison of leaf C, N, P content (**A–C**) and ratio (**D–F**) of different functional groups of desert saline shrubs. Note: I: euhalophyte; II: pseudohalophytes; III: secretohalophyte. Lowercase letters indicate significant differences between plants ($p < 0.05$).

The stoichiometric ratios of C/N, C/P, and N/P in the leaves of different functional groups were also different (Figure 2). The C/N ratios of pseudohalophytes (24.53 ± 1.08) and secretohalophytes (25.83 ± 0.59) were significantly higher than that of euhalophytes (18.29 ± 0.54) ($p < 0.05$), the C/P of euhalophytes (310.98 ± 10.31) was significantly lower than that of both pseudohalophytes (395.41 ± 14.87) and secretohalophytes (380.36 ± 9.37) ($p < 0.05$), and the N/P of euhalophytes (17.71 ± 0.57) and pseudohalophytes (16.74 ± 0.55) were significantly higher than that of secretohalophytes (15.08 ± 0.38) ($p < 0.05$).

3.2. Characteristics of Soil Factors of Desert Halophyte Shrubs

Except for the soil C content (8.75 mg g^{-1}) and EC (1.58) of pseudohalophytes that were significantly lower than those of other functional groups, other factors had no significant differences between the soils of different desert halophyte functional groups (Table 2). The correlation results showed that except for soil TN and EC, other soil indicators showed significant correlations at the overall level (Figure 3A). For euhalophyte soil, soil C showed a significant negative correlation with soil TN and pH, soil TN was positively correlated with pH and soil TP, and soil TP was positively correlated with pH and negatively correlated with EC (Figure 3B). For pseudohalophyte soil, soil C and soil TN and pH and EC were negatively correlated, and soil TN and soil TP and soil TP and pH were significantly positively correlated (Figure 3C). For secretohalophyte soil, soil C indicated a significant negative correlation with soil TP and pH, and soil TN was positively correlated with soil TP, whereas soil TP was significantly positively correlated with pH, and there was a significant negative correlation between pH and EC (Figure 3D).

Table 2. Multiple comparisons of soil factors of different functional groups of saline shrubs.

Plant Group	Soil C (mg g^{-1})	Soil TN (mg g^{-1})	Soil TP (mg g^{-1})	pH	EC (mS cm^{-1})
Euhalophyte	9.99 ± 0.22 [a]	2.54 ± 0.17 [a]	2.34 ± 0.19 [a]	8.90 ± 0.34 [a]	10.04 ± 1.30 [a]
Pseudohalophyte	8.75 ± 0.31 [b]	2.74 ± 0.32 [a]	2.28 ± 0.33 [a]	8.90 ± 0.50 [a]	1.58 ± 0.40 [b]
Secretohalophyte	9.82 ± 0.23 [a]	2.24 ± 0.11 [a]	2.71 ± 0.19 [a]	8.80 ± 0.30 [a]	14.29 ± 1.86 [a]
Mean	9.71 ± 0.14 [a]	2.47 ± 0.10 [a]	2.48 ± 0.13 [a]	8.86 ± 0.21 [a]	10.14 ± 0.96 [a]

Note: Lowercase letters indicate significant differences in the same soil factors between different functional groups of saline shrubs ($p < 0.05$).

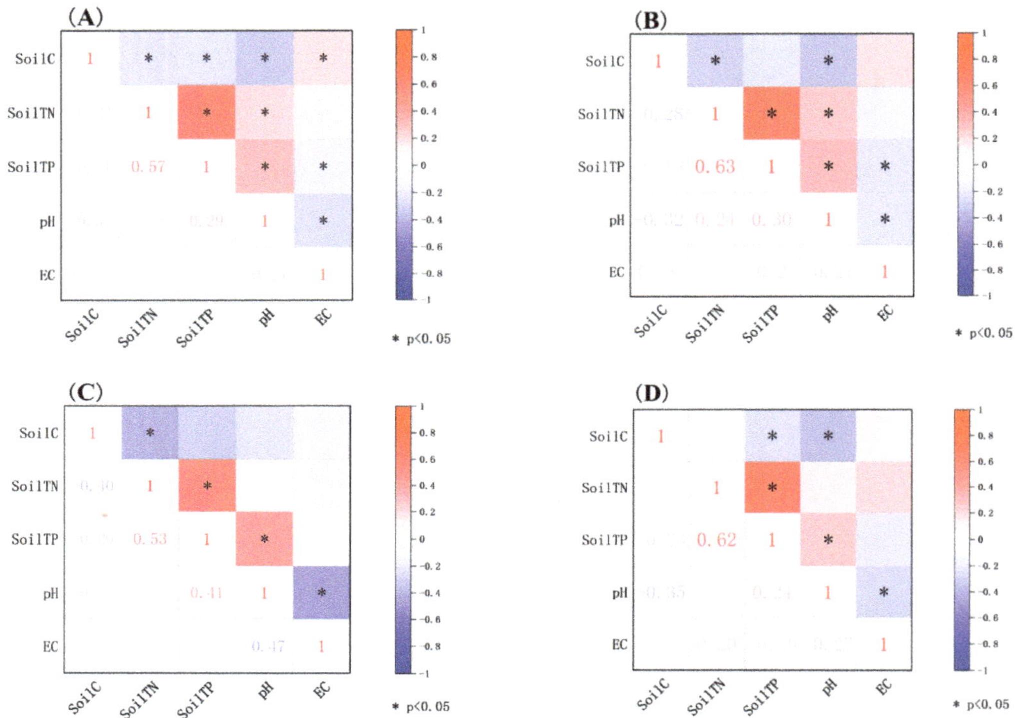

Figure 3. Correlation heat map of soil factors of desert halophytes in different functional groups. Note: (**A**) whole soil; (**B**) euhalophyte soil; (**C**) pseudohalophyte soil; (**D**) secretohalophyte soil.

3.3. Relationship between Ecological Stoichiometric Characteristics of Desert Halophyte Shrub Leaves and Soil Factors

To explore the correlation between the ecological stoichiometric characteristics of the leaves of different functional groups of halophyte shrubs and their soil factors, a redundancy analysis was performed (Figure 4). Data of all plants in the study area are shown in Figure 4A. The total explanation rates of the C, N, and P ecological stoichiometric characteristics of whole halophyte shrub leaves on the first and second ordination axes were 73.81% and 18.52%, respectively, and the cumulative interpretation amount reached 92.33%. Among all soil factors, the largest absolute value of the soil C cosine value shows that soil C has the greatest impact on plants. Soil C and EC had the longest arrow connection.

Therefore, the changes in the ecological stoichiometric characteristics of plant leaves can be well explained by soil C content and EC. Soil C was proportional to leaf N, P content, and N/P, and inversely proportional to C/N, C/P, and C, among which soil C had the greatest correlation with N; in addition, EC and P were positively correlated. Soil TN and pH were proportional to C/P and N/P, and inversely proportional to P; soil TP was inversely proportional to N and P and proportional to C, and the correlation with P was significantly greater than that with the other leaf ecological stoichiometric characteristics. There were some differences in the effects of soil factors on the ecological stoichiometry characteristics of plant leaves. The order of importance of the influence of each soil factor from the ecological stoichiometry characteristics of plant leaves was soil C (57.8%) > EC (23.2%) > soil TP (7.4%) > pH (5.9%) > soil TN (5.8%) (Table 3). Among all the factors, soil C and EC had very significant effects on the ecological stoichiometry of halophyte leaves ($p < 0.01$), and their interpretation rates for the ecological stoichiometry characteristics of halophyte leaves were 57.8% and 23.2%, respectively. The effect of soil TP and pH on the ecological stoichiometry characteristics of plant leaves was significant ($p < 0.05$), while the effect of soil TN on the ecological stoichiometry characteristics of plant leaves was not significant ($p = 0.05$) (Table 3).

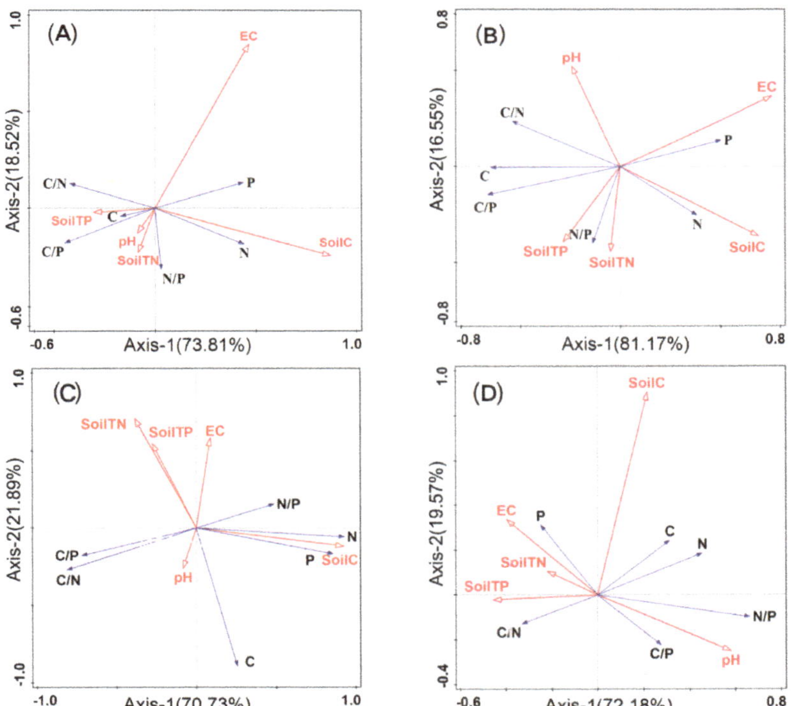

Figure 4. Redundancy analysis two−dimensional ordination diagram of the relationship between leaf C, N, P stoichiometric characteristics and soil factors. Note: (**A**) whole plants; (**B**) euhalophytes; (**C**) pseudohalophytes; (**D**) secretohalophytes.

The leaf C, N, and P stoichiometry of euhalophytes (Figure 4B), pseudohalophytes (Figure 4C), and secretohalophytes (Figure 4D) were explained by 81.17%, 71.73%, and 72.18%, respectively, in the first ordination axis, but the soil factors had different effects on the leaf C, N, and P stoichiometric characteristics. Meanwhile, EC had the greatest impact on euhalophytes, and soil C had the greatest impact on pseudohalophytes and secretohalophytes. According to the angle in the diagram (Figure 4B–D), it can be observed that soil C was significantly negatively correlated with the C/N, C/P, and C content of euhalophytes, and positively correlated with the N content. Soil C was positively correlated with the C, N, and P of pseudohalophytes and secretohalophytes, and negatively correlated with the C/N and C/P of secretohalophytes. The soil TN was significantly negatively correlated with the P content in euhalophytes, N/P and C/P in secretohalophytes, and C, N, and P in pseudohalophytes. The soil TN was positively correlated with the N/P of euhalophytes, and the P content and C/N of secretohalophytes. The soil TP was significantly positively correlated with the C content, C/P, N/P, and C/N of euhalophytes. However, it was significantly negatively correlated with the P content of euhalophytes, C and P of pseudohalophytes, and N and N/P of secretohalophytes. There was a positive correlation between the pH, C, and C/N of euhalophytes, C content of pseudohalophytes, and N, C/P, and N/P of halophytes, but pH was only significantly negatively correlated with P of secretohalophytes. EC was significantly negatively correlated with the C content and C/P of euhalophytes and pseudohalophytes, and the C/P and N/P of secretohalophytes. EC was positively correlated with the phosphorus content in both euhalophytes and secretohalophytes.

Table 3. Results of the importance ranking and significance tests of the overall soil factors' variables interpretation.

Name	Importance Rank	Contribution %	Pseudo-F	*p*
Soil C	1	57.8	25.6	0.002
Soil EC	2	23.2	10.7	0.002
Soil TP	3	7.4	3.5	0.02
pH	4	5.9	2.8	0.048
Soil TN	5	5.8	2.7	0.05

4. Discussion

4.1. Stoichiometric Characteristic of C, N, and P Elements in Leaves of Desert Halophyte Shrubs

The characteristics of plants' C, N, and P not only reflect the characteristics of plants, but also reflect the long-term adaptation and response to their environment [22,23]. The C content (409.31 ± 3.26 mg g^{-1}) of the whole halophyte shrub leaves was lower than that of the global terrestrial plants (464.00 ± 32.10 mg g^{-1}) [23] and the Sonoran Desert plants (434.8 ± 1.8 mg g^{-1}) [2], indicating that the C reserve capacity was weak in this study, which was related to the strong diurnal temperature difference and low annual precipitation in the study area. The contents of N (19.9 mg g^{-1}) and P (1.2 mg g^{-1}) in the leaves of halophyte shrubs in this study were lower than in the China desert plants (24.4 and 1.7 mg g^{-1}) [24], China desert halophytes (28.1 and 1.9 mg g^{-1}) [25], China sand plants (34.1, (2.5 mg g^{-1}) [26], and Sonoran Desert plants (25.9 and 1.5 mg g^{-1}) [2] (Table S3). Scarce rainfall and intense evaporation result in extremely low soil moisture and nutrient contents in the study area, which ultimately leads to inadequate nutrient availability to desert halophyte shrubs [13]. Under conditions of water and nutrient stress, desert halophyte shrubs may be limited in their ability to take up nitrogen and phosphorus nutrients, which reduces nutrient cycling and metabolic activity, resulting in reduced nutrient feedback mechanisms between plants and the soil. As a result, plant nutrient contents are also expected to decrease. Desert halophyte shrubs adapt to their environment through a number of physiological regulators, such as slowing plant growth to reduce metabolic rates and reducing resource requirements to cope with the barren environment [27].

There were significant differences in the C, N, and P stoichiometric characteristics among the leaves of desert halophyte shrub groups in the study area, indicating that there were some differences in the resource utilization efficiency and adaptation strategies in response to adverse environments [19,28]. As the main element of plant dry matter, C is also the substrate and energy source of various physiological metabolic processes of plants [25]. The C content in the leaves of halophyte shrubs was significantly different among different halophyte species. The leaves of some halophytes are scaly, forming a special stem leaf callus similar to the stem. The tissue is composed of large epidermis, porous cells, and water storage and vascular tissues. Salinity also improves the function of these tissues, making the saline plants better able to adapt to extreme drought conditions, which helps plants to enhance photosynthesis and improve the ability of plant leaves to synthesize carbohydrates [28]. The low C content in the leaves of euhalophytes may be related to their salt tolerance. The fleshy development of euhalophyte leaves can dilute excess salt in the body to alleviate physiological water shortages [3]. N and P are two important limiting elements in plant growth and development. Their effectiveness is one of the main factors regulating the plant litter decomposition rate and ecosystem nutrient balance, affecting the entire growth process of plants [29]. Euhalophytes have strong and unique salt tolerance mechanisms. Under salt stress conditions, the N and P contents in the leaves are significantly increased. In addition, non-protein N accumulation in euhalophytes is higher under salt stress conditions, which plays a crucial role in osmotic regulation [25,30]. Pseudohalophytes and secretohalophytes have different salt tolerance mechanisms to euhalophytes. The low N content in secretohalophytes is mainly due to the formation of salt glands or salt vesicles on their leaves, which can secrete excessive salt ions in plants [30]. The low P content in pseudohalophytes is due to the suberization of the root cortex. The main component of suberization is water-insoluble fatty substances, which are impermeable, making it difficult for salts dissolved in soil water to enter plants [31].

C/N and C/P ratios are important physiological indicators of plants. Generally, plants with higher C/N and C/P ratios have relatively lower growth rates [24] and higher carbon sequestration efficiencies [32]. Secretohalophytes and pseudohalophytes had the highest C/N and C/P ratios, and they had higher carbon sequestration advantages in nutrient-poor habitats than euhalophytes [13]. Overall, C/N and C/P were high, which can be attributed to the extreme drought, high temperature, and strong evaporation in the study area [16,30]. To resist harsh environments, plants reduce their growth rate, resulting in relatively high C/N and C/P ratios in the leaves [32]. Leaf N/P is considered an important indicator of plant and soil nutrient limitation [33]. The N/P of euhalophytes and pseudohalophytes was greater than 16, and that of secretohalophytes was less than 16. This indicates that the growth of euhalophytes and pseudohalophytes is more susceptible to P limitation, whereas the growth of secretohalophytes is more susceptible to both N and P limitations. Overall, the N/P ratio was much higher than in other regions [34,35], and greater than 16, indicating that P is the main limiting element for desert plant growth in the study area. Nutrient deficiency and water shortage lead to insufficient nutrient supply for desert halophytes [25]. Under limited water and nutrient conditions, the ability of desert halophytes to assimilate P may be limited, thereby reducing nutrient cycling and metabolic activity, and ultimately leading to a decrease in nutrient feedback mechanisms between plants and soil [30]. In addition, less precipitation and greater evaporation makes it difficult for desert halophytes to obtain phosphorus from the soil, thereby making it more vulnerable to P limitation in this area.

4.2. Characteristics and Relationship of Soil Factors

Soil C, N, and P contents can indicate soil nutrient storage, nutrient cycling, and balance and are important indicators of soil organic matter composition and quality [36]. The soil C content (8.75 mg g^{-1}) of pseudohalophytes was the lowest among the halophyte groups, and the overall C content was low (9.71 mg g^{-1}), which was far lower than the national level (11.12 mg g^{-1}) [37]. It is possible that the soil productivity of desert halophyte shrubs in Xinjiang is low, which is related to the nutrient conditions of the soil itself and the lower feedback of plant litter to the soil in the study area [30]. There was no significant difference in soil N and P contents among different halophyte functional groups, but the overall N and P contents were high (2.47 and 2.48 mg g^{-1}); they were significantly higher than the average global level (1.06 and 0.65 mg g^{-1}) [37] because the N and P contents increased under salt stress [4], which is consistent with the results of Wang et al. (2017) [25]. EC can represent some physical information about soil, such as salt state, nutrient content, and water content [38,39]. The pH was significantly negatively correlated with EC in the soil of all three halophytes, and EC was related to the salt content. The larger the EC, the higher the content of soluble salt ions [38]. The soil EC (1.58) of pseudohalophytes was the lowest among the halophytes, indicating that the soil soluble salt ion content was the lowest.

4.3. Response of Desert Halophyte Shrubs' Stoichiometry to Soil Factors

Plant growth and development are closely related to their environment. Changes in the element content and the stoichiometric ratio in plants reflect the response and adaptation of plants to the environment [40–42]. Many studies have found that soil is the main nutrient source for plant growth and development, and its nutrient characteristics have an important impact on plant leaf N and P contents [43,44]. Soil nutrients significantly affect the ecological stoichiometric characteristics of plant leaves [45]. Based on the redundancy analysis of the relationship between the stoichiometric characteristics of different functional groups of salinized shrubs and soil factors (Figure 4), there was a significant correlation between the stoichiometric characteristics of the different functional groups of the halophyte shrub and soil factors. Overall, the growth and development of desert saline shrubs were limited by soil C and EC. However, for the different functional groups of halophyte shrubs, the responses of leaf C, N, and P ecological stoichiometric characteristics to soil factors were different. EC represents soil salinity, and among them, it had the greatest impact on euhalophytes. It was found that the protein content of euhalophytes was very high under drought and salt stress, indicating that the plant has a higher nitrogen uptake or metabolism capacity [3]. In addition, it was shown that euhalophytes also absorb large amounts of salts and store them in their bodies during growth, and in addition to accumulating salts, they grow certain contents whose growth is positively correlated with external salinity [3,6]. For pseudohalophytes and secretohalophytes, soil C is one of the greater factors affecting their nutrient characteristics. Soil C is mainly derived from the content of soil organic matter and the decomposition of apoplastic matter, and is strongly influenced by plants, water and heat, and parent material. Due to drought and salinity stress, plants need to increase the proportion of C-rich tissues (e.g., lignin) in their own bodies to protect the plant body from damage to adapt to the drought environment [3,46].

Only leaf N had a significant positive correlation with soil TN in euhalophytes (Figure 4B), indicating that soil TN has a limiting effect on euhalophyte growth. The soil TN level determines the absorption of N by euhalophyte leaves to a certain extent [47,48]. The C and P contents in the leaves of secretohalophytes were positively correlated with soil C and soil TP, respectively (Figure 4C,D), indicating that secretohalophytes were limited by soil C and P. However, leaf C was only positively correlated with soil C in pseudohalophytes, indicating that the growth and development of pseudohalophytes were limited by soil C [49]. This may be because pseudohalophytes increase their salt tolerance through the compensatory growth of roots [3], and reduced C content is transported up leaves by pseudohalophytes; thus, leaf C was positively correlated with soil C. There was no significant difference in soil N and P contents among the different functional groups of halophyte shrubs, but the utilization and consumption strategies of nutrient elements in different functional groups of halophyte shrubs were different [30]. This indicates that the ecological stoichiometric characteristics of desert saline shrub leaves are not directly determined by the characteristics of soil nutrient contents, but are related to their own genetic characteristics, which reflects the unique adaptation mechanism of different desert halophyte shrubs to habitats [50], consistent with the results of Luo et al. (2017) [51] and Song et al. (2020) [52]. Under soil moisture- and nutrient-scarce conditions, Xinjiang desert halophyte shrubs have formed their own unique stoichiometric characteristics and physiological ecology, reflecting the relatively stable adaptability of desert saline shrubs to extreme environments [16]. The environmental indicators selected in this study are limited, and more environmental indicators will be combined in future studies to conduct a comprehensive analysis of the relationship between halophytes and environmental factors.

5. Conclusions

It is of great importance to reveal the nutritional characteristics and ecological adaptation mechanisms of desert halophytes to study the element accumulation and utilization characteristics of different functional groups and their adaptability to harsh soil from the perspective of plant nutrition. In this study, C content of the whole halophyte shrub leaves was lower than that of the global terrestrial plants and the Sonoran desert plants. The contents of N and P in the leaves of halophyte shrubs were lower than in the China desert plants, China desert halophytes, China sand plants, and Sonoran desert plants. Through the analysis of the stoichiometric characteristics of C, N, and P in the leaves of euhalophytes, secretohalophytes, and pseudohalophytes, and their correlations with soil factors, it was found that there were significant differences in the leaf contents and ratios. This indicates that the N and P nutrient limitations and nutrient resource utilization efficiencies of halophytes in different functional groups were different. The redundancy analysis results showed that the C, N, and P contents in the leaves of euhalophytes were most affected by EC, whereas the leaves of pseudohalophytes and secretohalophytes were most affected by soil C content, indicating that different functional groups of halophyte shrubs had different responses to soil factors. In this study, euhalophytes and pseudohalophytes can be applied with phosphate fertilizer due to P limitation, and salt-secreting halophytes can be appropriately applied with nitrogen fertilizer and phosphate fertilizer due to the common limitation of N and P. Desert halophytes show unique advantages, especially in saline alkali habitats. Desert halophytes can not only prevent salt accumulation in the tilth, but also improve soil structure and fertility, and promote good plant soil nutrient cycling in desert ecosystems. Therefore, in terms of desert halophytes' management, we need to pay attention to their adaptation to salt and differences in nutrient utilization so as to better use halophytes to improve the soil conditions in desert areas.

Supplementary Materials: The following supporting information can be downloaded at: https://www.mdpi.com/article/10.3390/f13122121/s1, Table S1: Geography information and climate characteristics of the sampling sites in the deserts of Xinjiang, China; Table S2: List of species sampled in the deserts of Xinjiang, China; Table S3: Leaf C, N, P content and C/N, C/P, N/P ratio of plants in various regions.

Author Contributions: Y.L. carried out the fieldwork and wrote the first draft of the manuscript. C.L. wrote the first draft of the manuscript. L.G. assisted with revising the draft manuscript. C.M. carried out the fieldwork. All authors contributed to the article and approved the submitted version. All authors have read and agreed to the published version of the manuscript.

Funding: Third Xinjiang Scientific Expedition Program (No. 2021xjkk0903), China Postdoctoral Science Foundation (No. 2022M722667), and Department of Education of Xinjiang Uygur Autonomous Region, Dr. Tianchi Program Project (No. TCBS202123).

Data Availability Statement: The available data can be obtained in the supplementary files, but anyone who uses the data must get the consent of the authors.

Acknowledgments: We thank Yanming Gong for their assistance with field sampling and laboratory work.

Conflicts of Interest: The authors declare that the research was conducted in the absence of any commercial or financial relationships that could be construed as a potential conflict of interest.

References

1. Sterner, R.W.; Elser, J.J. *Ecological Stoichiometry: Biology of Elements from Molecules to the Biosphere*; Princeton University Press: Princeton, NJ, USA, 2002.
2. Castellanos, A.E.; Llano-Sotelo, J.M.; Machado-Encinas, L.I.; Lopez-Pina, J.E.: Romo-Leon, J.R.; Sardans, J.; Penuelas, J. Foliar C, N, and P stoichiometry characterize successful plant ecological strategies in the Sonoran Desert. *Plant Ecol.* **2018**, *219*, 775–788. [CrossRef]
3. Zhao, K.F.; Li, F.Z.; Zhang, F.S. *Halophytes in China*, 2nd ed.; Science Press: Beijing, China, 2013.

4. He, M.Z.; Dijkstra, F.A.; Zhang, K.; Li, X.R.; Tan, H.J.; Gao, Y.H.; Li, G. Leaf nitrogen and phosphorus of temperate desert plants in response to climate and soil nutrient availability. *Sci. Rep.* **2014**, *4*, 6932. [CrossRef] [PubMed]

5. Chatterton, N.J.; Goodin, J.R.; Duncan, C. Nitrogen metabolism in *Atriplex polycarpa* as affected by substrate nitrogen and NaCl salinity. *Agron. J.* **1971**, *63*, 271–274. [CrossRef]

6. Zhang, K.; Li, C.J.; Li, Z.S.; Zhang, F.H.; Zhao, Z.Y.; Tian, C.Y. Characteristics of mineral elements in shoots of three annual halophytes in a saline desert, Northern Xinjiang. *J. Arid. Land* **2013**, *5*, 244–254. [CrossRef]

7. He, M.Z.; Dijkstra, F.A.; Zhang, K.; Tan, H.J.; Zhao, Y.; Li, X.R. Influence of life form, taxonomy, climate, and soil properties on shoot and root concentrations of 11 elements in herbaceous plants in a temperate desert. *Plant Soil* **2016**, *398*, 339–350. [CrossRef]

8. Zhang, X.M.; Wang, Y.D.; Zhao, Y.; Xu, X.W.; Lei, J.Q.; Hill, R.L. Litter decomposition and nutrient dynamics of three woody halophytes in the Taklimakan desert highway shelterbelt. *Arid. Land Res. Manag.* **2017**, *31*, 335–351. [CrossRef]

9. Liu, R.; Cieraad, E.; Li, Y. Summer rain pulses may stimulate a CO_2 release rather than absorption in desert halophyte communities. *Plant Soil* **2013**, *373*, 799–811. [CrossRef]

10. Amatangelo, K.L.; Vitousek, P.M. Stoichiometry of ferns in Hawaii: Implications for nutrient cycling. *Oecologia* **2008**, *157*, 619–627. [CrossRef]

11. Luo, Y.; Peng, Q.W.; Li, K.H.; Gong, Y.M.; Liu, Y.Y.; Han, W.X. Patterns of nitrogen and phosphorus stoichiometry among leaf, stem and root of desert plants and responses to climate and soil factors in Xinjiang, China. *Catena* **2021**, *199*, 105100. [CrossRef]

12. Zhang, X.L.; Zhou, J.H.; Guan, T.Y.; Cai, W.T.; Jiang, L.H.; Lai, L.M.; Gao, N.N.; Zheng, Y.R. Spatial variation in leaf nutrient traits of dominant desert riparian plant species in an arid inland river basin of China. *Ecol. Evol.* **2019**, *9*, 1523–1531. [CrossRef]

13. Wang, L.; Zhao, G.; Li, M.; Zhang, M.; Zhang, L.; Zhang, X.; Xu, S. C:N:P stoichiometry and leaf traits of halophytes in an arid saline environment, northwest China. *PLoS ONE* **2015**, *10*, e0119935. [CrossRef] [PubMed]

14. Yuan, K.Y.; Xu, H.L.; Zhang, G.P.; Yan, J.J. Aridity and high salinity, rather than soil nutrients, regulate nitrogen and phosphorus stoichiometry in desert plants from the individual to the community level. *Forests* **2022**, *13*, 890. [CrossRef]

15. Luo, Y.; Gong, L.; Zhu, M.L.; An, S.Q. Leaf and soil ecological stoichiometric characteristics of four shrub species in desert area of upper reaches of Tarim River. *Acta Ecol. Sin.* **2017**, *37*, 8326–8335. (In Chinese) [CrossRef]

16. Sun, M.M.; Zhai, B.C.; Chen, Q.W.; Li, G.Q.; Du, S. Response of density-related fine root production to soil and leaf traits in coniferous and broad-leaved plantations in the semiarid loess hilly region of China. *J. For. Res.* **2022**, *33*, 1071–1082. [CrossRef]

17. Zheng, Z.Y.; Ma, Z.G.; Li, M.X.; Xia, J.J. Regional water budgets and hydroclimatic trend variations in Xinjiang from 1951 to 2000. *Clim. Chang.* **2017**, *144*, 447–460. [CrossRef]

18. Zhang, W.T.; Wu, H.Q.; Gu, H.B.; Feng, G.L.; Wang, Z.; Sheng, J.D. Variability of soil salinity at multiple spatio-temporal scales and the related driving factors in the oasis areas of Xinjiang, China. *Pedoshere* **2015**, *24*, 753–762. [CrossRef]

19. Xi, J.B.; Zhang, F.S.; Tian, C.Y. *Halophytes in Xinjiang*; Science Press: Beijing, China, 2006.

20. Zhang, X.; Guan, T.; Zhou, J.; Cai, W.; Gao, N.; Du, H.; Zheng, Y. Community characteristics and leaf stoichiometric traits of desert ecosystems regulated by precipitation and soil in an arid area of China. *Int. J. Environ. Res. Public Health* **2018**, *15*, 109. [CrossRef]

21. Sardans, J.; Grau, O.; Chen, H.Y.H.; Janssens, I.A.; Ciais, P.; Piao, S.L.; Penuelas, J. Changes in nutrient concentrations of leaves and roots in response to global change factors. *Glob. Chang. Biol.* **2017**, *23*, 3849–3856. [CrossRef]

22. Hou, X.Y. *Vegetation Geography and Chemical Constituents of Dominant Plants in China*; Science Press: Beijing, China, 1982.

23. Elser, J.J.; Fagan, W.F.; Denno, R.F.; Dobberfuhl, D.R.; Folarin, A.; Huberty, A.A.; Interlandi, S.; Kilham, S.S.; McCauley, E.; Schulz, K.L.; et al. Nutritional constraints in terrestrial and freshwater food webs. *Nature* **2000**, *408*, 578–580. [CrossRef]

24. Li, Y.L.; Mao, W.; Zhao, X.Y.; Zhang, T.H. Leaf nitrogen and phosphorus stoichiometry in typical desert and desertified regions, north China. *Environ. Sci.* **2010**, *31*, 1716–1725. [CrossRef]

25. Wang, L.L.; Wang, L.; He, W.L.; An, L.Z.; Xu, S.J. Nutrient resorption or accumulation of desert plants with contrasting sodium regulation strategies. *Sci. Rep.* **2017**, *7*, 17035. [CrossRef] [PubMed]

26. Zhang, L.F.; Wang, L.L.; He, W.L.; Zhang, X.F.; An, L.Z.; Xu, S.J. Patterns of leaf N:P stoichiometry along climatic gradients in sandy region, north of China. *J. Plant Ecol.* **2018**, *11*, 218–225. [CrossRef]

27. Li, J.L.; Sun, X.; Zheng, Y.; Lü, P.P.; Wang, Y.L.; Guo, L.D. Diversity and community of culturable endophytic fungi from stems and roots of desert halophytes in northwest China. *MycoKeys* **2020**, *62*, 75–95. [CrossRef] [PubMed]

28. Meng, X.Q.; Zhou, J.; Sui, N. Mechanisms of salt tolerance in halophytes: Current understanding and recent advances. *Open Life Sci.* **2018**, *13*, 149–154. [CrossRef]

29. Yang, Y.; Liu, B.R.; An, S.S. Ecological stoichiometry in leaves, roots, litters and soil among different plant communities in a desertified region of Northern China. *Catena* **2018**, *166*, 328–338. [CrossRef]

30. Luo, Y.; Chen, Y.; Peng, Q.W.; Li, K.H.; Mohammat, A.; Han, W.X. Nitrogen and phosphorus resorption of desert plants with various degree of propensity to salt in response to drought and saline stress. *Ecol. Indic.* **2021**, *125*, 107488. [CrossRef]

31. Li, A.; Guo, D.; Wang, Z.Q.; Liu, H.Y. Nitrogen and phosphorus allocation in leaves, twigs, and fine roots across 49 temperate, subtropical and tropical tree species: A hierarchical pattern. *Funct. Ecol.* **2010**, *24*, 224–232. [CrossRef]

32. Gong, Y.M.; Lv, G.H.; Guo, Z.J.; Chen, Y.; Cao, J. Influence of aridity and salinity on plant nutrients scales up from species to community level in a desert ecosystem. *Sci. Rep.* **2017**, *7*, 6811. [CrossRef]

33. Güsewell, S. N:P ratios in terrestrial plants: Variation and functional significance. *New Phytol.* **2004**, *164*, 243–266. [CrossRef]

34. Tong, R.; Zhou, B.Z.; Jiang, L.N.; Ge, X.G.; Cao, Y.H.; Shi, J.X. Leaf litter carbon, nitrogen and phosphorus stoichiometry of Chinese fir (*Cunninghamia lanceolata*) across China. *Glob. Ecol. Conserv.* **2021**, *27*, e01542. [CrossRef]

35. Han, W.X.; Fang, J.Y.; Guo, D.L.; Zhang, Y. Leaf nitrogen and phosphorus stoichiometry across 753 terrestrial plant species in China. *New Phytol.* **2005**, *168*, 377–385. [CrossRef] [PubMed]
36. Gao, R.; Ai, N.; Liu, G.Q.; Liu, C.H.; Zhang, Z.Y. Soil C:N:P stoichiometric characteristics and soil quality evaluation under different restoration modes in the loess region of Northern Shanxi Province. *Forests* **2022**, *13*, 913. [CrossRef]
37. Li, Y.Q.; Zhao, X.Y.; Zhang, F.X.; Awada, T.; Wang, S.K.; Zhao, H.L.; Zhang, T.H.; Li, Y.L. Accumulation of soil organic carbon during natural restoration of desertified grassland in China's Horqin Sandy Land. *J. Arid. Land* **2015**, *7*, 328–340. [CrossRef]
38. Yang, Y.H.; Ji, C.J.; Ma, W.H.; Wang, S.F.; Wang, S.P.; Han, W.X.; Mohammat, A.; Robinson, D.; Smith, P. Significant soil acidification across northern China's grasslands during 1980s–2000s. *Glob. Chang. Biol.* **2012**, *18*, 2293–2300. [CrossRef]
39. Liu, W.; Li, R.J.; Han, T.T.; Cai, W.; Fu, Z.W.; Lu, Y.T. Salt stress reduces root meristem size by nitric oxide-mediated modulation of auxin accumulation and signaling in arabidopsis. *Plant Physiol.* **2015**, *168*, 343–U607. [CrossRef]
40. Yang, X.; Tang, Z.; Ji, C.; Liu, H.; Ma, W.; Mohhamot, A.; Zheng, C. Scaling of nitrogen and phosphorus across plant organs in shrubland biomes across Northern China. *Sci. Rep.* **2014**, *4*, 5448. [CrossRef]
41. Zheng, S.X.; Shangguan, Z.P. Spatial patterns of leaf nutrient traits of the plants in the Loess Plateau of China. *Trees* **2007**, *21*, 357–370. [CrossRef]
42. He, J.S.; Fang, J.Y.; Wang, Z.H.; Guo, D.L.; Flynn, D.F.B.; Geng, Z. Stoichiometry and large-scale patterns of leaf carbon and nitrogen in the grassland biomes of China. *Oecologia* **2006**, *149*, 115–122. [CrossRef]
43. Zhang, L.; Guo, H.; Bao, A.K. Unique salt-secreting structure of halophytes-salt vesicles. *Plant Physiol.* **2019**, *55*, 232–240. [CrossRef]
44. Cho, Y.; Sudduth, K.A.; Chung, S.O. Soil physical property estimation from soil strength and apparent electrical conductivity sensor data. *Biosyst. Eng.* **2016**, *152*, 68–78. [CrossRef]
45. An, H.; Tang, Z.S.; Keesstra, S.; Shangguan, Z.P. Impact of desertification on soil and plant nutrient stoichiometry in a desert grassland. *Sci. Rep.* **2019**, *9*, 9422. [CrossRef] [PubMed]
46. Rong, Q.Q.; Liu, J.T.; Cai, Y.P.; Lu, Z.H.; Zhao, Z.Z.; Yue, W.C.; Xia, J.B. Leaf carbon, nitrogen and phosphorus stoichiometry of Tamarix chinensis Lour. in the Laizhou Bay coastal wetland, China. *Ecol. Eng.* **2015**, *76*, 57–65. [CrossRef]
47. Flowers, T.J.; Torke, P.F.; Yeo, A.R. The mechanism of salt tolerance in halophytes. *Annu. Rev. Plant Physiol.* **1977**, *28*, 89–121. [CrossRef]
48. Flowers, T.J.; Colmer, T.D. Salinity tolerance in halophytes. *New Phytol.* **2008**, *179*, 945–963. [CrossRef] [PubMed]
49. Yi, L.P.; Ma, J.; Li, Y. Soil salt and nutrient concentration in the rhizosphere of desert halophytes. *Acta Ecol. Sin.* **2007**, *27*, 3565–3571. [CrossRef]
50. Didi, D.A.; Su, S.P.; Sam, F.E.; Tiika, R.J.; Zhang, X. Effect of plant growth regulators on osmotic regulatory substances and antioxidant enzyme activity of Nitraria tangutorum. *Plants* **2022**, *11*, 2559. [CrossRef]
51. Sardans, J.; Peñuelas, J. The role of plants in the effects of global change on nutrient availability and stoichiometry in the plant-soil system. *Plant Physiol.* **2012**, *160*, 1741–1761. [CrossRef] [PubMed]
52. Song, Z.P.; Liu, Y.H.; Su, H.X.; Hou, J.H. N-P utilization of Acer mono leaves at different life history stages across altitudinal gradients. *Ecol. Evol.* **2020**, *10*, 851–862. [CrossRef] [PubMed]

Article

Carbon Allocation of *Quercus mongolica* Fisch. ex Ledeb. across Different Life Stages Differed by Tree and Shrub Growth Forms at the Driest Site of Its Distribution

Yang Qi, Hongyan Liu *, Wenqi He, Jingyu Dai, Liang Shi and Zhaopeng Song

MOE Laboratory for Earth Surface Processes, College of Urban and Environmental Sciences, Peking University, Beijing 100871, China
* Correspondence: lhy@urban.pku.edu.cn

Abstract: There are less than 10% of woody species that can have both tree and shrub growth forms globally. At the xeric timberline, we observed the tree-to-shrub shift of the *Quercus mongolica* Fisch. ex Ledeb.. Few studies have explored the underlined mechanism of this morphological transition of tree-to-shrub in arid regions. To examine whether the tree-to-shrub shift affects carbohydrate allocation and to verify the effect of life stage on non-structural carbohydrate (NSC) storage, we measured the concentration of soluble sugar and starch of *Q. mongolica* in the seedlings, saplings, and adult trees by selecting two sites with either tree or shrub growth forms of *Q. mongolica* at the driest area of its distribution. Accordingly, there was no significant difference in the radial growth with different growth forms ($p > 0.05$). The results showed that the effects of growth form on NSC concentrations are significant in the seedling and sapling stages, but become less pronounced as *Q. mongolica* grows. The results of the linear mixed model showed that life stage has a significant effect on soluble sugar concentration of tree-form ($p < 0.05$), starch and TNC concentration of shrub-form ($p < 0.05$). Compared with a shrub form without seedling stage, a tree form needs to accumulate more soluble sugar from seedling stage to adapt to arid environment. Saplings and adult shrubs store more starch, especially in thick roots, in preparation for sprout regeneration. Our study shows that the same species with tree and shrub forms embody differentiated carbohydrate allocation strategies, suggesting that shrub form can better adapt to a drier habitat, and the tree-to-shrub shift can benefit the expansion of woody species distribution in dryland.

Keywords: non-structural carbon; *Quercus mongolica*; growth-reproduction-storage tradeoffs; xeric timberline; acclimation

Citation: Qi, Y.; Liu, H.; He, W.; Dai, J.; Shi, L.; Song, Z. Carbon Allocation of *Quercus mongolica* Fisch. ex Ledeb. across Different Life Stages Differed by Tree and Shrub Growth Forms at the Driest Site of Its Distribution. *Forests* **2022**, *13*, 1745. https://doi.org/10.3390/f13111745

Academic Editors: Xiao-Dong Yang, Nai-Cheng Wu and Xue-Wei Gong

Received: 3 September 2022
Accepted: 5 October 2022
Published: 23 October 2022

Publisher's Note: MDPI stays neutral with regard to jurisdictional claims in published maps and institutional affiliations.

1. Introduction

Trees adapt their size and shape to match their growing environment [1]. According to a global plant trait database, approximately 9.2% of woody species have both shrub and tree growth forms [2]. Compared with trees with a single stem, shrubs are reduced in tree height and have basal sprouting stems [2]. In more disturbed environments, multiple potential growth forms should allow woody plants to acclimate better and reproduce faster than those with less morphological plasticity [3].

Tree-to-shrub shifts with the reduction of tree height ensure the safety of water transportation. As taller trees are generally at greater risk of hydraulic failure due to embolism in areas where water becomes progressively more limiting, the same species will tend to grow shorter in arid areas [4]. Tree-to-shrub shifts also ensure hydraulic safety and carbon assimilation, avoiding forest dieback [5].

Tree-to-shrub shifts may also represent a change in the way of reproduction, from seed reproduction to resprouting regeneration. Resprouting regeneration avoids the costs associated with sexual reproduction, such as the production of seeds and accessories [6]. Reproductive allocation (RA) is considered to participate in resource trade-offs regarding vegetative growth and defense [7,8]. Most angiosperms have the ability to regenerate through resprouting. Resprouting regeneration accelerates the regeneration process, speeds up tree replacement, and improves the ability to avoid mortality and withstand drought [9,10]. In regions with frequent disturbance, the community tends to have a high proportion of species that mostly resprout [11,12]. A global multispecies study has also revealed that growth form is the factor that has the greatest impact on seed masses, the seeds produced by shrubs being significantly smaller than those produced by trees [13].

Few studies have examined whether this morphological transition of tree-to-shrub in arid regions also alters the physiological carbon allocation strategy of trees, which plays a key role in ecosystem dynamics and plant acclimation to changing environmental conditions [14]. A mechanistic understanding of how plants, particularly long-living organisms, such as trees, allocate and remobilize stored carbohydrates is still very poor. Previous studies have shown that carbon allocation in adult trees involves at least three trade-offs between storage, growth, and reproduction [15,16]. Nonstructural carbohydrates (NSC) play a central role in plant functioning because they are building blocks and energy carriers for plant metabolic processes. Although, on an annual basis, net carbon flux to storage may be small relative to allocation of respiration and growth, recent studies tend to suggest that storage represents a sink that can compete with other sinks like growth [17]. Storage is used in plants for maintenance respiration, growth resumption, foliage building in the spring, and protection tree physiological integrity against environmental stresses, such as frost [18,19], defoliation [20,21], shade [22,23], insect attacks, and wounds [15,24]. In arid regions of our concern, soluble sugars play important roles in cavitation induction, signaling, and repairment, as the xylem of trees undergoes diurnal and seasonal cavitation and repair [25–27]. We therefore hypothesize that growth form affects carbon allocation.

Differences in tree life stages have long been ignored in the studies of NSC dynamics, although the intensity of survival stress commonly changes during tree ontogeny. Seedlings have shallow roots and can only absorb water from shallow soils, while large trees can use water from deeper soils [28]. Considering operability, field control experiments mostly use tree seedlings as the research objects [29]. If the results of the field control experiment are simply applied to adult trees, it may lead to inaccurate conclusions [30]. Simultaneous measurements of the NSC concentrations of each organ at different life stages are required. We further hypothesize that the NSC concentration at different life stages differs.

To test the above hypotheses, we selected the natural *Quercus mongolica* Fisch. ex Ledeb. forest in the Saihanwula Nature Reserve in Inner Mongolia, China. The site is at the xeric timberline, where the driest site of *Q. mongolica* distribution and forests are threatened by long-term water limitation [5]. *Q. mongolica*, a drought-resistant and cold-resistant deciduous tree species [31], is naturally secondary and dominant in the temperate forests of northern China. From shady to sunny slopes, tree-to-shrub growth form shifts typically occur at the xeric timberline [32]. According to the water balance calculation, the soil available water of the shady slopes is 253 mm per year, much more than that of the sunny slopes (about 0 mm per year) [32]. We simultaneously measured the NSC concentrations of each organ in three stages of life: seedling, sapling, and adult tree. Carbon allocation to growth, storage, and reproduction function was assessed by quantifying the differences in radial growth, nonstructural carbohydrates concentrations, and number of resprouts, respectively [7,20].

2. Materials and Methods

2.1. Study Area and Site Features

The Saihanwula National Nature Reserve (43°59′–44°27′ N, 118°18′–118°55′ E) is located in the southern part of the Greater Khingan Mountains. It is located in the semiarid region, with cold winters and little snowfall; summer is hot with sufficient sunlight. The annual average temperature is 2 °C, and the average annual precipitation is 400 mm [29]. From shady to sunny slopes, tree-to-shrub growth form shifts typically occur (Figure S1, Table S1) [32].

The average soil thickness of the shady slopes was measured as 48.6 cm and that of the sunny slopes was 24.3 cm. We measured the soil water content and soil bulk density of two different slopes aspects using the ring knife method, and the results showed that the surface soil water content of the shady slopes was significantly higher than that of the sunny slopes ($p < 0.05$) (Figure S2a). The soil bulk density at 0–10 cm and 10–20 cm on the sunny slopes was significantly higher than that on the shady slopes ($p < 0.05$) (Figure S2b).

2.2. Field Survey

We conducted site surveys and sample collection in July and August 2019. In this study, different life stages of *Q. mongolica* were distinguished, including adult trees, saplings, and seedlings. Multiple plant organs were sampled separately. First, a 10×10 m plot was set up, and trees with a diameter at breast height ≥ 10 cm were defined as adult trees [33]. Trees with a diameter at breast height ranging from 1 to 9.9 cm were defined as saplings, and those with unlignified stem diameters < 1 cm and tree heights < 30 cm were defined as seedlings. We investigated 12 plots on the sunny slopes and 15 plots on the shady slopes while collecting samples. The heights and diameters at breast height (1.3 m above the ground) of every tree were measured in each plot. Sprout regeneration seedlings are differentiated from seed regeneration seedlings by emergence of the stem from a lateral root of the parent tree running in the upper soil horizon (Figure 1). We distinguished seedlings in this study on that basis. The ratio of seedling from seed germination was calculated by digging around each whole plant and removing the superficial soil temporarily to confirm the contact of the root sprout to its parent tree.

Figure 1. Sprout regeneration sapling and seedling. (**a**), The lateral roots of sapling are connected to the parent tree. (**b**), The roots of the seedling are much older than the seedling itself.

2.3. Sample Collection

To compare stem growth rates across different growth forms, we sampled two tree ring cores from 10 adult trees per growth form in the study site using increment borers with an inner diameter of 5 mm parallel to the contour on the opposite sides of the tree trunk at breast height (1.3 m). After air drying, the cores were polished using progressively fine sandpaper until tree ring details were clearly visible. All polished tree ring samples were dated using the cross-dating technique [34]. Tree ring widths were measured using a width meter with an accuracy of 0.001 mm using LINTAB. The quality of cross-dating was then validated using the COFECHA program [35]. Some cores were redated until the series intercorrelated up to 0.6 to ensure that the measured ring widths were reliable. Based on the ring widths, the BAI was calculated as:

$$\mathrm{BAI} = \pi \left(r_t{}^2 - r_{t-1}{}^2 \right)$$

where r_t is the stem radius at year t and r_{t-1} is the stem radius in the year before year t [36].

Considering that the tree canopy position has no significant effect on the branch NSC concentration in previous studies [37], this study did not sample according to the canopy position. The first-level branches in the middle of the canopy were randomly selected as standard branches. Branches with good growth were cut off by branch shears, and then the leaves on the branches were picked off (attempting to choose leaves without insect eggs). An increment borer was used to drill 4 tree cores at the breast height of each sample tree. During each sampling, the drilling position was slightly moved, and the sampling was in a Z-shape to avoid overlapping with the previous sampling position and decrease the experimental error. Two root samples from each tree were excavated using iron picks, shovels, and branch shears to cut from the soil layer between the root and the farthest end of the canopy at a depth of 5–30 cm. After washing the soil on the root surface, they were divided into thick roots (>5 mm) and fine roots (<2 mm), according to diameters.

The saplings are dug up with the roots, and the sprouted root saplings are collected from the roots of the parent tree. We took the seed regeneration seedlings to the shady slopes and the sprouted seedlings to the sunny slopes. Since the seedlings are small, the whole plant is brought back to the laboratory and decomposed into leaves, stems, and roots (fine roots and thick roots). A total of 6 seedlings, 4 saplings, and 5 adult tree form samples were collected on the shady slope. A total of 11 seedlings, 9 saplings, and 8 adult shrub form samples were collected on the sunny slope.

2.4. Measurements of NSC Concentration

All samples were deactivated using a microwave oven at high heat (600 W) to denature the enzyme, and then dried to a constant weight in a drying oven at 65 °C. In this study, the concentrations of starch and soluble sugar were determined by the anthracene copper concentrated sulfuric acid method [38–41]. Total nonstructural carbohydrates (TNCs) were defined as the sum of soluble sugar and starch concentrations.

One gram of purified anthrone was weighed and dissolved in 1000 mL of dilute sulfuric acid solution to obtain an anthrone reagent. It was prepared with 100 μg/mL glucose standard solution for drawing a standard curve. The grinded plant tissue weighed about 0.05 g and the actual weight was recorded and extracted repeatedly with 80% ethanol. After collecting the supernatant, anthrone reagent was added to measure the absorbance with a spectrophotometer (UV-1800 PC, Shanghai MAPADA Instruments, Shanghai, China) at 620 nm wavelength. With the sugar concentration in the filtrate analyzed from the standard curve (or calculated by a linear regression formula), the percentage of sugar in the sample was calculated, the unit of sugar concentration being a percent of sugar per gram of the sample dry weight.

Perchloric acid and distilled water were added to the residue after extraction of soluble sugar for repeated extraction. Anthrone reagent was added to the supernatant and measured on a spectrophotometer to calculate the absorbance. The starch concentration was read from the standard curve.

2.5. Statistical Analysis

Multivariate analysis of variance (MANOVA) was used to test whether there were significant differences in different growth forms. A mixed linear model was used to analyze whether each factor had a significant effect on the NSC concentration. The NSC concentration was used as the dependent variable, the growth forms, life stages, organs, and their interactions were used as fixed factors, and different individuals were used as random factors. All statistical analyses and figure graphing were performed in R version 4.1.2 and Origin version 2020b.

3. Results

3.1. Growth Features of Tree and Shrub Form Quercus mongolica

The results of the quadratic survey showed that 1/3 of seedlings for the tree form were produced by seed germination, while all the seedlings for the shrub form were sprout production (Table 1).

Table 1. Sample plot overview. Diameter at breast height (DBH), tree height, and adult tree age of *Quercus mongolica* Fisch. ex Ledeb. are shown at different growth stages (seedling, sapling, adult trees) of different growth forms.

Growth Form	Seedling		Sapling		Adult Tree/Shrub		
	Height (m)	Seedlings from Seed Germination	Height (m)	DBH (cm)	Height (m)	DBH (cm)	Age (Year)
Tree	0.14 ± 0.06 [a]	33.3%	3.10 ± 2.25 [a]	5.60 ± 2.49 [a]	7.21 ± 1.81 [a]	11.65 ± 1.32 [a]	33.8 ± 16.6 [a]
Shrub	0.16 ± 0.08 [a]	0	2.41 ± 1.55 [b]	5.13 ± 2.90 [a]	4.26 ± 1.23 [b]	11.33 ± 2.13 [a]	48.1 ± 12.7 [a]

Different lowercase letters indicate significant differences between growth forms ($p < 0.05$).

The tree heights of saplings and adult trees on the shady slope were significantly higher than those on the sunny slope ($p < 0.05$). DBH of saplings and adult trees, as well as height of seedlings, did not differ significantly between the two slopes ($p > 0.05$). There was no significant difference in the age of the two growth forms of *Q. mongolica* ($p > 0.05$).

3.2. Carbon Allocation for Reproduction and Growth

The basal number of sprouts of shrub form *Q. mongolica* was significantly higher than that of tree form adult trees ($p < 0.001$) (Figure 2a). Most trees of tree form only had single main trunk, while those of shrub form mostly grew with multiple stems. The ages of different growth forms are relatively similar (Table 1). Although the shrub form showed less growth variances relative to tree form (Figure 2b), the growth difference between them was not significant ($p > 0.05$, Figure 2c). There is no significant interannual differences in multi-year BAI between tree and shrub growth forms ($p > 0.05$, Figure 2d).

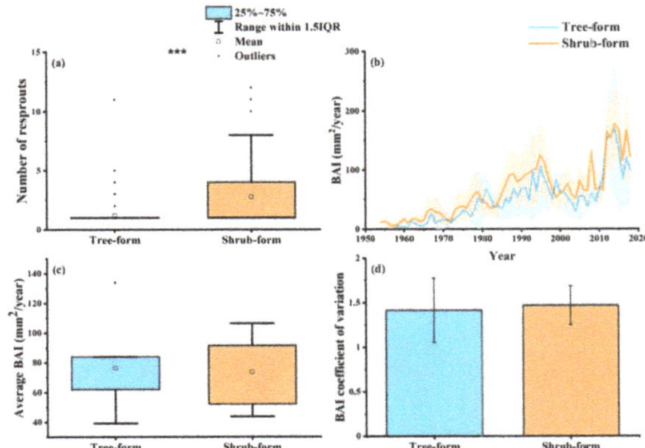

Figure 2. Carbon allocation reflected in growth and reproduction. (**a**), Box plots show the resprout number of different growth form has significant difference (***, $p < 0.001$). (**b**), Interannual variation in the basal area increment (BAI) of *Q. mongolica*. Shaded areas are the variations in BAI of average BAI standard variation. (**c**), Box plots show multi-year average BAI of two growth forms have no significant difference ($p > 0.05$). (**d**), BAI coefficients of variations have no significant difference between the two growth forms ($p > 0.05$). The coefficient of variation (C.V) is calculated as the ratio of the BAI standard deviation to the mean.

3.3. Carbon Allocation for NSC Storage

The results of the mixed linear model show that the differences in both growth form and organ significantly affected the soluble sugar, starch, and total nonstructural carbohydrates (TNCs) ($p < 0.01$, Table 2). Life stage significantly affected soluble sugar concentration ($p < 0.05$) but had no significant effect on starch and TNC concentrations ($p > 0.05$). Different life stages and sites showed significant interactions in starch and TNC ($p < 0.05$). Growth form and organ showed significant interactions on soluble sugar, starch, and TNC ($p < 0.01$).

Table 2. Results of mixed linear models for factors affecting NSC concentration. The NSC concentration was used as the dependent variable, life stage, growth form, and organ. Their interactions were used as fixed factors, and different individuals were used as random factors.

Fixed Factors	Soluble Sugar	Starch	TNC
Growth form	37.7754 ***	7.9883 **	27.0347 ***
Life stage	5.1772 *	1.0088	0.0049
Organ	12.2682 ***	13.7526 ***	14.6334 ***
Growth form × Life stage	2.3192	3.4001 *	3.9453 *
Growth form × Organ	5.9409 **	4.9803 **	6.7068 ***

Values indicate the F values, and stars indicate the significance level (*, $p < 0.05$; **, $p < 0.01$; ***, $p < 0.001$).

NSC concentrations in *Q. mongolica* with different growth forms were affected differently by life stage and organ (Table 3). The soluble sugar concentration of tree form *Q. mongolica* was significantly affected by life stage ($p < 0.05$). The starch concentration and TNCs of shrub form *Q. mongolica* were significantly affected by life stage ($p < 0.01$). NSC concentrations with different growth forms were significantly affected by different organs ($p < 0.001$).

Table 3. Mixed linear model results of tree and shrub growth forms for factors affecting NSC concentration. In each growth form, the NSC concentration was used as the dependent variable, life stage, and organs were used as fixed factors. Different individuals were used as random factors.

Fixed Factors	Tree Form			Shrub Form		
	Soluble Sugar	Starch	TNC	Soluble Sugar	Starch	TNC
Life stage	5.00 *	1.07	4.00	3.17	19.01 ***	12.65 **
Organ	12.83 ***	10.85 ***	13.17 ***	10.67 ***	7.55 ***	8.21 ***

Values indicate the F values, and stars indicate the significance level (*, $p < 0.05$; **, $p < 0.01$; ***, $p < 0.001$).

Results of MANOVA analysis showed that, for *Q. mongolica* of both tree and shrub forms, the concentrations of soluble sugar and starch were different in each life stage (Figure 3). There were no significant differences in starch and soluble sugar concentrations in the same organ of tree form *Q. mongolica* at different life stages ($p > 0.05$, Figure 3a,c). The changes in NSCs in shrub form *Q. mongolica* were completely different. The concentration of soluble sugar in adult tree stems (1.4%) was significantly lower than that in seedling (2.9%) and sapling (3.2%) stems ($p < 0.05$, Figure 3b). Meanwhile, the starch concentrations in the leaves (5.8%) and stems (6.6%) of the seedling stage were significantly lower than those of the saplings and the adult trees ($p < 0.05$, Figure 3d). The starch concentration in the thick roots (13.7%) of the adult trees was significantly higher than those of the seedlings (7.7%) and saplings (7.9%) ($p < 0.05$, Figure 3d).

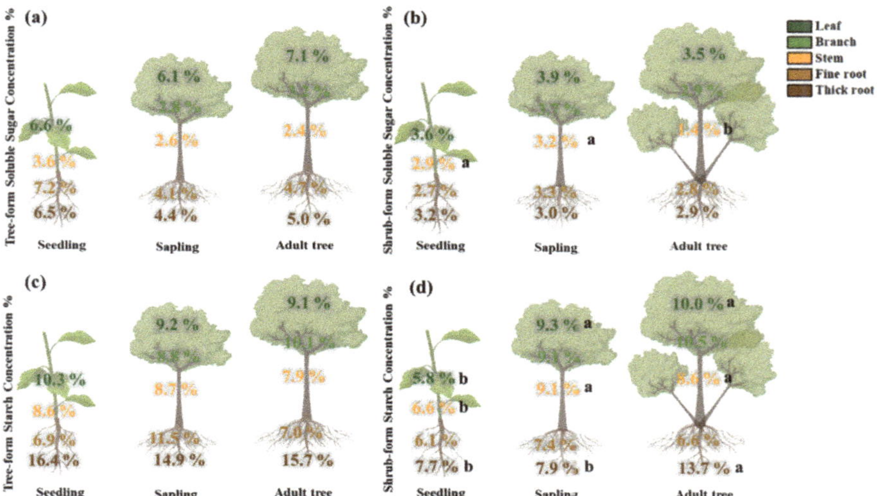

Figure 3. Effects of the interaction between life stages and organs on NSC concentration. (**a**), Soluble sugar concentration for tree form. (**b**), Soluble sugar concentration for shrub form. (**c**), Starch concentration for tree form. (**d**), Starch concentration for shrub form. Different lowercase letters indicate significant differences between life stages in the same organ ($p < 0.05$).

In terms of NSC concentration in different organs, except for the tree form seedlings, the soluble sugar concentration in leaves (3.5%–7.7%) was the highest at each life stage, and the soluble sugar concentration in stems (1.4%–3.6%) was relatively low (Figure 3a,b). Except for the shrub form saplings, the starch concentration in the thick roots (7.7%–16.4%) was always the highest, and the starch concentration in the fine roots (6.1%–11.5%) was relatively low (Figure 3c,d).

3.4. Effects of Different Life Stages on NSC Storage

The results of a linear mixed model for different life stages showed that growth form significantly affected the soluble sugar, starch, and TNC contents of seedlings ($p < 0.01$, Table 4). However, there was no significant effect on the NSC concentration of saplings ($p > 0.05$), though there was a significant effect on the soluble sugar concentration in adult trees ($p < 0.01$, Table 4).

Table 4. Results for mixed linear models of NSC storage between different life stages. In each life stage, the NSC concentration was used as the dependent variable, growth forms, organs. Their interactions were used as fixed factors, and different individuals were used as random factors.

Fixed Factors	Seedling			Sapling			Adult Tree		
	Soluble Sugar	Starch	TNC	Soluble Sugar	Starch	TNC	Soluble Sugar	Starch	TNC
Growth form	16.53 **	32.10 **	42.17 **	5.30	1.75	2.60	12.38 **	0.11	2.62
Organ	3.93 *	9.14 **	7.04 **	9.00 ***	1.81	2.83 *	6.07 ***	7.81 ***	9.48 ***
Growth form × Organ	2.30	3.63 *	3.17 *	6.22 **	7.05 ***	6.55 ***	1.84	0.41	0.71

Values indicate the F values, and stars indicate the significance level (*, $p < 0.05$; **, $p < 0.01$; ***, $p < 0.001$).

The MANOVA results showed that at the seedling stage of the shrub form, the soluble sugar (3.0% decrease), starch (4.4% decrease), and TNC (7.5% decrease) concentrations in leaves, and the concentrations of starch and TNC in stems (2.1% decrease) and thick roots (8.7% decrease) were significantly lower than those of the tree form (Figure 4). The soluble sugar concentration in fine roots (4.5% decrease) of the shrub form was significantly lower (Figure 4). In the sapling stage, the NSC concentrations of shrub form in thick roots (soluble sugar decreased 1.3% and starch decreased 7.0%) and soluble sugar concentration in the leaves (2.2% decrease) were significantly lower than those of tree form (Figure 5). The NSC concentrations in branches (soluble sugar increased 1.0% and starch increased 0.4%) and stems (soluble sugar increased 0.5% and starch increased 0.4%) were higher than those of the tree form, but the difference was not significant (Figure 5). For adult trees, the concentration of soluble sugar in leaves (3.6% decrease), stems (0.9% decrease), and fine roots (1.9% decrease) of the shrub form were lower, and the TNC concentration in fine roots (2.3% decrease) was lower (Figure 6). While starch (0.4% increase) and TNC concentrations (0.2% increase) in branches were higher than those of treeform, the difference was not significant (Figure 6).

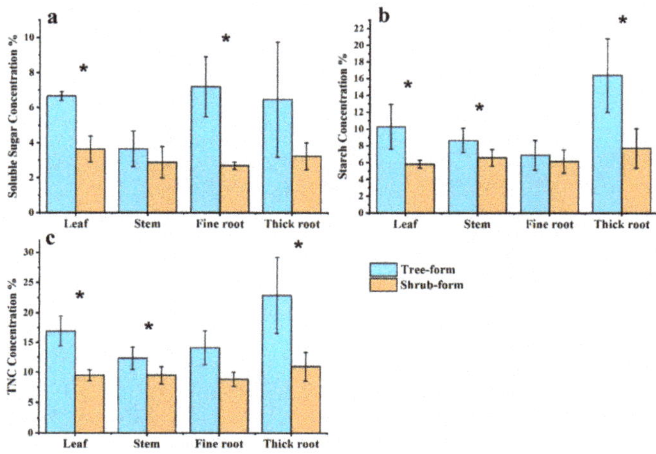

Figure 4. Effects of growth form on NSC concentration in seedlings. Changes of (**a**), soluble sugar, (**b**), starch and (**c**), TNC concentration of each organ. Asterisks indicate that MANOVA shows significant differences between different growth form trees (*, $p < 0.05$).

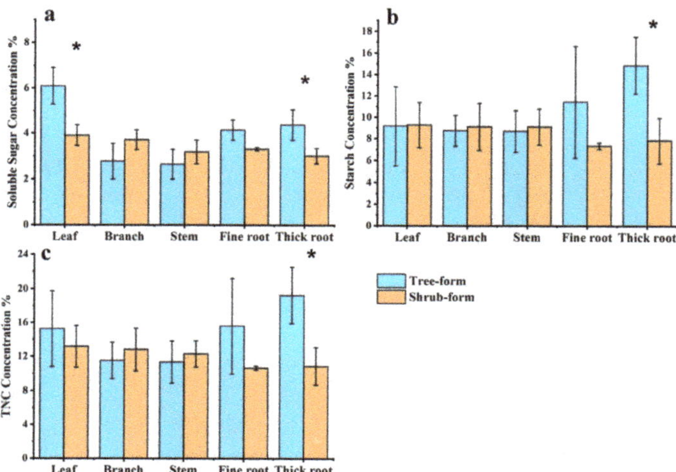

Figure 5. Effects of growth form on NSC concentration in saplings. Changes of (**a**), soluble sugar, (**b**), starch and (**c**), TNC concentration of each organ. Asterisks indicate that MANOVA shows significant differences between the two growth forms (*, $p < 0.05$).

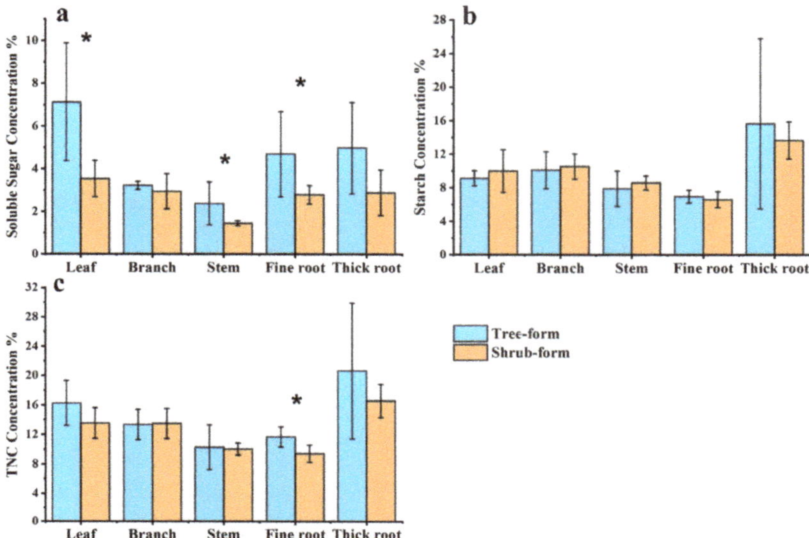

Figure 6. Effects of growth form on NSC concentration in adult trees/shrubs. Changes of (**a**), soluble sugar, (**b**), starch and (**c**), TNC concentration of each organ. MANOVA shows significant differences between the two growth forms (*, $p < 0.05$).

Different organs had a significant effect on the NSC concentrations of seedlings and adult trees ($p < 0.05$) and had a significant effect on the soluble sugar concentration of saplings ($p < 0.001$) (Table 4). Interactions between growth forms and organs on the NSC concentrations were significant in saplings and seedlings ($p < 0.05$), except for the soluble sugar of seedling (Table 4).

4. Discussion

4.1. Growth Form Affecting Carbon Allocation for Both Tree Growth and Reproduction

Our results showed that the effects of tree and shrub growth forms were significant for the seedlings and sapling of *Q. mongolica* and became less pronounced as it grew up. Due to the smaller carbon pool of seedlings, the carbon in seedlings relies on constant carbon assimilation and was therefore more prone to dramatic fluctuations in response to environmental stress [42]. Tree-to-shrub significantly decreased starch accumulation in thick roots at the seedling and sapling stages and in leaves and stems at the seedling stage (Figures 4 and 5). Meanwhile, changes in the growth form from tree to shrub at the sapling and adult tree stages resulted in increased starch concentrations in leaves, branches, and stems, but the increase was not significant (Figures 5 and 6). This distribution method might facilitate the rapid transportation and utilization of soluble sugars and starches [29].

Some previous studies suggested that the primary function of NSC storage was to obtain higher storage at the expense of reduced growth and sacrifice of seed reproduction to ensure the survival of the species in extreme environments [20,43]. Although we could not sample and compare the differences in the quantity and quality of seeds produced in this experiment, evidence for seedling regeneration patterns and the number of resprouts indicated that we observed a change in the regeneration mode, leading to a change in regeneration modes from seedling to sprout production (Figure 2).

By comparing the tree ring BAI, the growth difference between those was not significant (Figure 2). There was no significant difference in the interannual growth variations of trees with different growth forms. It was suggested that there was a trade-off between reserve storage and the production of new tissues, and the growth form and plant size was generally used to quantify the trade-off [44]; however, this trade-off was not found in our study. Therefore, we speculated that *Q. mongolica* in an arid environment would change NSC storage patterns and reduce investment in seed production to ensure vegetative growth.

4.2. Divergent Carbon Allocation at Different Life Stages

The NSC concentration in the different life stages of *Q. mongolica* were distinct, and the soluble sugar concentration was significantly different with the life stages (Table 2). Soluble sugar concentration in seedlings is relatively high. Compared with adult trees, seedlings did not have lignified stems, and the absence of non-storing heartwood and older sapwood, therefore they could store more soluble sugars than adult trees [30]. Additionally, soluble sugars were also higher in seedling roots. The NSC concentration in the seedling stem could be three to four times that of an adult tree [45,46].

Previous studies suggested that the NSC concentration either increased or decreased gradually with growth [16,47]. Peaks in starch and soluble sugar levels could also occur at intermediate stages of life [48]. However, there was no such consistent pattern between saplings and adult trees in our study, and the NSC concentration could not be simply scaled according to allometric relationships. The concentration of NSCs was more likely to fluctuate during plant growth, with allocations adjusted according to growth needs at distinct stages. Our results indicated that tree form *Q. mongolica* significantly changed the soluble sugar concentration with tree growth, while the starch and TNC concentrations of shrub form changed significantly. Specifically, the concentration of soluble sugar in the stems of shrub form adults had significantly decreased, and the concentration of starch in the thick roots had significantly increased. Starch concentrations in leaves and stems of shrub form seedlings were significantly lower than those of saplings and adult trees. Compared with shrub form *Q. mongolica*, tree form *Q. mongolica* needed to accumulate more soluble sugar at the seedling stage to adapt to arid environment [49]. Shrub form saplings and adults stored more starch, especially in thick roots, in preparation for sprout regeneration, which could be explained by the classic paradigm that the availability of stored carbohydrate reserves was the major driver of resprouting [50].

It should be noted that our sampling was a one-time sampling, and the NSC concentration in plants often fluctuated with the seasons [26,51,52], so our results could only represent the nonstructural carbon concentration in *Q. mongolica* during our sampling period.

5. Conclusions

In summary, our study revealed changes in carbon allocation at different growth forms. We observed adaptation to dry climate through carbon allocation adjustments for growth, reproduction, and storage in the driest regions of the *Q. mongolica* distribution. Our study could shed light on the adaptation of trees to a dry climate through NSC allocation, which could benefit the improvement of vegetation dynamics models.

Supplementary Materials: The following supporting information can be downloaded at: https://www.mdpi.com/article/10.3390/f13111745/s1, Figure S1: Photos of the tree and shrub-form Quercus mongolica. a, Tree-form oaks on shady slopes. b, Shrub-form oaks on sunny slopes. Figure S2: Physical proper-ties of soil on shady and sunny slopes. a, Soil moisture content, b, soil bulk density measured by the ring knife method. Asterisks indicate significant differences at the same soil layer depth on different slopes (*, $p < 0.05$; **, $p < 0.01$). Table S1: Basic Information for the shady and sunny slope of Quercus mongolica.

Author Contributions: H.L. and Y.Q. conceived of the study idea. Y.Q. performed the analyses. H.L., Y.Q., W.H., J.D., L.S. and Z.S. interpreted the results and implications. H.L. supervised the research. Y.Q. and H.L. wrote the first draft of the manuscript. All authors revised the text and provided critical feedback. All authors have read and agreed to the published version of the manuscript.

Funding: This work was granted by National Key Research and Development Program of China (Grant No. 2022YFF0801803) and National Natural Science Foundation of China (No. 42161144008).

Data Availability Statement: The data that support the findings of this study are available from the corresponding author upon reasonable request.

Conflicts of Interest: The authors declare that they have no conflicts of interest.

References

1. Jucker, T.; Fischer, F.J.; Chave, J.; Coomes, D.A.; Caspersen, J.; Ali, A.; Loubota Panzou, G.J.; Feldpausch, T.R.; Falster, D.; Usoltsev, V.A.; et al. Tallo: A global tree allometry and crown architecture database. *Glob. Change Biol.* **2022**, *28*, 5254–5268. [CrossRef] [PubMed]
2. Scheffer, M.; Vergnon, R.; Cornelissen, J.H.C.; Hantson, S.; Holmgren, M.; van Nes, E.H.; Xu, C. Why trees and shrubs but rarely trubs? *Trends Ecol. Evol.* **2014**, *29*, 433–434. [CrossRef] [PubMed]
3. Vincent, G.; Harja, D. Exploring Ecological Significance of Tree Crown Plasticity through Three-dimensional Modelling. *Ann. Bot.* **2007**, *101*, 1221–1231. [CrossRef] [PubMed]
4. Stovall, A.E.L.; Shugart, H.; Yang, X. Tree height explains mortality risk during an intense drought. *Nat. Commun.* **2019**, *10*, 4385. [CrossRef]
5. Dai, J.; Liu, H.; Wang, Y.; Guo, Q.; Hu, T.; Quine, T.; Green, S.; Hartmann, H.; Xu, C.; Liu, X.; et al. Drought-modulated allometric patterns of trees in semi-arid forests. *Commun. Biol.* **2020**, *3*, 405. [CrossRef]
6. Barrett, S.C.H. Influences of clonality on plant sexual reproduction. *Proc. Natl. Acad. Sci. USA* **2015**, *112*, 8859–8866. [CrossRef]
7. Qiu, T.; Andrus, R.; Aravena, M.; Ascoli, D.; Bergeron, Y.; Berretti, R.; Berveiller, D.; Bogdziewicz, M.; Boivin, T.; Bonal, R.; et al. Limits to reproduction and seed size-number trade-offs that shape forest dominance and future recovery. *Nat. Commun.* **2022**, *13*, 2381. [CrossRef]
8. Wenk, E.H.; Falster, D.S. Quantifying and understanding reproductive allocation schedules in plants. *Ecol. Evol.* **2015**, *5*, 5521–5538. [CrossRef]
9. Xu, C.; Liu, H.; Zhou, M.; Xue, J.; Zhao, P.; Shi, L.; Shangguan, H. Enhanced sprout-regeneration offsets warming-induced forest mortality through shortening the generation time in semiarid birch forest. *Forest Ecol. Manag.* **2018**, *409*, 298–306. [CrossRef]
10. Zeppel, M.J.B.; Harrison, S.P.; Adams, H.D.; Kelley, D.I.; Li, G.; Tissue, D.T.; Dawson, T.E.; Fensham, R.; Medlyn, B.E.; Palmer, A.; et al. Drought and resprouting plants. *New Phytol.* **2015**, *206*, 583–589. [CrossRef]
11. Malanson, G.P.; Trabaud, L. Vigour of post-fire resprouting by *Quercus coccifera* L. *J. Ecol.* **1988**, *76*, 351–365. [CrossRef]
12. Bellingham, P.J.; Sparrow, A.D. Resprouting as a life history strategy in woody plant communities. *Oikos* **2000**, *89*, 409–416. [CrossRef]
13. Moles, A.T.; Ackerly, D.D.; Webb, C.O.; Tweddle, J.C.; Dickie, J.B.; Pitman, A.J.; Westoby, M. Factors that shape seed mass evolution. *Proc. Natl. Acad. Sci. USA* **2005**, *102*, 10540–10544. [CrossRef] [PubMed]
14. Frank, D.; Reichstein, M.; Bahn, M.; Thonicke, K.; Frank, D.; Mahecha, M.D.; Smith, P.; Velde, M.; Vicca, S.; Babst, F.; et al. Effects of climate extremes on the terrestrial carbon cycle: Concepts, processes and potential future impacts. *Global. Chang. Biol.* **2015**, *21*, 2861–2880. [CrossRef] [PubMed]
15. Lauder, J.D.; Moran, E.V.; Hart, S.C. Fight or flight? Potential tradeoffs between drought defense and reproduction in conifers. *Tree Physiol.* **2019**, *39*, 1071–1085. [CrossRef] [PubMed]
16. Genet, H.; Breda, N.; Dufrene, E. Age-related variation in carbon allocation at tree and stand scales in beech (*Fagus sylvatica* L.) and sessile oak (*Quercus petraea (Matt.) Liebl.*) using a chronosequence approach. *Tree Physiol.* **2010**, *30*, 177–192. [CrossRef]
17. Green, J.K.; Keenan, T.F. The limits of forest carbon sequestration. *Science* **2022**, *376*, 692–693. [CrossRef]
18. Long, R.W.; Dudley, T.L.; D'Antonio, C.M.; Grady, K.C.; Bush, S.E.; Hultine, K.R. Spenders versus savers: Climate-induced carbon allocation trade-offs in a recently introduced woody plant. *Funct. Ecol.* **2021**, *35*, 1640–1654. [CrossRef]
19. Yin, X.H.; Hao, G.Y.; Sterck, F. A trade-off between growth and hydraulic resilience against freezing leads to divergent adaptations among temperate tree species. *Funct. Ecol.* **2022**, *36*, 739–750. [CrossRef]
20. Wiley, E.; Casper, B.B.; Helliker, B.R. Recovery following defoliation involves shifts in allocation that favour storage and reproduction over radial growth in black oak. *J. Ecol.* **2017**, *105*, 412–424. [CrossRef]
21. Jacquet, J.S.; Bosc, A.; O'Grady, A.; Jactel, H. Combined effects of defoliation and water stress on pine growth and non-structural carbohydrates. *Tree Physiol.* **2014**, *34*, 367–376. [CrossRef] [PubMed]
22. Piper, F.I. Patterns of carbon storage in relation to shade tolerance in southern South American species. *Am. J. Bot.* **2015**, *102*, 1442–1452. [CrossRef] [PubMed]
23. Maguire, A.J.; Kobe, R.K. Drought and shade deplete nonstructural carbohydrate reserves in seedlings of five temperate tree species. *Ecol. Evol.* **2015**, *5*, 5711–5721. [CrossRef] [PubMed]
24. Ferrenberg, S.; Kane, J.M.; Langenhan, J.M. To grow or defend? Pine seedlings grow less but induce more defences when a key resource is limited. *Tree Physiol.* **2015**, *35*, 107–111. [CrossRef] [PubMed]
25. Morgan, J.M. Osmoregulation and water-stress in higher-plants. *Annu. Rev. Plant Physiol. Plant Mol. Biol.* **1984**, *35*, 299–319. [CrossRef]
26. Guo, J.S.; Gear, L.; Hultine, K.R.; Koch, G.W.; Ogle, K. Non-structural carbohydrate dynamics associated with antecedent stem water potential and air temperature in a dominant desert shrub. *Plant Cell Environ.* **2020**, *43*, 1467–1483. [CrossRef]

27. Hartmann, H.; Trumbore, S. Understanding the roles of nonstructural carbohydrates in forest trees—From what we can measure to what we want to know. *New Phytol.* **2016**, *211*, 386–403. [CrossRef]

28. Drake, P.L.; Mendham, D.S.; White, D.A.; Ogden, G.N. A comparison of growth, photosynthetic capacity and water stress in Eucalyptus globulus coppice regrowth and seedlings during early development. *Tree Physiol.* **2009**, *29*, 663–674. [CrossRef]

29. He, W.; Liu, H.; Qi, Y.; Liu, F.; Zhu, X. Patterns in nonstructural carbohydrate contents at the tree organ level in response to drought duration. *Global. Chang. Biol.* **2020**, *26*, 3627–3638. [CrossRef]

30. Hartmann, H.; Adams, H.D.; Hammond, W.M.; Hoch, G.; Landhäusser, S.M.; Wiley, E.; Zaehle, S. Identifying differences in carbohydrate dynamics of seedlings and mature trees to improve carbon allocation in models for trees and forests. *Environ. Exp. Bot.* **2018**, *152*, 7–18. [CrossRef]

31. Xu, X.; Wang, Z.; Rahbek, C.; Sanders, N.J.; Fang, J. Geographical variation in the importance of water and energy for oak diversity. *J. Biogeogr.* **2016**, *43*, 279–288. [CrossRef]

32. Dai, J.; Lu, S.; Qi, Y.; Liu, H. Tree-to-Shrub Shift Benefits the Survival of *Quercus mongolica* Fisch. ex Ledeb. at the Xeric Timberline. *Forests* **2022**, *13*, 244. [CrossRef]

33. Zhu, Y.; Comita, L.S.; Hubbell, S.P.; Ma, K. Conspecific and phylogenetic density-dependent survival differs across life stages in a tropical forest. *J. Ecol.* **2015**, *103*, 957–966. [CrossRef]

34. Schweingruber, F.H. *Tree Rings: Basics and Applications of Dendrochronology*; Springer Science & Business Media: Dordrecht, The Netherlands, 2012.

35. Holmes, R.L. Computer-assisted quality control in tree-ring dating and measurement. *Tree-Ring Bull.* **1983**, *43*, 69–78.

36. Monserud, R.A.; Sterba, H. A basal area increment model for individual trees growing in even-and uneven-aged forest stands in Austria. *For. Ecol. Manag.* **1996**, *80*, 57–80. [CrossRef]

37. Li, M.H.; Hoch, G.; Körner, C. Spatial variability of mobile carbohydrates within *Pinus cembra* trees at the alpine treeline. *Phyton* **2001**, *41*, 203–213.

38. Trevelyan, W.E.; Harrison, J.S. Studies on yeast metabolism. I. Fractionation and microdetermination of cell carbohydrates. *Biochem. J.* **1952**, *50*, 298–303. [CrossRef]

39. Osaki, M.H.U.S.; Shinano, T.; Tadano, T. Redistribution of carbon and nitrogen compounds from the shoot to the harvesting organs during maturation in field crops. *Soil Sci. Plant Nutr.* **1991**, *37*, 117–128. [CrossRef]

40. Quentin, A.G.; Pinkard, E.A.; Ryan, M.G.; Tissue, D.T.; Baggett, L.S.; Adams, H.D.; Maillard, P.; Marchand, J.; Landhäusser, S.M.; Lacointe, A.; et al. Non-structural carbohydrates in woody plants compared among laboratories. *Tree Physiol.* **2015**, *35*, 1146–1165. [CrossRef]

41. Ashwell, G. Colorimetric analysis of sugars. In *Methods in Enzymology*; Academic Press: Cambridge, MA, USA, 1957; Volume 3, pp. 73–105.

42. Niinemets, Ü. Responses of forest trees to single and multiple environmental stresses from seedlings to mature plants: Past stress history, stress interactions, tolerance and acclimation. *Forest Ecol. Manag.* **2010**, *260*, 1623–1639. [CrossRef]

43. Wiley, E.; Helliker, B. A re-evaluation of carbon storage in trees lends greater support for carbon limitation to growth. *N. Phytol.* **2012**, *195*, 285–289. [CrossRef] [PubMed]

44. Vesk, P.A. Plant size and resprouting ability: Trading tolerance and avoidance of damage? *J. Ecol.* **2006**, *94*, 1027–1034. [CrossRef]

45. Weber, R.; Schwendener, A.; Schmid, S.; Lambert, S.; Wiley, E.; Landhausser, S.M.; Hartmann, H.; Hoch, G. Living on next to nothing: Tree seedlings can survive weeks with very low carbohydrate concentrations. *New Phytol.* **2018**, *218*, 107–118. [CrossRef]

46. Hoch, G.; Richter, A.; Körner, C. Non-structural carbon compounds in temperate forest trees. *Plant Cell Environ.* **2003**, *26*, 1067–1081. [CrossRef]

47. Sala, A.; Hoch, G. Height-related growth declines in ponderosa pine are not due to carbon limitation. *Plant Cell Environ.* **2009**, *32*, 22–30. [CrossRef] [PubMed]

48. Martin, A.R.; Thomas, S.C. Size-dependent changes in leaf and wood chemical traits in two Caribbean rainforest trees. *Tree Physiol.* **2013**, *33*, 1338–1353. [CrossRef]

49. Qiu, T.; Sharma, S.; Woodall, C.W.; Clark, J.S. Niche Shifts From Trees to Fecundity to Recruitment That Determine Species Response to Climate Change. *Front. Ecol. Evol.* **2021**, *9*, 719141. [CrossRef]

50. Bowen, B.J.; Pate, J.S. The Significance of Root Starch in Post-fire Shoot Recovery of the Resprouter *Stirlingia latifolia* R. Br. (Proteaceae). *Ann. Bot.* **1993**, *72*, 7–16. [CrossRef]

51. Tixier, A.; Guzmán-Delgado, P.; Sperling, O.; Amico Roxas, A.; Laca, E.; Zwieniecki, M.A. Comparison of phenological traits, growth patterns, and seasonal dynamics of non-structural carbohydrate in Mediterranean tree crop species. *Sci. Rep.* **2020**, *10*, 347. [CrossRef]

52. Gilson, A.; Barthes, L.; Delpierre, N.; Dufrene, E.; Fresneau, C.; Bazot, S. Seasonal changes in carbon and nitrogen compound concentrations in a Quercus petraea chronosequence. *Tree Physiol.* **2014**, *34*, 716–729. [CrossRef]

forests

Article

Regulatory Control and the Effects of Condensation Water on Water Migration and Reverse Migration of *Halostachys caspica* (M.Bieb.) C.A.Mey. in Different Saline Habitats

Lu Qin [1,2,†], Xuemin He [3,4,5,†], Guanghui Lv [3,4,5,*] and Jianjun Yang [3,4,5]

1 College of Tourism, Xinjiang University, Urumqi 830046, China
2 Key Laboratory for Sustainable Development of Xinjiang's Historical and Cultural Tourism, Urumqi 830046, China
3 College of Ecology and Environment, Xinjiang University, Urumqi 830017, China
4 Key Laboratory of Oasis Ecology of Education Ministry, Urumqi 830017, China
5 Xinjiang Jinghe Observation and Research Station of Temperate Desert Ecosystem, Ministry of Education, Urumqi 830017, China
* Correspondence: ler@xju.edu.cn; Tel.: +86-0991-2111427
† These authors contributed equally to this work.

Citation: Qin, L.; He, X.; Lv, G.; Yang, J. Regulatory Control and the Effects of Condensation Water on Water Migration and Reverse Migration of *Halostachys caspica* (M.Bieb.) C.A.Mey. in Different Saline Habitats. *Forests* **2022**, *13*, 1442. https://doi.org/10.3390/f13091442

Academic Editor: Giovanna Battipaglia

Received: 23 July 2022
Accepted: 2 September 2022
Published: 8 September 2022

Publisher's Note: MDPI stays neutral with regard to jurisdictional claims in published maps and institutional affiliations.

Abstract: Condensation water has been a recent focus in ecological hydrology research. As one of the main water sources that maintains the food chain in arid regions, condensation water has a significant impact on water balance in arid environments and plays an important role in desert vegetation. This study takes drought desert areas and high-salinity habitats as its focus—selecting *Halostachys caspica* (M.Bieb.) C.A.Mey. and its community in mild, moderate, and severe salinity soil—analyzed the source of condensation water utilized by these plants, and calculated its percentage of contribution. I. Study results revealed: (1) Scale-like leaves can absorb condensation water and the order of condensation water contribution to plant growth in different salinity habitats are severe > mild > moderate, such that the average contribution rates were 11.13%, 7.10%, and 3.79%, respectively; (2) The migration path of water movement in these three communities are formed in two main ways: (a) rain and condensation water recharge the soil to compensate for groundwater, while some groundwater compensates for river water and partially returns to the atmosphere by soil evaporation and plant transpiration; and (b) rain and condensation water directly compensate for river water and plant roots absorb river water, groundwater, and soil water in order to grow; (3) in mild habitats, the water movement path in plants is as follows: shallow root → stem → branches → leaves and shallow root → deep root; (4) in moderate habitats, stems act as the bifurcation point and the path follows as: stem → branches → leaves and stem → shallow root → deep root; and (5) in severe habitats, the path is as follows: deep root → shallow root → stem → branches → leaves, and finally returning to the atmosphere. These results elucidate the contribution of condensation water on *Halostachys caspica* growth and the migration path through the *Halostachys caspica* body. Condensation water obtained by *Halostachys caspica* communities in different salinity habitats provides a theoretical basis and data supporting the need for future research of condensation water on plants at the physiological level in arid regions and provides reference for the protection of saline soil and its ecological environment in arid regions.

Keywords: condensation water; isotope; *Halostachys caspica*; salinity; moisture migration

1. Introduction

In arid regions, precipitation is limited and evaporation is intense, and any supplementary water has a positive impact on the ecosystems. Condensation water and precipitation are two water sources in desert regions that play an important role in desert ecosystems [1–3]. In the desert, water is scarce and apart from precipitation, condensation water is an important, vital source of water. In drought conditions, there is less soil water content and fewer perennial plants. Despite the small volume of condensation water, it plays an important role in the local water balance [4–6], and especially in drought years its importance is more obvious [5,7].

In the abundant precipitation region, the amount of condensation is trivial. However, in the arid and semi-arid regions, condensation water plays a supplementary role (e.g., in the desert along the Atlantic coast) [8–10]. Maphangwaa et al. [11] found that atmospheric water vapor is the main water source that lichen absorb, and it is atmospheric water vapor, not precipitation, that determines the amount of lichen richness and overall distribution. Thus, conducting research on condensate water and the water balance in arid regions should be a focus of present and future studies.

In a former study conducted by Tao and Zhang [12], it was demonstrated that the tippy of desert moss crust can significantly reduce and delay evaporation in the crust, which prolongs plant hydration time. The greater the amount of precipitation, the more obvious the slowing effect is observed in the tippy of desert moss crust. This slowing effect is also helpful as it allows for the utilization of condensation and precipitation, enhancing the moss crust's ability to adapt to drought conditions. In a separate study, Temina and Kidron [13] discovered that the duration of condensate water determines the distribution of lichen on rocks in the Negev desert. Cheng et al. [14] found in the Alpine Sandy Desert that biological soil crust is conducive to moisture absorption and condensate water and, with the development of the crust, the content of water vapor increases. As an important source of water, condensation water plays a significant role in the survival of vegetation in arid areas [15,16]. Therefore, when conducting research on water balance in arid regions, condensate water cannot be ignored and the importance of water vapor on desert vegetation has been a recent focus of present-day studies.

There are two aspects of plants' use for condensation water. One is indirect utilization, that is, when the dew congeals on the surface of the plant or when fog is intercepted by the canopy and as a result, small water droplets form on a big plant's leaves and water drips down to the soil surface where it is absorbed by the root system of shallow root plants. This use of condensation water allows scholars to focus on the issues of water sources for root systems, such as groundwater, soil natural water, precipitation, fog, dew, and more. Thus, the contribution of various sources of water supply for a specific plant can be calculated, mainly for water sources that contain different δD and $\delta^{18}O$ isotopes to distinguish various sources. For example, Goebel and Lascano [17] conducted a quantitative analysis of the water use conditions of cotton, including the utilization of condensation water by measuring the different sources of water containing δD and $\delta^{18}O$.

Additionally, when plants absorb water, it affects the direction of moisture migration through the plant body, but research on this characteristic is rare. Previous research focused on the quantity of water in terms of soil condensation, plant canopy condensation, occurrence regularity, and impact factors. There are very few studies on the quantity of condensation water utilized by plants and moisture migration path through the plant body [18], and the signature of ^{18}O isotope tracer (the phenomenon of the enrichment of ^{18}O in the photosynthetic organs, secondary branches, and trunk xylem) showed that the photosynthetic organs of desert woody plants are able to absorb canopy dew and transfer it to the trunk xylem. In high humidity conditions, assimilating branches of *Haloxylon ammodendron* (C.A.Mey.) Bunge actively absorbed canopy dew and transferred the canopy dew down to the secondary shoots through the reverse water potential (Ψ) gradient between photosynthetic organs and secondary branches (Ψ Photosynthetic organs > Ψ Secondary branches), and the photosynthetic organs can transport the excess dew to the trunk stem via reverse water potential gradient, which is conducive to the continuous absorption and utilization for canopy dew [18]. Leaves of desert plants usually carnify and degenerate to form spiny or scaly parts, while succulent leaves can effectively store water but reduce water transpiration by reducing the amount of leaf area exposed to the air [19]. *Halostachys caspica* is a model organism that belongs to the saline-xerophytic-juicy subshrub and is an important windproof, dune-fixing shrub, whose branchlets contain succulent juice and has scale-like leaves. Studies have shown that irregular leaf surface can capture small water droplets, and the leaves of *Halostachys caspica* have an irregular surface; thus, we can infer that *Halostachys caspica* could absorb and utilize condensate water.

In this study, we test the following three hypotheses: (1) The leaf of *Halostachys caspica* has the ability to absorb condensation water; (2) The contribution of condensation water to plant growth in *Halostachys caspica* habitats with salinity differences; and (3) The different ways in which condensation water is utilized by *Halostachys caspica*. This study selected the typical desert shrub species, *Halostachys caspica*, as the research organism, which is found in the Ebinur Lake wetland national nature reserve. The treatments were divided into three different salinity habitats: mild, moderate, and severe salinity soils. We conducted a field investigation and analyzed isotope composition to determine the source of condensation water that plants used, as well as its proportion, and investigated the migratory path of condensation water in the plant body. The findings of this study will help to further our understanding of the contribution of condensation water and the migration path in the *Halostachys caspica* body under different salinity conditions and to probe the migration path of obtained condensation water in the *Halostachys caspica* body. The results help to further understand the contribution and the migration path in the *Halostachys caspica* body to the obtained condensation water by the *Halostachys caspica* communities under a different salinity habitat. It is of great significance to elucidate the positive effects of condensation water on plant growth in desert ecosystems in order to provide a theoretical basis and data support for propelling the future research of condensation water on plant physiological level in arid regions and to provide reference for the protection of saline soil and its ecological environment in arid regions.

2. Materials and Methods

2.1. The Study Region

The study region is located on the northwest edge of the Junggar Basin ($82°36'$–$83°50$ E, $44°30'$–$45°09'$ N) of the Xinjiang Uygur Autonomous Region. The climate is very dry, there is scarce rainfall, copious amounts of sunlight, strong winds, dust storms, hot summers, and cold winters, which is typical of climates belonging to temperate continental arid regions [20]. Influenced by landform characteristics and climactic conditions, the distribution of vegetation within the Ebinur Lake basin is primarily dominated by two plant flora native to Asia and Mongolia. These two plants had a clear transition into these areas and represent most desert plant species in Xinjiang. *Halostachys caspica* is one of the dominant species in the study area and it is a saline, xerophytism, succulent subshrub that is mainly distributed in the salt-alkali beach area, river valley, and by the salt lake.

Salinization reverse succession was the main method utilized in this study. In the study area, we selected communities for mild, moderate, and severe salinity (Figure 1), and their soil salt content were $0.500 \pm 0.275\%$, $1.428 \pm 0.286\%$, and $2.022 \pm 0.329\%$, respectively. The mild salinity community is adjacent to the river, located at the north of the river with a vertical distance from the river of 50 m. The moderate salinity community is far away from the river bank, located at the north of the river with a vertical distance from the river of 3700 m. The severe salinity community is far away from the river bank, located south of the river with a vertical distance from the river of 700 m. Three well-growing *Halostachys caspica* were selected for isotope sampling in each habitat.

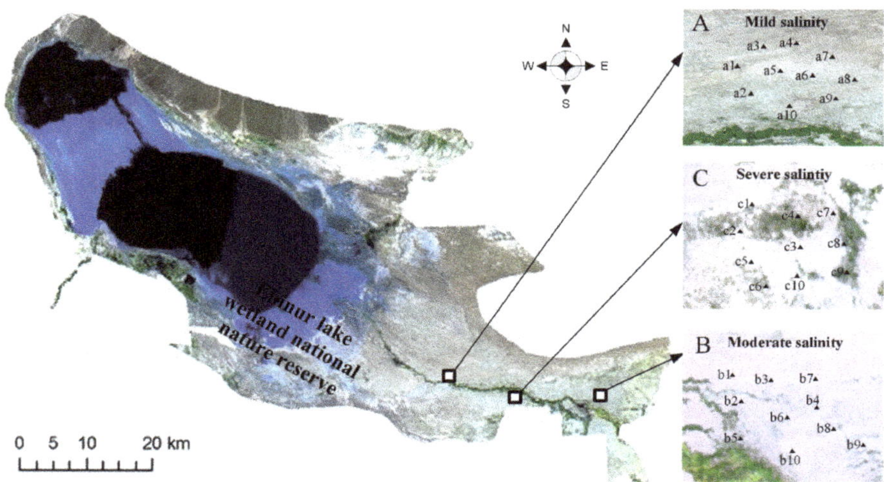

Figure 1. Schemes of study sample regions.

2.2. Experimental Method and Samples Collection

This article had set seven water sources, which were: topsoil (0–5 cm), shallow soil (5–20 cm), deep soil (20–30 cm), rain water, river water, ground water, and condensation water.

Plant samples were collected before sunrise and three species of plants were randomly selected as representatives of either the mild, moderate, or severe salinity habitats; there was a total of nine samples. We used loppers to acquire leaves. Suberification was utilized after gathering mature branches, stems, shallow roots (5–20 cm), and deep roots (20–50 cm) in order to remove the phloem but leave in the xylem. We then quickly put samples into 50 mL isotope sampling bottles and sealed them with parafilm membrane. Samples were transported back to the lab, then placed in the fridge to freeze samples at −20 °C until they were analyzed for isotope composition.

Soil samples were collected before sunrise from each of the three salinity habitats. Samples were gathered under the canopy of plant samples, close to the root. Samples were collected from the surface soil (0–5 cm), shallow soil (5–20 cm), and deep soil (20–30 cm). We then placed the samples into 50-mL isotope sampling bottles and sealed them with parafilm membrane. Samples were transported back to the lab, then placed in the fridge to freeze samples at −20 °C until they were analyzed for isotope composition.

Natural water samples were collected from the underground water source near the study point. To ensure the samples were true representatives, we collected deep water samples from the Aqikesu River. Rain and condensation water samples were collected two days before plant and soil sampling, then placed in a 50 mL isotope sampling bottle and sealed with parafilm membrane. Samples were transported back to the lab, then placed in the fridge to be kept at 4 °C until they were analyzed for isotope composition.

Using a handheld weather instrument (Kestrel 4500 NV), we were able to determine the air temperature, dew-point temperature, relative humidity of the atmosphere, wind speed, and air pressure at 1.5 m above the surface when all of the samples were collected.

2.3. Plant Physiological Determination

2.3.1. Presence and Absence Condensation Water Treatment Setting

In the mild salinity, moderate salinity, and severe salinity communities, some leaves with the same water potential were selected in the observation plot approximately 0.5 h before sunset, and they were divided into three groups, namely: (1) Before treatment (B), the initial water potential of branches were measured immediately after taking down the branches; (2) presence of condensation water treatment (W1), that is, branches were not bagged in the natural state; (3) absence of condensation water treatment (W0), sealed with plastic bags (so that the surface of branches was not affected by night condensation water).

2.3.2. Water Potentials

Approximately 0.5 h after sunrise the next day, the branches were removed to check the presence and absence of condensation water, and the water potential of the branches was immediately measured, with 9 repetitions in each group.

2.3.3. Fresh Leaf Water Absorption per Unit Area

In mild salinity, moderate salinity, and severe salinity communities, six leaves with and without condensate treatment were picked, the radius (r) and length (l) of the leaves were measured, and the weight $m1$ was weighed. Then, the leaves were placed in clean water and left for approximately 24 h. After the leaves fully absorbed water, they were removed from the water. The water on the surface of the leaves was completely absorbed by filter paper, and $m2$ was weighed again. The water absorption per unit area of fresh leaves was calculated by the following formula:

$$\text{Fresh leaf water absorption per unit area} = (m2 - m1)/(\pi r^2 l) \tag{1}$$

2.3.4. Transpiration Rate (Tr) and Water Use Efficiency (WUE)

The transpiration rate and the net photosynthetic rate (Pn) of *Halostachys caspica* under the presence and absence condensation water treatment were measured by cluster leaf chamber of Li-6400XT portable photosynthetic apparatus (Li-Cor, Lincoln, Nebraska, USA), and the WUE was calculated by the following formula:

$$\text{WUE} = \text{Pn}/\text{Tr} \tag{2}$$

In the presence and absence condensation water treatment, 30 repetitions are determined for each treatment.

2.4. Water Extraction and Isotope Determination

We used the Liquid Water Isotope Analyzer (LWIA, DLT–100, Los Gatos Research Inc., Mountain View, CA, USA) to determine the hydrogen and oxygen isotopic composition of the samples. Plant water and soil water were extracted using a cryogenic vacuum distillation line [21] and the extracted water samples were stored in sealed glass vials at 2C. Then, the hydrogen and oxygen isotopic compositions of the samples were determined by an isotope ratio infrared spectroscopy (IRIS) analyzer—the Liquid Water Isotope Analyzer (LWIA, DLT–100, Los Gatos Research Inc., Mountain View, CA, USA), and the determination of each sample was repeated 12 times. Analytical precision of individual measurement were ±0.1‰ for δD and ±0.25‰ for δ^{18}O. The isotopic composition can be expressed as:

$$\delta X = \left(\frac{R_{sample}}{R_{standard}} - 1 \right) \times 1000‰ \tag{3}$$

where X is D or ^{18}O, R_{sample} and $R_{standard}$ are the hydrogen or oxygen isotopic composition (^{2}H/^{1}H or ^{18}O/^{16}O molar ratio) of the sample and the standard water (Standard Mean Ocean Water, SMOW), respectively.

2.5. Data Analysis

A one-way analysis of variance (ANOVA) was utilized to analyze the data. Data manipulation and visual representations were conducted using Excel 2003 software and Sigmaplot 10.0. All of the statistical analyses were run with a significance level of 0.05 using StatView 5.0 (SAS Institute, Inc., Cary, NC, USA). There were seven sources of water in this study, according to the upper and lower limits method Proposed by Phillips and Gregg [22]. IsoSource 1.3.1 software was used to analyze the water source of plant and soil samples.

3. Result and Analysis

3.1. Plant Physiological Characteristics in Different Salinity Habitats

3.1.1. Water Potentials

As can be seen from the Figure 2, night condensation water can improve the water potential of plant leaves in those three communities, but without condensation water supplement, it will reduce the water potential of plant leaves, indicating that the leaves of *Halostachys caspica* can absorb condensation water and improve the water potential of plant leaves.

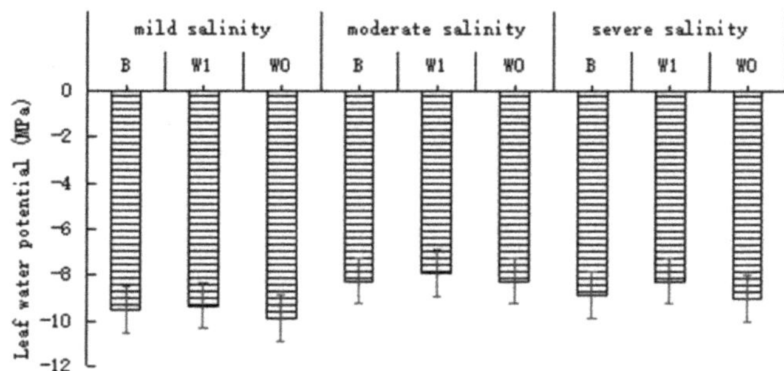

Figure 2. Comparison of *Halostachys caspica* leaves water potentials under presence and absence of water treatment in different salinity habitats. Before treatment (B), presence of condensation water treatment (W1), absence of condensation water treatment (W0).

3.1.2. Fresh Leaf Water Absorption per Unit Area

Using the results of fresh leaf water absorption per unit area in mild salinity, moderate salinity, and severe salinity communities under the presence and absence condensation water treatment, the following figure can be drawn.

In mild salinity, moderate salinity, and severe salinity communities, the fresh leaf water absorption per unit area of leaves of absence condensation water is higher than that presence condensation water (Figure 3), which indicates that the short-term lack of water on the surface of leaves can stimulate the water absorption per unit area. In addition, under the condition of absence condensation water treatment, the water absorption per unit area of leaves of *Halostachys caspica* in the mild salinity, moderate salinity, and severe salinity communities showed the following order: moderate salinity community > severe salinity community > mild salinity community, which indicated that if the leaves were temporarily short of water, and once there was enough water in the air, and the salinization degree increased, the plant leaves growing on them would have stronger water absorption capacity.

Under the presence and absence condensation water treatment, the transpiration rates of three communities are as follows: severe salinity community > moderate salinity > mild salinity community (Figure 4), which indicated that the ecological effect of condensation water on severe salinity community was higher than that of mild salinity and moderate salinity community. In those three salinity habitats, the transpiration rate of plants treated of presence condensation water was higher than that absence condensation water, which indicated that condensation water can improve the transpiration rate of plants and had certain ecological effects on plants. In addition, it was found that the condensation water formed at night can supplement the evaporated water to a certain extent in the three communities [23].

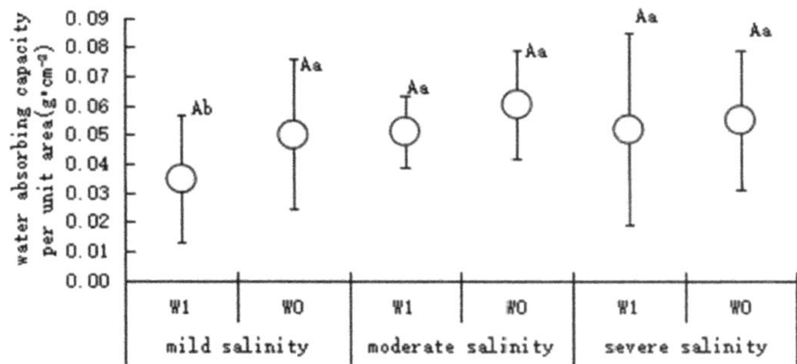

Figure 3. Fresh leaf water absorption per unit area under presence and absence of condensation water treatment. The same lowercase letter indicates that there is no significant difference among different treatments in the same habitat ($p > 0.05$). The same capital letter indicates that there is no significant difference between different habitat of the same treatment ($p > 0.05$), presence of condensation water treatment (W1), absence of condensation water treatment (W0), the same below.

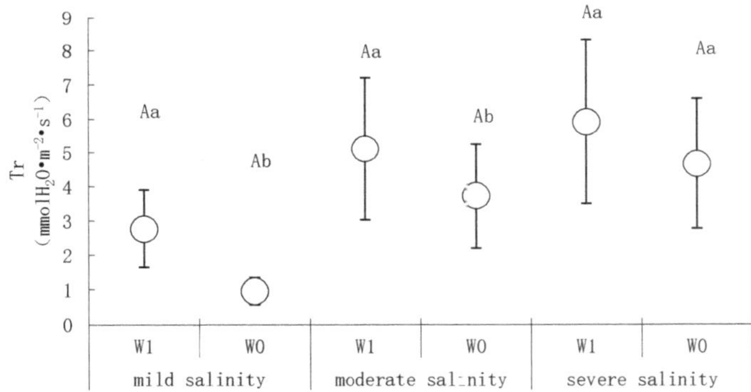

Figure 4. Tr of *Halostachys caspica* under the presence and absence condensation water treatment.

The instantaneous water use efficiency (WUE) of plants is a comprehensive physiological index to evaluate the adaptability of plants to the environment.

Daily mean value of the WUE of moderate and mild salinity communities under the presence condensation water treatment higher than the absence condensation water (Figure 5), which may be temporary water balance is out of balance in plants at night without condensation water supplement, which reduces the water use efficiency. In the severe salinity, the daily average WUE of *Halostachys caspica* under the presence condensation water treatment is lower than that of absence treatment. This is because the severe salinity community is far from the water source, lives in a low water environment for a long time, has a survival strategy, has a strong self-regulating mechanism for water stress to improve water utilization ability to cope with water deficit [24], which is similar to Yu's research result for maize [25].

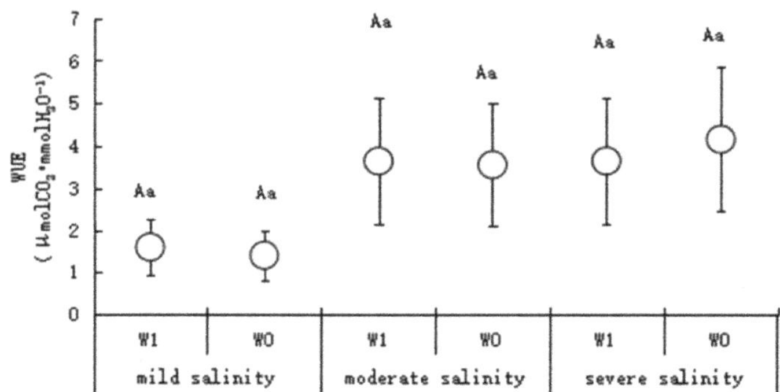

Figure 5. Comparison of *Halostachys caspica* WUE under presence and absence of condensation water treatment in different salinity habitats.

3.1.3. δD and δ^{18}O Composition in Plant Structures and the δD-δ^{18}O Relationship in Different Salinity Habitats

Plant leaves enriched with δD and δ^{18}O isotopes have partially negative δD values ranging from -40.0844 ± 1.6106 to $-40.0844 \pm 1.6706\%$; when there are partially positive δ^{18}O values, they range from 4.9758 ± 0.686 to $12.469 \pm 1.653\%$ (Figure 6). The max value was observed in the severe salinity community, while the minimum value was observed in the light salinity community, which indicates that *Halostachys caspica* growing in severe salinity soil were better at evapotranspiration. Additionally, compared to aerial parts of the plant, such as branches and stems, the shallow roots from moderate salinity soil had lower δD and δ^{18}O values. This may be because soil evaporation has allowed for the transpiration and evaporation at the roots of the plant. Moreover, transpiration and evaporation in aerial parts of the plant strengthen water delivery from the roots to aerial parts of plant, which suggests that there is a regulatory effect to the plant from soil moisture through a hydraulic redistribution effect [26].

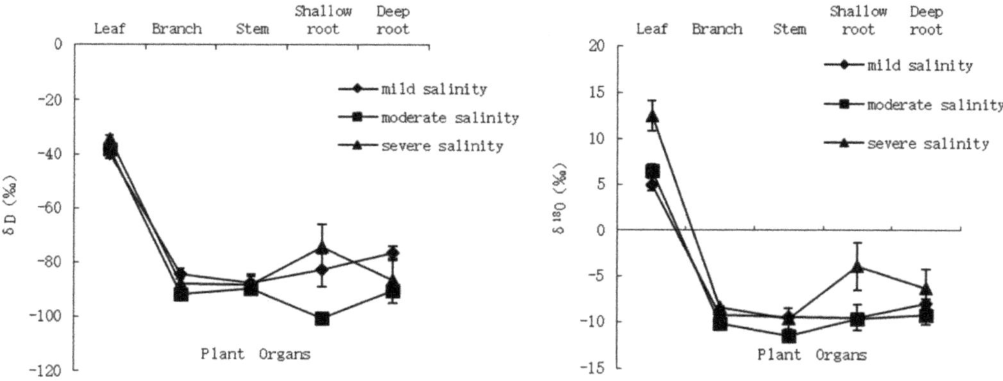

Figure 6. Distribution of δD and δ^{18}O composition of structures in different salinity habitats.

The three salinity habitats were ranked as mild > moderate > severe, which indicates that the degree of soil salinity enhances the evapotranspiration ability of *Halostachys caspica* (Figure 7). This in turn causes an isotope fractionation effect. It could also be a survival strategy for plant growth in saline soil is to take excess salt from the plant of body as evapotranspiration increases.

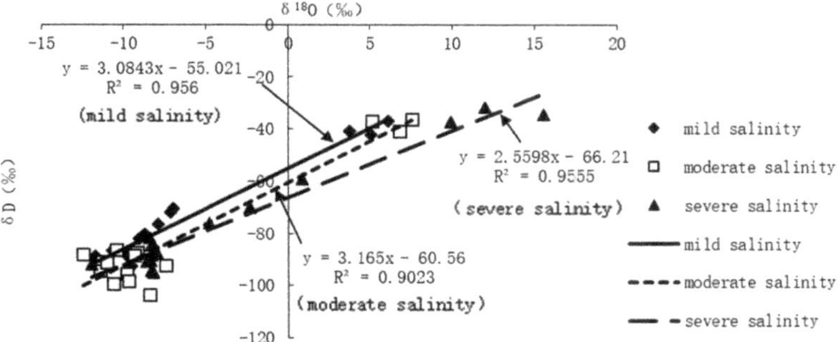

Figure 7. δD-$\delta^{18}O$ relationship in water samples from plant structures found in different salinity habitats. Fitting equations for each habitat are noted in the figure.

3.2. Hydrogen and Oxygen Isotope Composition and Soil Profiles of Halostachys Caspica Communities

Results revealed that there are significant differences in δD and $\delta^{18}O$ values based on the salinity habitat (Figure 8). The values were (-70.524 ± 4.326)–(-46.432 ± 3.638)‰ and (-5.197 ± 0.906)–(1.987 ± 0.626)‰, respectively. The higher the composition of δD and $\delta^{18}O$ isotopes, the more enriched the soil surface layer and with the increasing soil depth, the δD and $\delta^{18}O$ values decrease gradually.

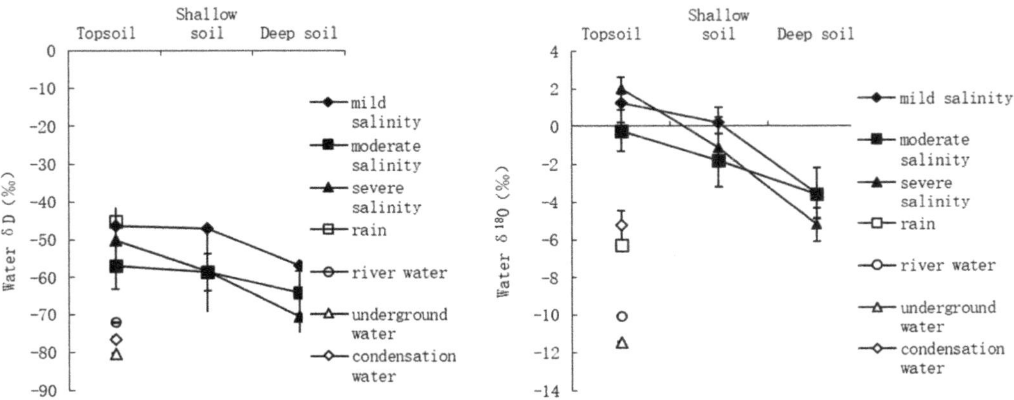

Figure 8. δD and $\delta^{18}O$ of soil water in different soil and natural water of salinity habitats.

All of the soil layers were enriched with δD in the mild salinity habitat. The topsoil in severe salinity habitats had slightly higher $\delta^{18}O$ values than the mild salinity, while $\delta^{18}O$ values in mild salinity soil water were higher than moderate and severe, which indicates that there is soil evaporation and an isotope fractionation effect in mild salinity, although the volatility of δD and $\delta^{18}O$ values are at a minimum. Meanwhile, soil water hydrogen and oxygen stable isotopes are lacking in deep soil of severe salinity habitats.

Results revealed that closer to the surface, $\delta^{18}O$ values for soil water are partially positive. Additionally, $\Delta\, \delta D$ values for soil water in the different salinity habitats were between the values for rain and the other three sources of water (river water, condensation water, and groundwater). Lastly, the $\delta^{18}O$ values were all partially negative for all water sources.

3.3. The Contribution of Condensation Water to Plant Growth

Due to strong fractionation in the leaves, there were no results for leaf water source, so we analyzed the water source for branches, stems, and shallow roots. Water sources were topsoil, shallow soil, deep soil, rain water, river water, groundwater, and condensation water (Table 1).

Table 1. Sources of water in different salinity habitats (mean $\pm$ SE).

Habitat	Contribution of Water Source (%)				
	Soil Water	**Rain**	**River Water**	**Ground Water**	**Condensation Water**
mild salinity	11.375 ± 2.071	8.525 ± 1.280	27.025 ± 0.330	46.1 ± 4.594	7.05 ± 1.100
moderate salinity	6.775 ± 2.507	4.575 ± 1.661	18.5 ± 5.750	66.4 ± 11.253	3.775 ± 1.375
severe salinity	27.9 ± 9.484	11.925 ± 2.258	20.1 ± 3.955	28.975 ± 10.150	11.125 ± 2.478

Results revealed that the main contributors to plant growth in mild salinity habitats were soil water (11.375%), rain (8.53%), river (27.03%), groundwater (46.1%), and condensation water (7.1%). In the moderate salinity habitats, the main contributors were 1.8% (Topsoil), 2.18% (Shallow soil), 2.8% (Deep soil), 4.58% (Rain), 18.5% (River water), 66.4% (Ground water), and 3.78% (Condensation water). In the severe salinity habitat, the main contributors were 27.91% (Soil water), 11.93% (Rain), 20.1% (River water), 28.98% (Ground water), and 11.13% (Condensation water).

To summarize, results revealed that groundwater is the main contributor to plant growth in *Halostachys caspica* with the maximum contribution in the moderate salinity community (66.4%) followed by the mild salinity community (28.89%) and the minimum contribution in the severe salinity community (28.98%). Additionally, there were multiple water absorption strategies exhibited in *Halostachys caspica* growth, which ensures that when there are water restrictions other strategies can instead be employed. Lastly, the contribution of condensation water to plant growth in severe salinity habitats was the largest, with an average value of 11.13% and a maximum of 82%. In the mild salinity habitat, the average value was 7.1% and the maximum value was 56%. In the moderate salinity habitat, the average value was 3.79% and the maximum value was 36%.

3.4. Water Migration Path in the Plant Body

Generally speaking, when water is transported in a mature plant, hydrogen and oxygen stable isotope fractionation will not occur. Only the leaves or high salinity plant will have fractionation, thus, one can use hydrogen and oxygen stable isotope technology to quantify these compositions in the vast majority of land plants. The $\delta^{18}O$ isotope is present at the earliest water source, thus, this value will gradually increase over time and the moisture migration path in the body of *Halostachys caspica* in different salinity habitats can be determined (Figure 9a–c).

In all three habitats, rain and condensation water directly contributed to the river water and supplied the soil with water through soil infiltration. Groundwater also supplies the river with water and through soil evaporation and transpiration into the atmosphere, plant roots are able to absorb river water, groundwater, and soil water, which all contribute to plant growth.

All of the plant structures, except leaves, had $\delta^{18}O$ values similar to soil water values. Yang [27] also observed this in *Haloxylon ammodendron* at the study area and speculated that in addition to rain, river water, groundwater, and soil water, *Haloxylon ammodendron* may take advantage of condensation water through strong evaporation through their leaves as a result of day and night temperature differences. Moreover, Yang [27] calculated that the contributions of condensation water to plant growth were 7.1% (mild salinity), 3.78% (moderate salinity), and 11.13% (severe salinity), which indicates that the contribution of condensed water to plants in different saline habitats is influenced by the soil properties of saline habitats.

In the plant body, water does not simply migrate from the roots to the leaves then return to the atmosphere through evaporation. Rather, in *Halostachys caspica* communities under mild salinity conditions, water migration flows as follows: shallow roots → stems → branches → leaves; there are also short water circuits from shallow roots to deep roots. In moderate salinity conditions, the stems acts as a bifurcation point and one path of water migrates from: stems → shallow and deep roots, while the other path flows from: stems → branches → leaves. In severe salinity conditions, water migrates from: stems → shallow and deep roots →branches → leaves, there is also a water loop from the stems to shallow roots.

In terms of soil water, there are different $\delta^{18}O$ compositions in topsoil and shallow soil water along the vertical section. Shallow soil water $\delta^{18}O$ composition showed a partially negative trend from bottom to top under the topsoil, which indicates that there is an ascending phenomenon. In mild and moderate salinity conditions, topsoil water and shallow soil water $\delta^{18}O$ composition under the topsoil showed a partially positive trend up to the surface. This may be because mild and moderate salinity habitats are near the river and, thus, the soil surface absorbs more precipitation [23].

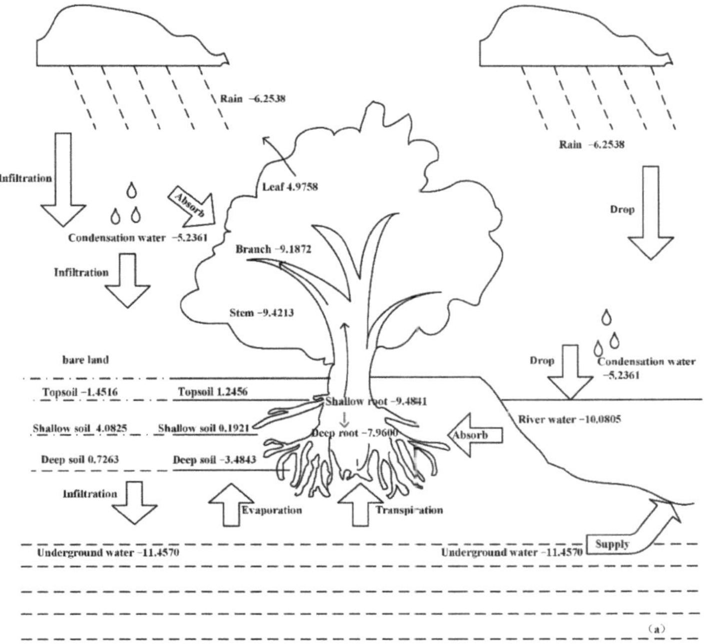

Figure 9. *Cont.*

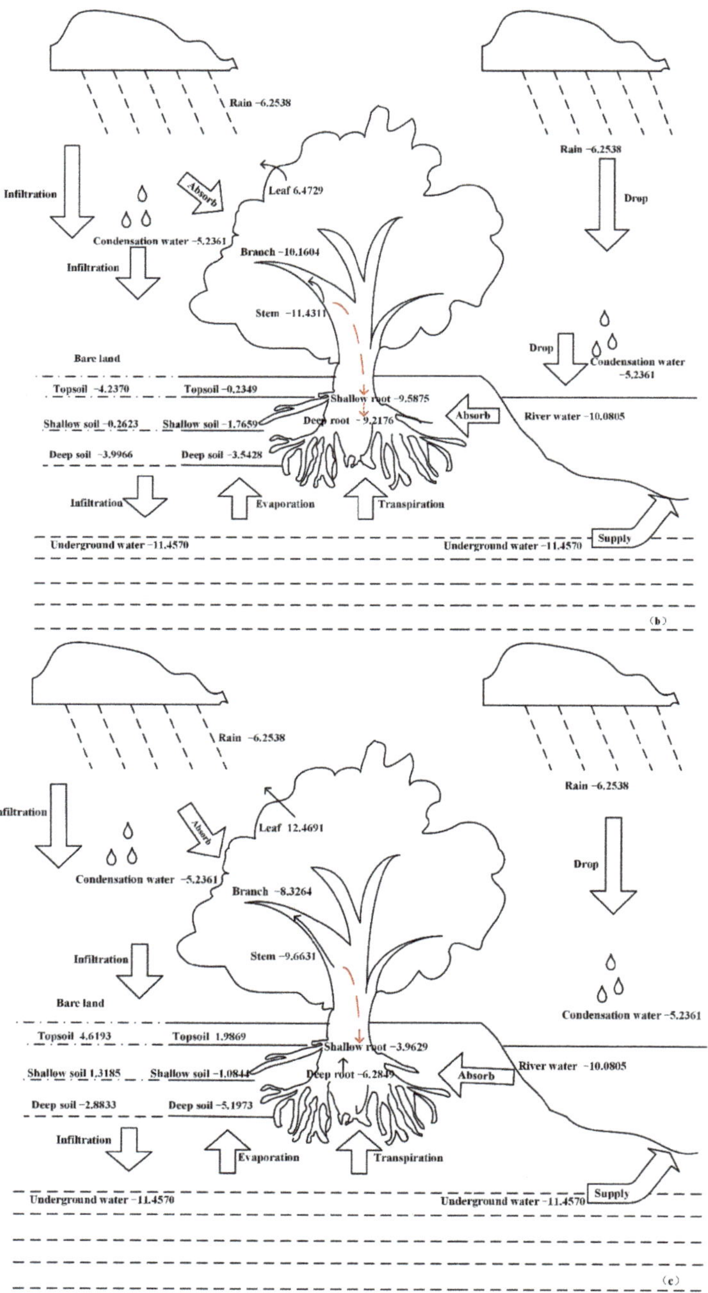

Figure 9. (**a**) Soil-vegetation-atmosphere water migration diagram in a mild salinity community. (**b**) Soil-vegetation-atmosphere water migration diagram in a moderate salinity community. (**c**) Soil-vegetation-atmosphere water migration diagram in a severe salinity community. Note: the unit of $\delta^{18}O$ is presented as percentages. The arrow means the direction of water transport.

4. Discussions

4.1. The Contribution of Condensation Water to the Water Needed for the Growth Halostachys Caspica

In the wild, the water that plants utilize come from precipitation, soil water, runoff (including melting snow and ice), and ground water [28]. In arid regions, there is little precipitation, so groundwater is the only water source available for plant survival, especially perennial plants [29,30]. Using hydrogen and oxygen stable isotopes to study the water sources of plants, we quantitatively analyzed the origin and direction of water and the selective absorption and water use in three ecological environments that differed in salinity levels. Groundwater was determined to be the main water source utilized for plant growth by *Halostachys caspica* in different salinity habitats found within the study area. The utilization of ground water in the mild, moderate, and severe salinity habitats accounted for 46.1%, 66.4%, and 28.98% of the total water use, respectively. Thereunto, the severe salinity plants relied on groundwater the least, followed by the mild habitat, while the moderate habitat was the highest, such that groundwater accounted for more than half of all water sources. Therefore, it is clear that the change of groundwater level has had the greatest influence on plant growth in the moderate salinity area.

Although groundwater is the main water source, soil water is probably the most important source of water for these plants. Due to the nature of the soil (i.e., particle size and porosity), there are differences in the composition of hydrogen and oxygen stable isotopes. In general, surface soil water is able to contact surface moisture, thus, there is relatively large evaporation intensity to produce hydrogen and oxygen stable isotopic fractionation; this is based on established theories that state there is "light" molecular priority to evaporate ($^1H^1H^{16}O$) compared to deeper soil water levels. Surface soil water is more enriched with heavier hydrogen and oxygen stable isotopes, so there are significant differences in the levels of hydrogen and oxygen stable isotopes. Soil profiles taken at different depths show that soil water hydrogen and oxygen stable isotopic composition are indeed different, such that there is relatively stable isotopic content up to the deepest layer [26,31]. Moreover, non-moving water was a result of evaporation close to the surface area [32].

The results of this study also revealed that the closer to the soil surface, soil water $\delta^{18}O$ levels are partially positive, which is similar to the findings of other studies [33]. This is because the light isotope of oxygen evaporates, which leads to ^{18}O relative enrichment. As for δD values, soil water in the three salinity habitats had $\Delta\delta D$ values between those of rain and the other three natural water bodies (river water, condensation water and groundwater). This indicates that soil water comes primarily from rain, river water, condensation water and groundwater. Furthermore, soil water $\delta^{18}O$ values were all partially negative relative to the natural water bodies and were similar to condensation water ($-5.23614 \pm 0.72364\%$). This further illustrates that soil water in the study area comes from condensation water. These results are similar to the findings of Zhu and Jiang [34], who conducted a study in the northern Loess Plateau. Namely, they found that atmospheric condensation water contributed between 0–20 cm to soil water, and that it is difficult to contribute to the soil water at the 30 cm soil layer and below.

Some scholars employ the use of thermal pulse technology, isotope tracer techniques, fluorescent tracer methods, thermal ratio, botanical anatomy, pressure chamber, and physiological measurements [35] and found that plant leaves absorb water to ease water deficit in the body of the plant through the leaf stomata, bristles, fissure, drainage organs, and other structures on the surface of leaves [6,36–38]. In a study conducted by Burgess et al. [39], isotopic tracer techniques were used on the leaves of *Sequoia sempervirens* located in the United States and found that their main source of water comes from the fog formed overnight. Ellsworth and Williams [40] conducted a study on 16 species of drought and semi-arid shrubs, and one species of mesophytic herbage kept under control conditions. They found that due to transpiration processes, there was a δD and $\delta^{18}O$ enrichment found in leaf water, which is likely the source of water for young stems. Zhuang and Zhao [38] also conducted a study on whether the leaves of desert coat plants, *Bassia dasyphylla*, and glabrous plants,

Agriophyllum squarrosum (Linn.) Moq., absorbed condensation water. Results revealed that leaves of desert plants can absorb condensation water and through the process of photosynthesis, water retention and growth all respond to the presence of condensation water. Meanwhile, glabrous plants could not absorb condensation water. Vitarelli et al. [41] also conducted research on croton plants and demonstrated that the trichoid structure is the key structure through which plants absorb atmospheric water. Yan et al. [42] used a molecular ecology approach to research the intrinsic mechanism of the leaves of *Tamarix ramosissima*, a desert woody plant that absorbs condensation water from the canopy. Yang et al. [43] studied short-life desert plants and found that the leaf and stem can absorb condensation water, and increasing amounts of condensation water can significantly affect population dynamics. Cen and Liu [44] conducted research on the effects of simulated condensation water on the physiological characteristics of the leaf surface structure of *Leymus chinensis* and *Agropyron cristatum* under drought conditions and found that the aboveground biomass and root biomass increased with the presence of condensation water, and that condensation water can both protect and repair damage on plant leaf surface structures following stress induced by drought conditions. In this study, the leaf of *Halostachys caspica* could absorb the condensation water, this is because that hydrophilic polysaccharide compounds are found in cell walls of leaf epidermal cells, eutrophic cells and vascular bundle cell for photosynthetic organs of plants in arid region, these polysaccharides connect an extracellular network within the photosynthetic organs, and accept the moisture absorbed by the cuticle and transport it to the xylem. The microstructure on the surface of photosynthetic organs and the hydrophilic compounds inside which are the material basis of the canopy could absorb the condensation water [18]. In addition, In photosynthetic organs, the aquaporins (AQPs) regulating the plasma membrane and vacuolar membrane plays an important role in the process of transporting the condensation water in the canopy among cells through the symplastic pathway in desert woody plants, which is the molecular mechanism that absorbs the condensation water in the canopy [18].

The contribution of condensation water on plant growth in *Halostachys caspica* should not be overlooked, and the degrees of condensation water use vary under different salinity conditions. The degrees of condensation water use in mild, moderate, and severe salinity habitats were 7.1%, 3.78%, and 11.13%, respectively. This is roughly equivalent to the contribution of rain water in these habitats, which was 8.53%, 2.18%, and 11.93%, respectively. The degrees of condensation water use was relatively high. This may be because the surface soil salinity level was also high, which allows for greater absorption of condensation water, which infiltrated the soil where plant roots were then able to absorb and use the water for growth. Furthermore, salt crusts can reduce soil moisture evaporation, allowing more soil condensation water to infiltrate it. It was also found in this study that the degree of isotopic fractionation was positively correlated with salt tolerance; namely, the salt tolerance of plants lead to the fractionation of hydrogen isotopes [40].

4.2. The Water Migration Path in Halostachys Caspica

Under normal conditions, there are two paths of water migration in plants. First, roots transport water through the root system so that the cells in the soil absorb moisture from the soil through osmosis and root hair cells absorb soil moisture from the soil to the root hair cell through osmosis, which then transfer moisture to the plant root catheter through the intercellular osmotic differential and this in turn transfers moisture to each plant structure on the ground. Second, plant cells above the ground absorb moisture when guard cells on the plant leaves open, which allow a small amount of water vapor from the atmosphere to be absorbed by the plant tissue.

The water migration paths in the three saline habitats were different and all had specific water movement patterns. In the mild habitat, the water movement path in plants followed as: shallow root → stem → branches → leaves and shallow root → deep root. In moderate habitats, stems acted as the bifurcation point where the water movement path followed as: stem→ branches → leaves and stem → shallow root → deep root. In severe habitats, the water movement path went from stem to shallow root. These water movement paths occur in the plants' xylem, not the epidermal cells, this may be because the photosynthetic organs of *Halostachys caspica* have the ability to absorb the condensation water in the canopy and transport the water to the xylem [18]. Another discovery was made in the study of the *Haloxylon ammodendron*. When the photosynthetic organs of the *Haloxylon ammodendron* absorb the water from the canopy, whose water potential keeps increasing, and when the water potential of photosynthetic organ increased to a certain degree, which may establish the reverse water potential gradient, namely Ψ Photosynthetic organs > Ψ Secondary branches, the photosynthetic organs can transfer excess water to the main stem via a reverse water potential gradient, and this is conducive to the continuous absorption and utilization of the condensation water in the canopy, it can be seen that the reverse water potential gradient is the energy structure of plant photosynthetic organs capable of absorbing the condensation water in the canopy [18].

It is clear that the water movement path in the mild habitat begins in the shallow root, while the path in the moderate and severe habitats begin in the stems. Thus, water migration paths start in different parts of the plant depending on the degree of salinity in the habitat. This may be related to the salt and water potential differences in plant structures. Previous studies have shown that hydrogen isotopic fractionation in drought-resistant, salt-tolerant, woody plants is likely to occur when water is being absorbed in the roots [45]. Thus, it seems this will also affect condensation water use by *Halostachys caspica*.

To summarize, condensation water absorption and utilization in the study plants took two forms: (1) Direct absorption where leaves directly absorbed condensation water, which can add moisture to the plant surface, reduce the surface temperature of leaves, and reduce the water loss of surface evaporation [46], or replenish the evaporation consumption, as well as go to the plant body for growth; and (2) Indirect absorption, where atmospheric condensation in the soil constitutes the soil water, while the atmospheric condensation water replenishes the river and groundwater that plant roots absorb in addition to soil water. These results reflect the findings of Goebel and Lascano [17], who measured the δD and $\delta\ ^{18}O$ composition of different water sources of cotton and quantitatively analyzed the water content, including condensation water. Chen et al. [47] also conducted a study on 50 plant species in arid and semi-arid regions in Central Ningxia Province and found that the plant leaves could absorb water and even had the ability to take advantage of small amounts of rainfall.

5. Conclusions

In conclusion, this study revealed that: (1) Scale-like leaves can actively absorb condensation water, the condensation water absorbed by leaves can supplement the water consumed by evaporation to a certain extent, and the contribution of condensation water to plant growth in severe salinity habitats is the greatest (11.13%); (2) The migration path of water movement in the three habitats followed two main paths: (a) rainwater and condensation water were recharged through soil to compensate for groundwater, while some groundwater compensated for river water, and these were in part returned to the atmosphere by soil evaporation and plant transpiration; and (b) rainwater and condensation water directly compensated for the river, such that plant roots not only absorbed river water but also groundwater and soil water to assist with plant growth; (3) In mild salinity habitats, water movement paths in plants followed as: shallow root → stem → branches → leaves; and shallow root → deep root; in moderate habitats, stems acted as the bifurcation point and the path of water followed as: stem → branches → leaves, as well as: stem → shallow root → deep root; in severe habitats, the water movement path followed as: deep root → shallow root → stem → branches → leaves and finally returning to the atmosphere; there is also a water circuit from stem to shallow root. Although there are clear condensation water movement paths, this remains to be studied further, specifically in the xylem of the plant.

Author Contributions: Conceptualization, L.Q., X.H. and G.L.; investigation: L.Q., X.H. and J.Y.; methodology, L.Q. and X.H.; software, L.Q. and X.H.; writing—original draft, L.Q., X.H. and J.Y.; writing—review and editing, L.Q. and X.H.; supervision, G.L.; funding acquisition, G.L., L.Q. and X.H. All authors have read and agreed to the published version of the manuscript.

Funding: This work was funded by the National Natural Science Foundation of China (Nos. 41571034, 32101360, 31760168, 31660120) and the Doctor Starts Project of Xinjiang University (202115120003).

Institutional Review Board Statement: Not applicable.

Informed Consent Statement: Not applicable.

Data Availability Statement: Not applicable.

Acknowledgments: We thank Jing Cao, Xiao Ying, Ting-Quan Wang in Key Laboratory of Oasis Ecology of Xinjiang University for their indispensable help in fieldwork.

Conflicts of Interest: The authors declare no conflict of interest.

References

1. Uclés, O.; Villagarcía, L.; Moro, M.J.; Canton, Y.; Domingo, F. Role of dewfall in the water balance of a semiarid coastal steppe ecosystem. *Hydrol. Process.* **2014**, *28*, 2271–2280. [CrossRef]
2. Fang, J.; Ding, Y.J. An experiental observation of the relationship between sandy soil condensation water and micrometerorogical factors in the arid desert region. *J. Desert Res.* **2015**, *35*, 1200–1205.
3. Pina, A.L.C.B.; Zandavalli, R.B.; Oliveira, R.S.; Martins, F.R. Dew absorption by the leaf trichomes of Combretum leprosum in the Brazilian semiarid region. *Funct. Plant Biol.* **2016**, *43*, 851–861. [CrossRef] [PubMed]
4. Agam, N.; Berliner, P.R. Dew formation and water vapor adsorption in semi-arid environments—A review. *J. Arid. Environ.* **2006**, *65*, 572–590. [CrossRef]
5. Hill, A.J.; Dawson, T.E.; Shelef, O.; Rachmilevitch, S. The role of dew in Negev Desert plants. *Oecologia* **2015**, *178*, 317–327. [CrossRef]
6. Wang, X.H.; Xiao, H.L.; Ren, J.; Cheng, Y.B.; Yang, Q. An ultrasonic humidification fluorescent tracing method for detecting unsaturated atmospheric water absorption by the aerial parts of desert plants. *J. Arid. Land* **2016**, *8*, 272–283. [CrossRef]
7. Kalthoff, N.; Fiebig-Wittmaack, M.; Meißner, C.; Kohler, M.; Uriarte, M.; Bischoff-Gauß, I.; Gonzales, E. The energy balance, evapo-transpiration and nocturnal dew deposition of an arid valley in the Andes. *J. Arid. Environ.* **2006**, *65*, 420–443. [CrossRef]
8. Jiang, J.; Wang, K.F.; Zhang, W.J. A study on the Cogulation water in the Sandy soil and its role in water balance. *Arid. Land Res.* **1993**, *10*, 1–9.
9. Agam, N. Diurnal Water Content Changes in the Bare Soil of a Coastal Desert. *J. Hydrol.* **2004**, *5*, 922–933.
10. Zhang, Q.; Wang, S.; Yang, F.L.; Yue, P.; Yao, T.; Wang, W.Y. Characteristics of Dew Formation and Distribution, and Its Contribution to the Surface Water Budget in a Semi-arid Region in China. *Bound.-Layer Meteorol.* **2015**, *154*, 317–331. [CrossRef]
11. Maphangwaa, K.W.; Musilb, C.F.; Raittc, L.; Zedda, L. Differential interception and evaporation of fog, dew and water vapor and elemental accumulation by lichens explain their relative abundance in a coastal desert. *J. Arid. Environ.* **2012**, *82*, 71–80. [CrossRef]
12. Tao, Y.; Zhang, Y.M. Effects of leaf hair points on dew deposition and rainfall evaporation rates in moss crusts dominated by Syntrichia caninervis, Gurbantunggut Desert, northwestern China. *Acta Ecol. Sin.* **2012**, *32*, 7–16. [CrossRef]
13. Temina, M.; Kidron, G.J. The effect of dew on flint and limestone lichen communities in the Negev Desert. *Flora-Morphol. Distrib. Funct. Ecol. Plants* **2015**, *213*, 77–84. [CrossRef]
14. Cheng, L.; Jia, X.H.; Wu, B.; Li, Y.S.; Zhao, X.B.; Zhou, H. Effects of biological soil crusts on the characteristics of hygroscopic and condensate water deposition in alpine sandy lands. *Acta Ecol. Sin.* **2018**, *38*, 5037–5046.
15. Zhuang, Y.L.; Zhao, W.Z. The ecological role of dew in assisting seed germination of the annual desert plant species in a desert environment, northwestern China. *J. Arid. Land* **2016**, *8*, 264–271. [CrossRef]
16. Arabnejad, H.; Mirzaei, F.; Noory, H. Greenhouse cultivation feasibility using condensation irrigation (studied plant: Basil). *Agric. Water Manag.* **2021**, *245*, 106526. [CrossRef]
17. Goebel, T.S.; Lascano, R.J. Time for cotton to uptake water of a known isotopic signature as measured in leaf petioles. *Agric. Sci.* **2014**, *5*, 170–177. [CrossRef]
18. Gong, X.W. *A Probe into the Mechanisms of Canopy Dew Uptake by Photosynthetic Organs of Desert Trees: Based on Molecular, Cellular and Physiological Perspectives*; Xin Jiang University: Urumqi, China, 2017.
19. Liu, J.Q. Characiter of Morphology, anatomy and Physiology of water about extreme xerophytes. *Acta Ecol. Sin.* **1983**, *1*, 15–20.
20. Chen, S.H.J.; Hou, P.; Li, W.H. *Comprehensive Scientific Investigation of the Ebinur Lake Wetland Nature Reserve*; Xinjiang Science and Technology Press: Urumqi, China, 2006.
21. Ehleringer, J.R.; Roden, J.; Dawson, T.E. Assessing ecosystemlevel water relations through stable isotope ratio analyses. In *Methods in Ecosystem Science*; Sala, O.E., Ed.; Springer: New York, NY, USA, 2000; pp. 181–198.
22. Phillips, D.L.; Gregg, J.W. Source partitioning using stable isotopes: Coping with too many sources. *Oecologia* **2003**, *136*, 261–269. [CrossRef]

23. Qin, L. Formation Mechanism of Condensation Water and Its Ecological Effects of *Halostachys caspica* Community in Different Salinity Habitats. Ph.D. Thesis, Xinjiang University, Urumqi, China, 2014.

24. Yang, X.D.; Anwar, E.; Zhou, J.; He, D.; Gao, Y.C.; Lv, G.H.; Cao, Y.E. Higher association and integration among functional traits in small tree than shrub in resisting drought stress in an arid desert. *Environ. Exp. Bot.* **2022**, *201*, 1–12. [CrossRef]

25. Yu, W.; Ji, R.P.; Feng, R.; Zhao, X.L.; Zhang, Y.S. Response of water stress on photosynthetic characteristics and water use efficiency of maize leaves in different growth stage. *Acta Ecol. Sin.* **2015**, *35*, 2902–2909.

26. Jia, G.D. Water Movement Mechanism of Plant-Soil System Using Stable Hydrogen and Oxygen Isotope Technology. Ph.D. Thesis, Beijing Forestry University, Beijing, China, 2013.

27. Yang, X.D. Hydraulic redistribution and mechanism of *Populus euphratica* and *Haloxylon persicum* in Ebinur Lake Wetland. Master's Thesis, Xinjiang University, Urumqi, China, 2011.

28. Bian, J.J.; Sun, Z.Y.; Zhou, A.G.; Yu, S.W. Advances in the D and ^{18}O isotopes of Water Source of Plants in Arid Areas. *Geol. Sci. Technol. Inf.* **2009**, *28*, 117–120.

29. Ehleringer, J.R.; Dawson, T.E. Water uptake by plants: Perspectives from stable isotopes composition. *Plant Cell Environ.* **1992**, *15*, 1073–1082. [CrossRef]

30. Li, W.H.; Zhou, H.H.; Yang, X.M.; Ding, H. Temporal and special distribution characteristics of aboveground biomass of grassland plant communities in an arid area. *Acta Prataculturae Sin.* **2010**, *19*, 186–195.

31. McCole, A.A.; Stern, L.A. Seasonal water use patterns of Juniperus ashei on the Edwards Plateau, Texas, based on stable isotopes in water. *J. Hydrol.* **2007**, *342*, 238–248. [CrossRef]

32. Liu, J.; Wei, W.; Zhang, L.; Wang, Y.; Duan, B.Q.; Liu, F.L. Application on Isotopes D and ^{18}Oof soil water in water movement of unsaturated Zone. *Investig. Sci. Technol.* **2012**, *5*, 38–43.

33. Zhao, L.J.; Xiao, H.L.; Cheng, G.D.; Song, Y.Y.; Zhao, L.; Li, C.Z.; Yang, Q. A Preliminary Study of Water Sources of Riparian Plants in the Lower Reaches of the Heihe Basin. *Acta Geosci. Sin.* **2008**, *26*, 709–718.

34. Zhu, Q.; Jiang, Z. Using stable isotopes to determine dew formation from atmospheric water vapor in soils in semiarid regions. *Arab. J. Geosci.* **2016**, *9*, 1–9. [CrossRef]

35. Li, H.B. It's Effects on the Moisture Characteristics of Plant in Semi-Arid Area. Ph.D. Thesis, Inner Mongolia Agricultural University, Hohhot, China, 2010.

36. Li, X.Y. Effects of gravel and sand mulches on dew deposition in the semiarid region of China. *J. Hydrol.* **2002**, *260*, 151–160. [CrossRef]

37. Ramírez, D.A.; Bellot, J.; Domingo, F.; Blasco, A. Can water responses in *Stipa tenacissima* L. during the summer season be promoted by nonrainfall water gains in soil? *Plant Soil* **2007**, *291*, 67–79. [CrossRef]

38. Zhuang, Y.L.; Zhao, W.Z. Experimental Study of effects of Artifical dew on Bassia dasyphylla and Agriophyllum squarrosum. *J. Desert Res.* **2010**, *30*, 1068–1074.

39. Burgess, S.S.O.; Awson, T.E.D. The contribution of fog to the water relations of Sequoia sempervirens (D.Don): Foliar uptake and prevention of dehydration. *Plant Cell Environ.* **2004**, *27*, 1023–1034. [CrossRef]

40. Ellsworth, P.Z.; Williams, D.G. Hydrogen isotope fraction during water uptake by woody xerophytes. *Plant Soil* **2007**, *291*, 93–107. [CrossRef]

41. Vitarelli, N.C.; Riina, R.; Cassino, M.F.; Meira, R.M.S.A. Trichome-like emergences in Croton, of Brazilian highland rock outcrops: Evidences for atmospheric water uptake. *Perspect. Plant Ecol. Evol. Syst.* **2016**, *22*, 23–35. [CrossRef]

42. Yan, X.; Zhou, M.X.; Dong, X.C.; Zou, S.B.; Xiao, H.L.; Ma, X.F. Molecular mechanisms of foliar water uptake in a desert tree. *AoB Plants* **2015**, *7*, 129. [CrossRef] [PubMed]

43. Yang, X.D.; Lv, G.H.; Ali, A.; Ran, Q.Y.; Gong, X.W.; Wang, F.; Liu, Z.D.; Qin, L.; Liu, W.G. Experimental variations in functional and demographic traits of Lappula semiglabra among dew amount treatments in an arid region. *Ecohydrology* **2017**, *10*, e1858. [CrossRef]

44. Cen, Y.; Liu, M.Z. Effects of dew on eco-physiological traits and leaf structures of Leymus chinensis and Agropyron cristatum grown under drought stress. *Chin. J. Plant Ecol.* **2017**, *41*, 1199–1207.

45. Lin, G.H. *Stable Isotope Ecology*; Higher Education Press: Beijing, China, 2013; p. 131.

46. Fang, J. Research progress on the ecological hydrological effect of condensation water. *J. Desert Res.* **2013**, *33*, 583–589.

47. Chen, L.; Yang, X.G.; Song, N.P.; Yang, M.X.; Xiao, X.P.; Wang, X. Leaf water uptake strategy of plants in the arid and semi-arid region of Ningxia. *J. Zhengjiang Univ. (Argic. Life Sci.)* **2013**, *39*, 565–574.

 *forests*

MDPI

Article

Spatial Scale Effects of Soil Respiration in Arid Desert Tugai Forest: Responses to Plant Functional Traits and Soil Abiotic Factors

Jinlong Wang [1,2,3], Xuemin He [1,2,3], Wen Ma [4], Zhoukang Li [1,2,3], Yudong Chen [1,2,3] and Guanghui Lv [1,2,3,*]

[1] College of Ecology and Environment, Xinjiang University, Urumqi 830017, China; wangjlxju@163.com (J.W.); he8669@163.com (X.H.); lzkeco@163.com (Z.L.); cyd666@stu.xju.edu.cn (Y.C.)
[2] Key Laboratory of Oasis Ecology of Education Ministry, Xinjiang University, Urumqi 830017, China
[3] Xinjiang Jinghe Observation and Research Station of Temperate Desert Ecosystem, Ministry of Education, Urumqi 830017, China
[4] College of Geography and Remote Sensing Sciences, Xinjiang University, Urumqi 830017, China; maww_08@126.com
[*] Correspondence: ler@xju.edu.cn; Tel.: +86-0991-2111427

Citation: Wang, J.; He, X.; Ma, W.; Li, Z.; Chen, Y.; Lv, G. Spatial Scale Effects of Soil Respiration in Arid Desert Tugai Forest: Responses to Plant Functional Traits and Soil Abiotic Factors. *Forests* **2022**, *13*, 1001. https://doi.org/10.3390/f13071001

Academic Editor: Mariangela Fotelli

Received: 1 June 2022
Accepted: 24 June 2022
Published: 25 June 2022

Abstract: Understanding the spatial variation law of soil respiration (Rs) and its influencing factors is very important when simulating and predicting the terrestrial carbon cycle process. However, there are still limitations in understanding how different sampling scales affect the spatial heterogeneity of Rs and whether the spatial scale effect will change with habitat types. Our objectives were to explore the effects of different sampling scales on the spatial variability of Rs and the relative importance of soil abiotic characteristics and plant traits in influencing the spatial variability of Rs. The Rs, soil properties, and plant traits were measured through field investigation and indoor analysis in the Tugai forest desert plant community in the Ebinur Lake Basin in northwest China. The Rs showed significant water gradient changes, with a coefficient of variation of 35.4%–58%. Plot types had significant effects on Rs, while the change of sampling scale did not lead to significant differences in Rs. At the plot scale, Rs spatial variation at the 5 m × 5 m sampling scale mainly depended on plant traits (leaf length, leaf thickness, leaf dry matter content, and leaf phosphorus content, $p < 0.05$), while Rs spatial variation at the 10 m × 10 m scale mainly depended on soil properties (soil total phosphorus, ammonium nitrogen, soil water content, and pH, $p < 0.05$). At the local scale, soil nutrients (soil available phosphorus and ammonium nitrogen) and plant traits (maximum plant height, leaf length, and phosphorus content) at the 5 m × 5 m scale jointly explained 49% of the spatial change of Rs. In contrast, soil microclimate (soil water content), soil nutrients (soil pH, available phosphorus, and nitrate nitrogen), and plant traits (leaf thickness) jointly explained 51% of the spatial variation of Rs at the 10 m × 10 m scale. These results demonstrate the potential to predict the spatial variability of Rs based on the combination of easily measured aboveground functional traits and soil properties, which provides new ideas and perspectives for further understanding the mechanism of Rs change in Tugai forests.

Keywords: desert ecosystem; ecosystem function; growing seasons; soil CO_2 emission; spatial variation

1. Introduction

Soil respiration (Rs) is an important ecological process in the global carbon balance and carbon cycle and has an important impact on global climate change [1]. Rs is often used as an indicator to evaluate soil biological activity, soil fertility, and even air permeability [2]. As the only output channel of the soil carbon pool and an important source of atmospheric CO_2, Rs flux has become a major research hotspot [3,4]. Although Rs is a critical part of the carbon cycle, the current understanding of Rs is relatively poor, and knowledge on the influencing factors of Rs and the variability of Rs among ecosystems remains limited.

In addition, the spatial heterogeneity of Rs and its influencing mechanism remain to be clarified [5]. Therefore, studying the Rs process and its influencing factors in depth is the key to exploring the regulation mechanism of regional carbon budgets, the carbon cycle, and carbon balance [6].

Rs is a process in which soil produces and releases CO_2 to the atmosphere [7]. Rs exhibits obvious spatial variation characteristics, mainly at the plot scale, regional scale, and global scale [7]. At the global scale, Rs has obvious zonal characteristics that are mainly affected by zonal changes in vegetation and climate [8,9]. At the regional scale, landscape [10,11], land use type [12,13], and vegetation type [14,15] affected Rs. At the plot scale, stand and canopy structure [16,17], root distribution [18], and the heterogeneity of major environmental factors and soil characteristics [19] lead to the spatial heterogeneity of Rs [20,21]. Currently, the dynamic chamber-infrared CO_2 analyzer (IRGA) method is mostly used to measure Rs in ecosystems with equipment such as LI-COR series instruments (including the 6400 and 8100 instruments: LI-COR Biosciences, Lincoln, NE, USA) [22]. Rs based on a certain number of chambers at the plot scale can be used to represent the average value of Rs in ecosystems [23]. Studies have shown that the spatial heterogeneity of Rs within a plot or between different plots is caused by the differences in soil water content (SWC) and soil physicochemical properties [22,24,25]. However, Rs is not only a physiological process that occurs in response to soil abiotic characteristics but is also a result of the combined action of several complex ecosystem processes [26]. Although soil temperature (ST) and SWC can reflect the temporal variability of Rs well, they do not fully explain the spatial heterogeneity of Rs within or between plots [27]. Therefore, strengthening the synchronous observation of Rs and biological factors will help to enhance understanding of the process and the mechanism of biological factors affecting Rs.

Biotic factors affect Rs by affecting soil microclimate and structure, litter quantity and quality, and root respiration [28]. During plant growth, Rs is mainly controlled by plant growth [29]. This is mainly manifested in the following ways: (1) the material basis of Rs is derived from plant photosynthesis [30,31]. High photosynthesis rates lead to high respiration rates [32], and the Rs rate is positively correlated with vegetation photosynthesis and total primary productivity (GPP) [33,34]. (2) The functional traits that drive carbon assimilation (leaf) and respiration (root) control root respiration [35]. Compared with leaves, root traits are more complex and are difficult to observe and measure [36]. Studies have shown that easily measured aboveground traits are good predictors of root traits [37]. Fine roots, specific root length, and root biomass are closely related to leaf area, height, or aboveground biomass [38]. In addition, plant functional traits show a pattern of coordination and adaptation along environmental gradients (e.g., soil fertility, water, or light) [37]. Therefore, differences in aboveground plant functional traits may lead to significant changes in Rs [30,39]. However, dynamic models of Rs that couple the combined effects of plant functional traits and soil abiotic factors (SWC, ST, and soil properties) are rarely reported [22,36].

Rs exhibits scale dependence in space and has different characteristics at different scales, thus showing a spatial scale effect [40–43]. Spatial scale is one of the basic problems in ecological research and is usually expressed by spatial extent and spatial grain [44]. Spatial extent describes the overall coverage area of the research object. Spatial grain refers to the sampling length and area represented by the smallest identifiable unit, such as quadrat or pixel [45–47]. The determination of spatial scale is closely related to the nature and requirements of the research problem, which determine the sampling points, the design density and workload of the grid, and the investment of funds. Therefore, the nested sampling design method can be used to explore the spatial scale effect of Rs by setting plots with different grain sizes in the same region [48,49], which can effectively reveal the distribution pattern and potential mechanism of Rs.

Tugai forest, a natural plant community composed of desert plants, is widely distributed along the lakeside and supply rivers of the Ebinur Lake Basin in northwest China [50]. As the only forest community that forms naturally in arid deserts, Tugai forests can not only endure drought but can also adapt to saline soil. Tugai forest plays an important role as a windbreak and in sand fixation, water conservation, and biodiversity maintenance [51,52]. This community is a valley forest dominated by arbor species and associated shrubs and herbs distributed along rivers in arid deserts. With the increase of distance from rivers, this type of community presents an obvious water gradient and community structure change [53], which provides an opportunity to identify the driving factors of Rs variability at different scales. Ecosystems such as the desert Tugai forest in the Ebinur Lake Basin are often highly sensitive to global climate change and play an important role in the carbon cycle in arid areas [54]. Thus, a study was designed to understand the variation pattern of Rs and its biotic and abiotic regulatory factors in the desert Tugai forest at different water gradients in the Ebinur Lake Basin. We assume that biotic and abiotic factors are different along the water gradient zone, and their regulatory mechanisms on Rs also differ. The objectives of this study were: (1) to quantify and compare the variability of Rs at different spatial scales, and (2) to evaluate the effects of biotic and abiotic factors on Rs at different spatial scales.

2. Materials and Methods

2.1. Study Site

The Ebinur Lake Wetland National Nature Reserve is located in the northwest of Xinjiang Uygur Autonomous Region, Bortala Mongolian Autonomous Prefecture (44°37′05″–45°10′35″ N, 82°30′47″–83°50′21″ E), and has a temperate continental arid climate. The average annual temperature is 6–8 °C, the annual precipitation is approximately 100 mm, the annual evaporation is approximately 1600 mm, and the groundwater depth is 1.8−2.7 m. The Ebinur Lake Wetland National Nature Reserve is mainly composed of eight soil types (Figure S1), dominated by silt and clay [51,52]. The main plant types are *Haloxylon ammodendron* (C.A.M.) Bge., *Populus euphratica* Oliv., *Tamarix ramosissima* Ldb., and *Phragmites australis* (Cav.) Trin. ex Steud.

2.2. Experimental Design

The water gradient zone is 8 km long, north of the Aqikesu River, and perpendicular to the riparian zone [55]. The climatic conditions in the gradient zone are relatively uniform, and there is little difference in topography and altitude. In August 2013, three 5 km long transects were established on the water gradient zone at intervals of 5 km [56]. According to the groundwater level and vegetation development, the gradient belt from southwest to northeast can be divided into riverbank habitat, transitional zone habitat, and desert margin habitat [19]. The riverbank habitat is located at the southern end of the gradient zone (0~1.50 km from the riparian zone), and the groundwater table is relatively shallow due to its proximity to the channel. The main plant species are *Populus euphratica*, *Halimodendron halodendron*, and *Phragmites australis*, and the soil nutrients and biomass are higher than in other regions [57]. The transitional zone habitat is located in the middle of the gradient zone (1.50~4.50 km from the riparian zone), and suitable water content maintains the symbiosis of 17 plants, including *Reaumuria soongorica* (Pall.) Maxim. and *Aeluropus pungens* (M. Bieb.) C. Koch. The desert margin habitat is at the northernmost end of the gradient zone (4.50~8.00 km from the riparian zone). The soil particle size is large, the organic matter content (1.35 g kg^{-1}) is low, and the soil water content (1.04%) is also low, making plant survival challenging. The species growing in this area mainly include *Haloxylon ammodendron*, *Nitraria tangutorum*, and *Suaeda salsa* [19].

In mid-July 2018, one of the three replicate transects was selected, and a 50 m × 50 m plot was established in each of the three habitats of this transect, namely plot 1, plot 2, and plot 3 (Figure 1). The nested sampling design method was used to systematically divide each plot according to the sampling scales of 10 m × 10 m and 5 m × 5 m, respectively [48,49]. For the vegetation survey, Rs monitoring, soil sampling, and plant sample collection, 25 grids were evenly set at each scale. First, the 50 m × 50 m plot was equally spaced into 25 grids of 10 m × 10 m. Then, the vertices of each grid were marked and fixed with PVC pipes as corner stakes, and the midpoint of four boundary lines of each 10 m × 10 m quadrat was measured by a ruler, and each quadrat was further divided into four grids of 5 m × 5 m. Finally, a total of 150 grids (150 grids = 25 grids × two sampling scales × three plots) were established in the three plots. The latitude, longitude, altitude, and other geographic information for each sample plot were recorded by handheld GPS, and each grid was numbered. Due to the negative effects of large-scale destructive soil sampling and expensive labor and time costs, it was impossible to test in all transects [58]. In addition, according to the findings of the previous research group, there were no significant differences in plant composition and soil properties among the three replicate transects (Table S1) [56]. Therefore, we hypothesized that our experimental design could reflect the response of Rs to habitat changes (water gradients) at different spatial scales. This non-repeated experimental design method has been widely used to study the spatial variability of Rs on the fine scale [58–61].

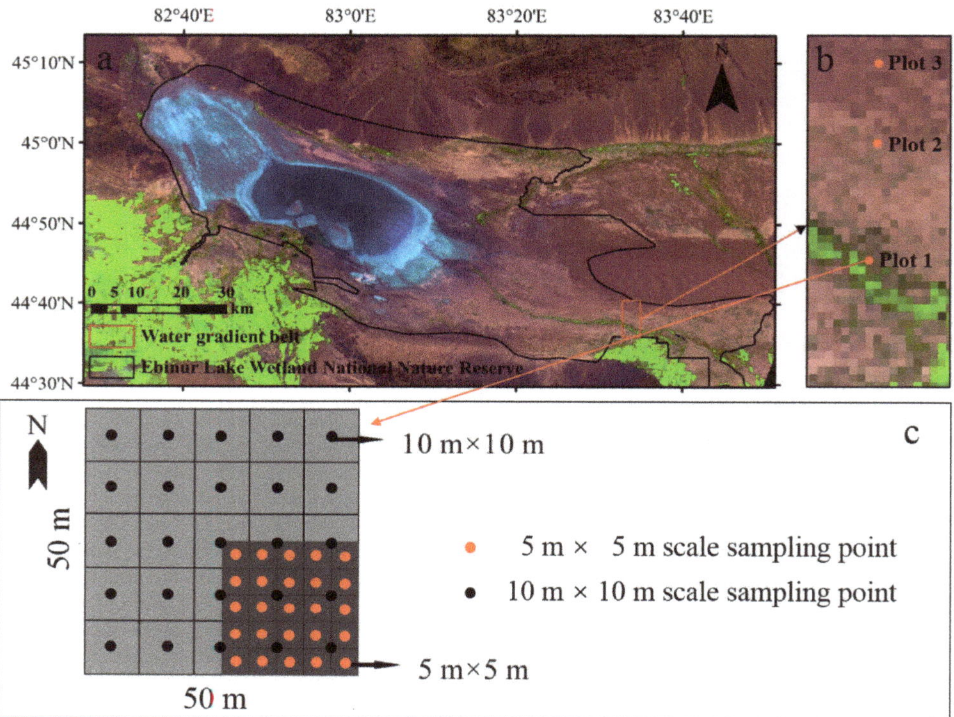

Figure 1. Locations of sampling sites. (**a**) The Ebinur Lake Wetland National Nature Reserve, (**b**) the study site along the water gradient zone, and (**c**) the quadrat distribution at different scales.

2.3. Rs and Soil Indicators

The center point of each 10 m × 10 m and 5 m × 5 m grid was taken as the Rs measurement point. To reduce the influence of human disturbance and plant photosynthesis on the Rs measurement results, a PVC collar (inner diameter: 20 cm, height: 15 cm) was placed one day in advance at the center of each grid to measure Rs (150 collars in total). The PVC collar was then vertically embedded into the soil until its surface was approximately 5 cm above ground level, and living plants inside the collar were removed. A period of sunny and windless weather was selected from mid-July to mid-August in 2018, and the Rs and ST were measured by an LI-8100 portable gas analysis system (LI-COR Inc., Lincoln, NE, USA) at 10:00–12:00 (local time) every day. The 10 m × 10 m sampling scale was measured before the 5 m × 5 m sampling scale. To eliminate the influence of above-ground plant respiration, the ground vegetation and large branches in the collar were all removed one day before the measurement [19]. Each PVC collar was measured three times, and the average value was taken as the Rs value at the measuring point [22,60]. Rs was measured and the 0–10 cm ST was measured by a thermocouple temperature probe with an LI-8100 instrument (LI-COR Inc., Lincoln, NE, USA).

After the surface litter was removed near each Rs monitoring point, soil samples of 0–10 cm were collected. Two soil samples were taken, and one soil sample was collected by aluminum box (the aluminum box was weighed in advance). After the aluminum box soil was collected, it was numbered and weighed to obtain its fresh weight. It was then brought back to the laboratory to determine the SWC by the drying method [61]. Another soil sample was dried, and debris was removed using a 2-mm sieve for later soil index determination. Soil pH was determined using a PHS-3C Precision pH meter (Shanghai Etorch Electro-Mechanical Technology Co., Ltd., Shanghai, China). The soil organic carbon (SOC), total nitrogen (TN), nitrate nitrogen (NN), ammonium nitrogen (AN), total phosphorus (TP), and available phosphorus (AP) were measured using the methods described by Wang et al. [19].

2.4. Community Survey and Plant Functional Trait Measurement

Community surveys were performed in each 5 m × 5 m and 10 m × 10 m grid throughout the period of Rs measurements. This process included recording the species names and measuring the plant number, coverage, and height of each species. According to the global functional characteristics measurement manual and combined with the vegetation characteristics in the study area, nine functional traits were selected, including leaf length (LL), leaf width (LW), leaf thickness (LT), maximum plant height (H_{max}), specific leaf area (SLA), leaf dry matter content (LDMC), leaf carbon content (LCC), leaf nitrogen content (LNC), and leaf phosphorus content (LPC) [62]. For each plant appearing in the grid, 30–50 fully developed, and healthy leaves were collected to measure the leaf traits [62]. LL (mm), LW (mm), and LT (mm) were measured by Vernier calipers with an accuracy of 0.01 mm. ImageJ was used to calculate leaf area after scanning all collected leaves [63]. The leaves were then placed in an oven at 80 °C for 48 h and immediately weighed with an electronic balance with an accuracy of 0.001 g. SLA (m^2 kg^{-1}) is defined as the ratio of leaf area (m^2) to leaf dry weight (kg). LDMC (%) is defined as the ratio of leaf dry weight (kg) to leaf fresh weight (kg). Finally, all dried plant leaves were ground and stored in self-sealing bags for the determination of leaf element contents. LPC, LNC, and LCC were determined by the molybdenum antimony resistance colorimetric method, H_2SO_4-H_2O_2-Kjeldahl method, and potassium dichromate dilution method, respectively [64].

2.5. Statistical Analyses

The coefficient of variation (CV) was calculated by standard deviation (SD)/mean value (mean) to represent the spatial variation of Rs and environmental factors. CV < 10% was considered weak variability, 10% < CV < 100% was considered moderate variability, and CV > 100% was considered strong variability [42]. Differences in Rs, soil properties, and plant traits between the two sampling scales were analyzed using independent-samples t-tests. Two-factor analysis of variance (ANOVA) was used to analyze the effects of plot type and sampling scale on Rs. Rs was converted into a natural logarithm before analysis to achieve normal distribution. GS+ 9.0 (Gamma Design Software, Plainwell, MI, USA) geostatistics software was used for semi-variance function model fitting and parameter calculation. The ratio $C/(C_0 + C)$ of partial sill and sill was used to reflect the spatial autocorrelation of Rs. $C/(C_0 + C)$ > 75%, 25%–75%, and <25% indicated strong, moderate, and weak spatial autocorrelation of Rs, respectively [65]. The ggpmisc package was used to conduct fitting regression analysis of Rs and the environmental factors (soil microclimate, soil nutrients, and plant traits) of the three plots. Due to the collinearity among environmental factors, the Hmisc package varclus function was used to cluster and evaluate the redundancy among environmental factors before the multiple regression analysis, and the factors with high correlation were removed [66].

The Vegan package varpart function was used to perform variance partitioning analysis (VPA) based on redundancy analysis (RDA) to assess the explanatory degree of different types of environmental factors (e.g., soil microclimate, soil nutrients, and plant traits) on the spatial change of Rs [67]. Before RDA analysis, to improve the normality and homogeneity of the variance of data, all environmental factors were log transformed. Then, the ordistep function was used for backward selection to delete the variables that were not significant in each explanatory set. Finally, 999 Monte Carlo permutation tests were performed on the RDA analysis results to quantitatively evaluate the contribution rates of different types of environmental factors to the spatial variation of Rs (i.e., the independent explanatory variables). To prevent overfitting, the adjusted R^2 was selected as the recognition criterion to evaluate the interpretation ability of the model.

3. Results

3.1. Changes in Rs and Environmental Factors

The mean Rs of plot 1 was the highest (5 m × 5 m: 0.29 ± 0.17 µmol m^{-2} s^{-1}; 10 m × 10 m: 0.28 ± 0.15 µmol m^{-2} s^{-1}), and that of plot 3 was the lowest (5 m × 5 m: 0.12 ± 0.05 µmol m^{-2} s^{-1}; 10 m × 10 m: 0.11 ± 0.04 µmol m^{-2} s^{-1}) (Figure 2a). Plot types had significant effects on Rs, but sampling scales and their interactions with plot types had no significant effects on Rs ($p > 0.05$, Table 1). At different sampling scales, ST, SWC, SSC, pH, TN, SOC, H$_{max}$, and LPC were not significantly different in the same plot ($p > 0.05$, Figure 2b–f), but were significantly different among the three plots ($p < 0.05$, Figure 2k,l,t). At the sampling scale of 5 m × 5 m, the AN, NN, TP, and AP of plot 1 were significantly higher than those of plot 2 and plot 3 ($p < 0.05$, Figure 2g–j), while there was no significant difference in AN, NN, TP, and AP between plot 2 and plot 3 ($p > 0.05$). At the sampling scale of 10 m × 10 m, except for AP and SLA, the variables showed significant differences among different plots ($p < 0.05$, Figure 2q). LW, LT, and LL were significantly different among different plots and sampling scales ($p < 0.05$, Figure 2m–o).

Table 1. Two-factor analysis of variance (ANOVA) for the effects of plot type and sampling scale on soil respiration.

Source of Variation	*df*	Sum Square	Mean Square	F Value	*p* Value
Plot type	2	0.76	0.38	32.08	<0.001
Sampling scale	1	0.03	0.03	2.87	0.09
Plot type: Sampling scale	2	0.03	0.01	1.17	0.31
Residuals	144	1.71	0.01	-	-

In addition to ST and pH, Rs and its environmental factors exhibited moderate spatial variations ($10\% \leq CV \leq 100\%$). Among them, the CV of Rs and environmental factors at different scales were not significantly different in three plots, but were different among different plots, showing the order of plot 1 > plot 2 > plot 3 (Table 2). At the sampling scale of 5 m × 5 m, Rs in plot 1, plot 2, and plot 3 had strong, medium, and weak spatial dependence, respectively. At the scale of 10 m × 10 m, the spatial dependence of Rs in plot 1 and plot 2 was stronger, while that in plot 3 was weaker (Table 3).

3.2. Multiple Regression Analysis of Rs and Environmental Factors in Specific Plot Types under Different Sampling Scales

The variables with higher correlations (LW, LC, LN, TN, SSC, and SOC) were removed because of Spearman's ρ2 > 0.7 (Figure S2), and all the other variables were entered into the final model. Under different sampling scales, the most significant environmental variables affecting the spatial variation of Rs in different plots could be explained by the multiple regression analysis model (Table S2). At the scale of 5 m × 5 m, LL ($0.02, p < 0.01$), LDMC

(2.98, $p < 0.01$), LPC (0.52, $p < 0.001$), and AP (0.01, $p < 0.05$) jointly explained 49% of Rs variation in plot 1. In plot 2, LPC had the most significant effect on Rs (0.12, $p < 0.05$). As the most significant variables affecting the Rs in plot 3, LL (-0.06, $p < 0.05$), LT (-1.23, $p < 0.05$), and AN (0.07, $p < 0.01$) jointly explained 42% of the total variation ($p < 0.005$). At the scale of 10 m × 10 m, LL (partial regression coefficient was 0.01, $p < 0.05$), SWC (0.04, $p < 0.001$), and pH (0.19, $p < 0.01$) were the most important variables affecting the Rs in plot 1, which could explain 70% of the spatial variation of the Rs in this plot ($p < 0.001$). The contribution of LPC, SWC, TP, and AN to the Rs in plot 2 was small but significant (-0.09–0.06, $p < 0.05$). In contrast, SWC had a unique effect on the spatial change of Rs in plot 3 (0.08, $p < 0.001$) that explained 36% of the total variation in this plot ($p < 0.001$).

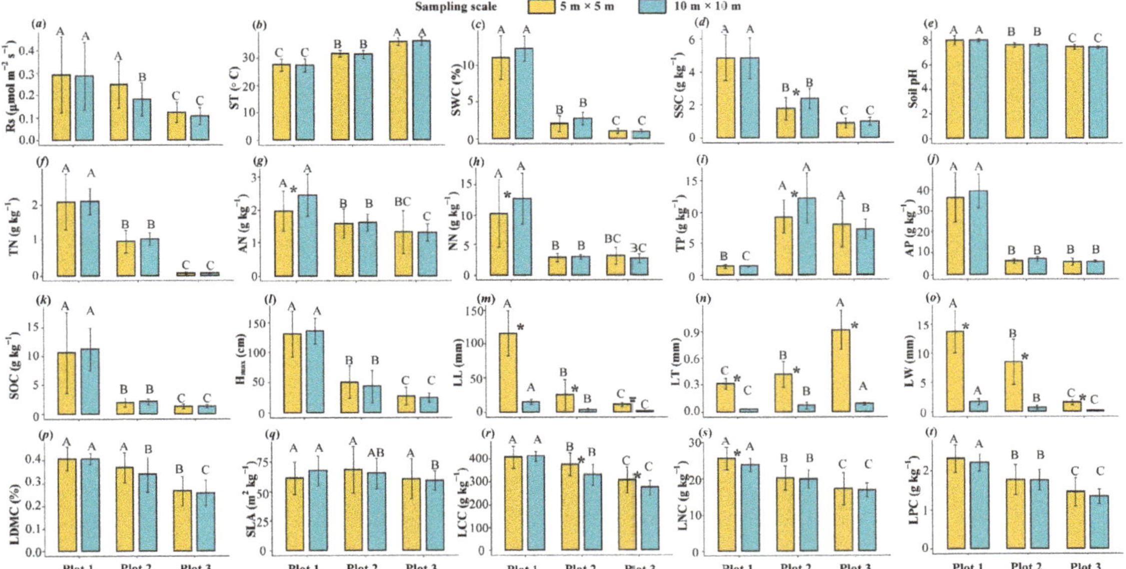

Figure 2. Comparison of soil respiration (a), soil properties (b–k), and plant traits (l–t) in different plot types and at different sampling scales. Uppercase letters indicate significant differences between plots ($p < 0.05$), while "*" indicate significant differences between sampling scales in specific plots ($p < 0.05$). SWC: Soil water content; ST: Soil temperature; pH: Soil pH; AP: Soil available phosphorus; TP: Soil total phosphorus; TN: Total nitrogen content; SOC: Soil organic carbon content; AN: Soil ammonium nitrogen; NN: Soil nitrate nitrogen; SSC: Soil salinity content; H_{max}: Maximum plant height; LL: Leaf length; LW: Leaf width; LT: Leaf thickness; LDMC: Leaf dry matter content; SLA: Specific leaf area; LPC: Leaf phosphorus content; LNC: Leaf nitrogen content; LCC: Leaf carbon content.

3.3. The Influencing Factors of Rs among Three Plot Types under Different Sampling Scales

Figure 3 displays the regression relationship between Rs and several environmental factors at different sampling scales. Among them, the relationships between Rs and SLA, TP, and AN were weak at the 5 m × 5 m scale ($p > 0.05$, Figure 3f,i), and the relationship between Rs and other factors was more obvious ($p < 0.05$, Figure 3). At the 10 m × 10 m scale, except for SLA, other influencing factors were significantly correlated with Rs ($p < 0.05$, Figure 3).

Table 2. Coefficient of variation (%) of soil respiration and environmental factors at different sampling scales.

Parameter	Plot 1		Plot 2		Plot 3	
	5 m × 5 m	10 m × 10 m	5 m × 5 m	10 m × 10 m	5 m × 5 m	10 m × 10 m
Soil respiration	58.0	52.8	41.9	41.7	37.0	35.4
Maximum plant height	29.3	16.0	53.8	62.4	55.4	31.9
Leaf length	29.0	24.5	88.3	81.5	29.3	25.0
Leaf width	26.8	32.3	46.3	37.1	29.6	18.1
Leaf thickness	20.0	14.4	34.9	55.3	24.1	17.0
Leaf dry matter content	12.8	6.1	17.7	22.4	23.9	21.4
Specific leaf area	21.9	19.0	28.6	19.6	27.8	13.5
Leaf carbon content	11.9	5.4	13.3	13.7	17.8	11.6
Leaf nitrogen content	12.3	7.7	16.2	12.7	26.0	11.2
Leaf phosphorus content	15.1	10.0	21.1	14.7	24.9	14.0
Soil organic carbon	27.1	14.1	52.7	31.0	35.6	25.2
Soil water content	8.0	8.8	3.5	4.3	3.7	4.1
Soil temperature	28.3	25.7	39.0	25.6	33.7	23.4
Soil available phosphorus	4.5	1.6	2.5	1.5	2.8	1.4
Soil salinity content	66.8	33.3	40.1	22.2	26.9	23.1
Soil pH	31.8	20.0	18.1	15.9	29.6	11.5
Soil total phosphorus	19.5	9.3	28.7	32.7	46.1	22.3
Total nitrogen content	37.5	17.0	33.9	17.1	48.8	27.2
Soil ammonium nitrogen	31.4	25.9	28.0	16.1	49.8	20.6
Soil nitrate nitrogen	55.4	33.1	22.3	10.9	43.6	25.8

Table 3. Parameters of the theoretical models for soil respiration at different sampling scales.

Plot	Sampling Scale	Variogram Model Type	Proportion ($C/(C_0 + C)$)	Range (m)	R^2
Plot 1	5 m × 5 m	Spherical	0.92	11.65	0.72
	10 m × 10 m	Spherical	0.98	15.45	0.94
Plot 2	5 m × 5 m	Linear	0.28	16.30	0.36
	10 m × 10 m	Gaussian	0.97	19.68	1.00
Plot 3	5 m × 5 m	Linear	<0.25	16.30	0.24
	10 m × 10 m	Linear	<0.25	32.60	0.64

The results of RDA and variation partitioning showed that plant traits (H_{max}, LL, and LP) at the 5 m × 5 m scale explained 17% of the spatial variation of Rs, and soil nutrients (AP and NN) and plant traits explained 49% of the spatial variation (Figure 4). At the sampling scale of 10 m × 10 m, soil microclimate (SWC), soil nutrients (pH, AP, and NN) and plant traits (LT) jointly explained 51% of the spatial variation of Rs. Among them, soil microclimate and soil nutrients accounted for a large proportion (32% and 16%, respectively), and plant traits explained 3% of the variation of Rs. The spatial change of Rs explained by soil nutrients and plant traits at both scales was significant ($p < 0.05$), while soil microclimate only explained a large part of the variation of Rs at the 10 m × 10 m scale ($p < 0.05$).

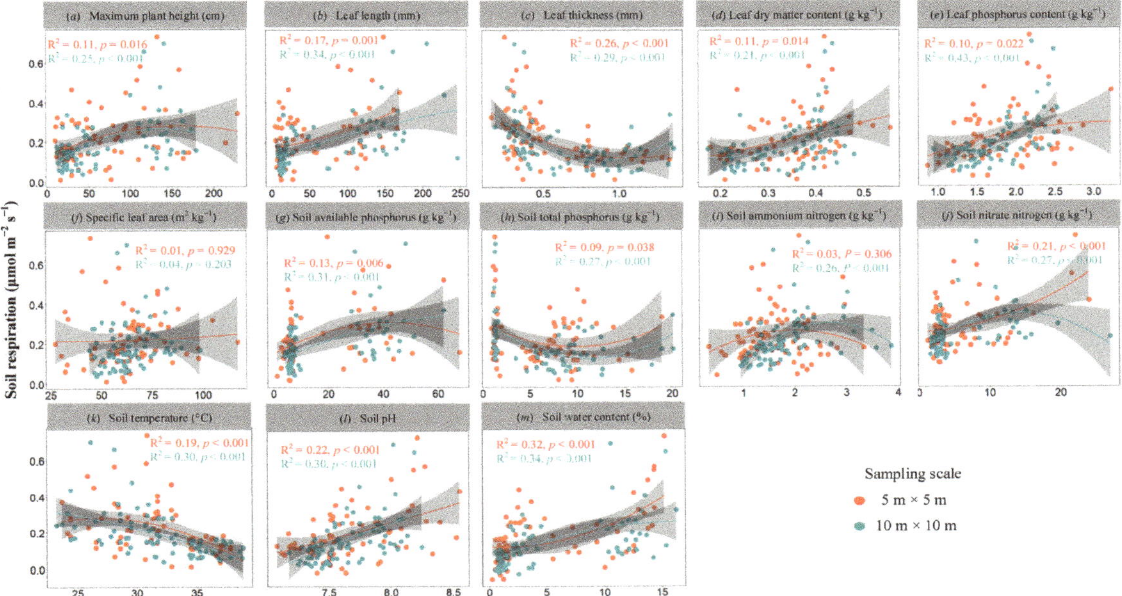

Figure 3. Relationship between soil respiration and plant traits (**a**–**f**) and soil properties (**g**–**m**) under different sampling scales.

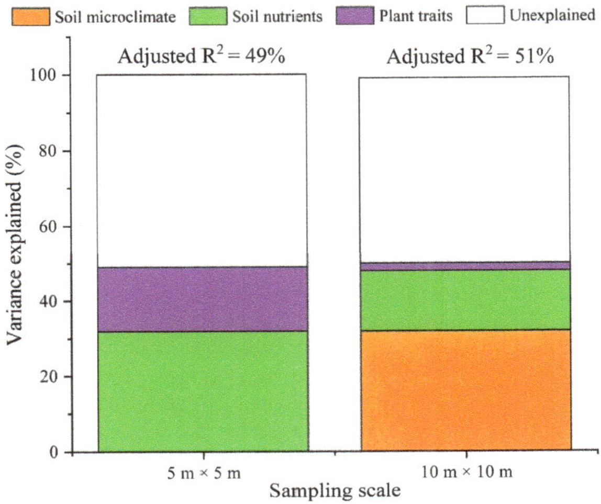

Figure 4. Variation partitioning for the effects of soil properties and plant traits on soil respiration.

4. Discussion

The results showed that the CV of Rs in plot 1 was the largest, and the CVs of soil properties and plant functional traits were greater than those in plot 2 and plot 3 (Table 2). The above results suggest that the biotic and abiotic factors in plot 1 had higher variability than those in plot 2 and plot 3, which might lead to stronger spatial variability of Rs by affecting plant and soil conditions [22,30]. At different spatial scales, Rs exhibits high variability [62–64]. Voroney and Russell [65] sampled a mature birch forest along a 40 m sampling line at 2–4 m intervals and found that the CV of Rs was between 16% and 45%. Rayment and Jarvis [66] reported that when the sampling scale was greater than 1 m, the spatial variation of Rs increased with the increase of sampling spacing, but the increase was not significant. Kosugi et al. [48] suggested that the CV of the Rs rate increased with the increase of spatial grain size (5 m × 5 m, 10 m × 10 m, and 50 m × 50 m). However, there was no difference in the variation coefficient of Rs between the 5 m × 5 m and 10 m × 10 m scales. Adachi et al. [67] measured the variation coefficients of Rs at 20 m × 20 m sampling scales of 1 hm^2 and 2 hm^2 and obtained CV values of 42.3%–43.7% and 39.6%–44.5%, respectively. In the present study, it was found that although Rs had obvious spatial variation, the variation coefficients of Rs in the same plot were consistent at two sampling scales (Figure 2 and Table 2). This indicates that relatively stable environmental factors at different sampling scales control the spatial variability of high or low Rs in the same plot [42,68]. The spatial autocorrelation ranges of Rs at the 5 m × 5 m and 10 m × 10 m sampling scales were 11.65–16.3 m and 15.45–32.60 m, respectively, which were consistent with the results of Kosugi et al. [48] (i.e., the spatial autocorrelation ranges of Rs were 7.9–14.1 m). These results indicate that it is important to evaluate the changes of Rs at different scales and analyze the key factors affecting the changes [19].

Rs includes autotrophic respiration (Ra) and heterotrophic respiration (Rh) [69]. Ra reflects carbon returned from roots and associated microbes to the atmosphere and regulates the distribution of photosynthetic carbon to underground tissues [29], namely root-derived respiration (i.e., root respiration and root exudates) [28]. Rh reflects the CO_2 released by the microbial decomposition of organic matter [69]. Therefore, Rs connects aboveground and underground carbon fractions [70] in terrestrial ecosystems and is significantly associated with a series of aboveground and underground plant traits and soil properties. However, it is difficult for researchers to obtain information on underground plant traits involving a certain spatial range. Because there is a strong correlation between changes in aboveground and underground traits [38], it is feasible to select easily measured aboveground functional traits to predict spatial variations in Rs [71]. The results showed that the key factors affecting Rs variability differed at different spatial scales. The effects of driving factors on Rs were significantly different, and plant functional traits could explain some variations in Rs (Table S2). This is because variations in aboveground plant traits can affect litter decomposition rates [72], soil microbial communities [73], and soil nutrient cycling [74,75], and they can directly and indirectly affect spatial changes in Rs [76]. When spatial scales change, the controlling factors of Rs variation at small scales may be replaced by factors that have a stronger impact at the large scale [68]. Martin and Bolstad [41] suggested that the spatial variation of Rs at the 0–1 m^2 scale was mainly affected by roots and litter, while at the scale of 1–100 m^2, it was mainly affected by root biomass, soil carbon/nitrogen content, and root nitrogen content. At the landscape scale, topography strongly affects soil moisture and soil properties and causes great changes in the spatial pattern of Rs. Shi and Jin [68] reported that the variation of Rs in a forest type in northeast China depended on forest structure parameters (mean diameter at breast height, maximum diameter at breast height, and total basal area) and soil physical and chemical properties (water-filled porosity, soil organic carbon content, and soil C:N). Soil physical and chemical properties (soil organic carbon content, soil C:N, and field capacity) control the variation of Rs among four forest types. A previous study has shown that the size of biomass controls the spatial distribution of Rs, while the effects of ST and SWC are relatively small [77]. However, biomass and plant diversity are related to the increase of root biomass, and root biomass regulates Rs in natural systems in the interaction with ST and SWC [36]. In this study, it was found that at the scale of 5 m × 5 m, the spatial variation of Rs in the three plots mainly depended on plant traits (LL, LT, LDMC, and LPC, $p < 0.05$), while at the scale of 10 m × 10 m, the spatial variation of Rs in the three plots mainly depended on soil properties (TP, AN, SWC, and pH, $p < 0.05$). The above results showed that the differences in biotic (plant functional traits) and abiotic factors (soil properties) increased correspondingly with the increase of the sampling scale, which may have been caused by the differences in soil physical and chemical properties and functional traits caused by plant community composition and vegetation patch distribution [23,28,78].

Xu and Qi [19] showed that the spatial change of Rs was related to root and microbial biomass, soil physical and chemical properties, ST, and SWC. As water decreases, the soil becomes barren, resulting in a gradual decrease in Rs. In the present study, soil nutrients and plant traits at two sampling scales jointly controlled the variation of Rs at the local scale ($p < 0.05$). The mean Rs of plot 3 was the lowest at any sampling scale (Figure 2a). This is because soil nutrients are the material basis for plant growth and soil microbial survival, and their content determines the growth and development of plants and the size of the microbial biomass [79]. The leaf is the site of plant photosynthesis, and the length, width, size, and thickness of leaves affect photosynthesis. Studies have shown that root respiration and photosynthesis have a strong coupling relationship [33], and the photosynthetic rate largely affects root activity. Plant leaves adapt to changes in the external environment, such as drought stress, by changing their physical forms [55], such as by changing the distribution of photosynthetic products to roots, leading to changes in soil autotrophic respiration [31,80]. Therefore, compared with plot 3, plot 1, with high photosynthesis and biomass, also had relatively strong Rs [57,81]. In future studies, the community functional parameters of plant functional traits and the influence mechanism of plant functional trait diversity on Rs should be focused on [19].

Studies have shown that SWC is very important among the various factors affecting the spatial variation of Rs [48,82]. The direct effect of SWC on Rs occurred through the physiological processes of roots and microorganisms, and the indirect effect on Rs occurred through the diffusion of substrate and oxygen [83,84]. A study by Cai et al. [17] in Kayanodaira Forest, Japan, found that the spatial variability of SRdaily (daily summed Rs) was well correlated with SWC, and SRdaily first increased and then decreased with the increase of SWC. Jiang et al. [62] showed that the spatial distribution of the Rs rate was significantly negatively correlated with SWC in subtropical evergreen broad-leaved forests in southern China. The results of the present study showed (Figure 3) that the variation of Rs at the local scale could be well explained by SWC (5 m × 5 m: $R^2 = 0.32$; 10 m × 10 m: $R^2 = 0.34$). In addition, compared with the sampling scale of 5 m × 5 m, the large spatial variation of Rs at the scale of 10 m × 10 m could be explained by SWC (Figure 4, $p < 0.05$). This is because SWC, at levels that are too low or too high, limits Rs [85], especially in arid or semi-arid areas. When SWC becomes a stress factor, it may replace temperature and become the main control factor of the Rs rate [19,86]. The effect of SWC on Rs is very complex and changeable. When the soil is dry, the soil metabolic activity increases with the increase of water content, and there is a positive correlation between them. When SWC exceeds a certain range (80% of saturated soil moisture), the Rs decreases with the increase of SWC, and there is a negative correlation between them [36,87]. These results indicated that the synergistic changes of plant functional traits and soil nutrients with SWC caused the spatial variation of the Rs rate at the local scale. Therefore, to improve the uncertainty of existing models for predicting Rs and better elucidate the potential mechanism of the spatial pattern of Rs in arid areas, it is necessary to quantify the spatial variation of Rs and its relationship with related environmental factors (plant traits, soil nutrients, and soil microclimate) at a fine scale (10 m × 10 m).

5. Conclusions

We have provided valid and reliable evidence that, with increasing sampling scale, the spatial variability of Rs was primarily influenced by soil properties, followed by plant traits, the relative importance of which depends on soil water conditions. In addition, the contribution of soil properties varies with plot types, and its influence in plot 1 (riverbank habitat) was greater than that in plot 3 (desert margin habitat). Among them, soil microclimate has a greater potential to enhance the spatial heterogeneity of Rs than soil nutrients. These results emphasize that the nested sampling design method can be used to quantify the relationship between Rs and related driving factors at different spatial scales in regions with large soil water changes, which can help inform the design of Rs field sampling schemes at local scales.

Supplementary Materials: The following supporting information can be downloaded at: https://www.mdpi.com/article/10.3390/f13071001/s1, Table S1: Comparison in soil properties of the measuring transect (transect 1) and the other two transects (transects 2–3); Figure S1: Soil types in the Ebinur Lake Wetland National Nature Reserve; Table S2: Multiple regression analysis of soil respiration and related influencing factors within the same plot. H_{max}: Maximum plant height; LL: Leaf length; LT: Leaf thickness; LDMC: Leaf dry matter content; SLA: Specific leaf area; LPC: Leaf phosphorus content; SWC: Soil water content; ST: Soil temperature; pH: Soil pH; AP: Soil available phosphorus; TP: Soil total phosphorus; AN: Soil ammonium nitrogen; NN: Soil nitrate nitrogen; NS, not significant; ND, not determined (removed by the stepwise regression analysis results). * $p < 0.05$, ** $p < 0.01$, *** $p < 0.001$; Figure S2: Cluster analysis of the measured environmental variables. H_{max}: Maximum plant height; LL: Leaf length; LW: Leaf width; LT: Leaf thickness; LDMC: Leaf dry matter content; SLA: Specific leaf area; LPC: Leaf phosphorus content; LNC: Leaf nitrogen content; LCC: Leaf carbon content; SWC: Soil water content; ST: Soil temperature; pH: Soil pH; AN: Soil ammonium nitrogen; AP: Soil available phosphorus; TP: Soil total phosphorus; TN: Total nitrogen content; NN: Soil nitrate nitrogen; SSC: Soil salinity content.

Author Contributions: Conceptualization, J.W. and G.L.; methodology, J.W.; software, J.W.; data curation, Z.L. and Y.C.; writing—original draft preparation, J.W.; writing—review and editing, X.H. and W.M.; supervision, G.L.; funding acquisition, G.L. and X.H. All authors have read and agreed to the published version of the manuscript.

Funding: This research was funded by the National Natural Science Foundation of China (31560131 and 31760168), Xinjiang Uygur Autonomous Region university scientific research project (XJEDU2020I002), Xinjiang Uygur Autonomous Region Innovation Environment Construction Project—Science and Technology Innovation Base Construction Project (PT2107), and the Xinjiang Uygur Autonomous Region Graduate Research and Innovation Project (XJ2020G012 and XJ2019G020).

Institutional Review Board Statement: Not applicable.

Informed Consent Statement: Not applicable.

Data Availability Statement: The data presented in this study are available on request from the corresponding author.

Acknowledgments: We are grateful to Xueni Zhang of Xinjiang University's College of Ecology and Environment for providing the data. We also thank the editor and the reviewers for their insightful and valuable suggestions, which greatly improved the quality of this manuscript.

Conflicts of Interest: The authors declare no conflict of interest.

References

1. Hashimoto, S.; Carvalhais, N.; Ito, A.; Migliavacca, M.; Nishina, K.; Reichstein, M. Global spatiotemporal distribution of soil respiration modeled using a global database. *Biogeosciences* **2015**, *12*, 4121–4132. [CrossRef]
2. Nunes, M.R.; Karlen, D.L.; Veum, K.S.; Moorman, T.B.; Cambardella, C.A. Biological soil health indicators respond to tillage intensity: A US meta-analysis. *Geoderma* **2020**, *369*, 114335. [CrossRef]
3. Pries, C.E.H.; Castanha, C.; Porras, R.C.; Torn, M.S. The whole-soil carbon flux in response to warming. *Science* **2017**, *355*, 1420–1423. [CrossRef] [PubMed]
4. Lei, J.; Guo, X.; Zeng, Y.; Zhou, J.; Gao, Q.; Yang, Y. Temporal changes in global soil respiration since 1987. *Nat. Commun.* **2021**, *12*, 403. [CrossRef] [PubMed]
5. Adachi, M.; Ito, A.; Yonemura, S.; Takeuchi, W. Estimation of global soil respiration by accounting for land-use changes derived from remote sensing data. *J. Environ. Manag.* **2017**, *200*, 97–104. [CrossRef]
6. Fóti, S.; Balogh, J.; Herbst, M.; Papp, M.; Koncz, P.; Bartha, S.; Zimmermann, Z.; Komoly, C.; Szabó, G.; Margóczi, K.; et al. Meta-analysis of field scale spatial variability of grassland soil CO_2 efflux: Interaction of biotic and abiotic drivers. *CATENA* **2016**, *143*, 78–89. [CrossRef]
7. Schlesinger, W.H.; Andrews, J.A. Soil respiration and the global carbon cycle. *Biogeochemistry* **2000**, *48*, 7–20. [CrossRef]
8. Hursh, A.; Ballantyne, A.; Cooper, L.; Maneta, M.; Kimball, J.; Watts, J. The sensitivity of soil respiration to soil temperature, moisture, and carbon supply at the global scale. *Glob. Chang. Biol.* **2016**, *23*, 2090–2103. [CrossRef]

9.   Tang, X.; Du, J.; Shi, Y.; Lei, N.; Chen, G.; Cao, L.; Pei, X. Global patterns of soil heterotrophic respiration—A meta-analysis of available dataset. *CATENA* **2020**, *191*, 104574. [CrossRef]
10. Stephan, E.; Groffman, P.; Vidon, P.; Stella, J.C.; Endreny, T. Interacting drivers and their tradeoffs for predicting denitrification potential across a strong urban to rural gradient within heterogeneous landscapes. *J. Environ. Manag.* **2021**, *294*, 113021. [CrossRef]
11. Riutta, T.; Kho, L.K.; Teh, Y.A.; Ewers, R.; Majalap, N.; Malhi, Y. Major and persistent shifts in below-ground carbon dynamics and soil respiration following logging in tropical forests. *Glob. Chang. Biol.* **2021**, *27*, 2225–2240. [CrossRef] [PubMed]
12. Ouyang, W.; Lai, X.; Li, X.; Liu, H.; Lin, C.; Hao, F. Soil respiration and carbon loss relationship with temperature and land use conversion in freeze-thaw agricultural area. *Sci. Total Environ.* **2015**, *533*, 215–222. [CrossRef] [PubMed]
13. Zhang, Y.; Zou, J.; Dang, S.; Osborne, B.; Ren, Y.; Ju, X. Topography modifies the effect of land-use change on soil respiration: A meta-analysis. *Ecosphere* **2021**, *12*, e03845. [CrossRef]
14. Han, G.; Yu, J.; Li, H.; Yang, L.; Wang, G.; Mao, P.; Gao, Y. Winter Soil Respiration from Different Vegetation Patches in the Yellow River Delta, China. *Environ. Manag.* **2012**, *50*, 39–49. [CrossRef]
15. Jian, J.; Steele, M.K.; Zhang, L.; Bailey, V.L.; Zheng, J.; Patel, K.F.; Bond-Lamberty, B.P. On the use of air temperature and precipitation as surrogate predictors in soil respiration modelling. *Eur. J. Soil Sci.* **2021**, *73*, e13149. [CrossRef]
16. Darenova, E.; Čater, M. Different Structure of Sessile Oak Stands Affects Soil Moisture and Soil CO_2 Efflux. *For. Sci.* **2018**, *64*, 340–348. [CrossRef]
17. Cai, Y.; Nishimura, T.; Ida, H.; Hirota, M. Spatial variation in soil respiration is determined by forest canopy structure through soil water content in a mature beech forest. *For. Ecol. Manag.* **2021**, *501*, 119673. [CrossRef]
18. Lee, J.-S. Relationship of root biomass and soil respiration in a stand of deciduous broadleaved trees—A case study in a maple tree. *J. Ecol. Environ.* **2018**, *42*, 19. [CrossRef]
19. Wang, J.; Teng, D.; He, X.; Qin, L.; Yang, X.; Lv, G. Spatial non-stationarity effects of driving factors on soil respiration in an arid desert region. *CATENA* **2021**, *207*, 105617. [CrossRef]
20. Tang, Y.-S.; Wang, L.; Jia, J.-W.; Fu, X.-H.; Le, Y.-Q.; Chen, X.-Z.; Sun, Y. Response of soil microbial community in Jiuduansha wetland to different successional stages and its implications for soil microbial respiration and carbon turnover. *Soil Biol. Biochem.* **2011**, *43*, 638–646. [CrossRef]
21. Yang, J.; Zhan, C.; Li, Y.; Zhou, D.; Yu, Y.; Yu, J. Effect of salinity on soil respiration in relation to dissolved organic carbon and microbial characteristics of a wetland in the Liaohe River estuary, Northeast China. *Sci. Total Environ.* **2018**, *642*, 946–953. [CrossRef] [PubMed]
22. Singh, R.; Singh, A.K.; Singh, S.; Srivastava, P.; Singh, H.; Raghubanshi, A.S. Geomorphologic heterogeneity influences dry-season soil CO_2 efflux by mediating soil biophysical variables in a tropical river valley. *Aquat. Sci.* **2019**, *81*, 43. [CrossRef]
23. Raich, J.W.; Schlesinger, W.H. The global carbon dioxide flux in soil respiration and its relationship to vegetation and climate. *Tellus B Chem. Phys. Meteorol.* **1992**, *44*, 81–99. [CrossRef]
24. Tang, J.; Baldocchi, D.D. Spatial-Temporal variation in soil respiration in an oak-grass savanna ecosystem in California and its partitioning into autotrophic and heterotrophic components. *Biogeochemistry* **2005**, *73*, 183–207. [CrossRef]
25. Xu, M.; Qi, Y. Soil-surface CO_2 efflux and its spatial and temporal variations in a young ponderosa pine plantation in northern California. *Glob. Chang. Biol.* **2001**, *7*, 667–677. [CrossRef]
26. Janssens, I.A.; Pilegaard, K. Large seasonal changes in Q 10 of soil respiration in a beech forest. *Glob. Chang. Biol.* **2003**, *9*, 911–918. [CrossRef]
27. Tang, J.; Baldocchi, D.; Qi, Y.; Xu, L. Assessing soil CO_2 efflux using continuous measurements of CO_2 profiles in soils with small solid-state sensors. *Agric. For. Meteorol.* **2003**, *118*, 207–220. [CrossRef]
28. Metcalfe, D.B.; Fisher, R.A.; Wardle, D.A. Plant communities as drivers of soil respiration: Pathways, mechanisms, and significance for global change. *Biogeosciences* **2011**, *8*, 2047–2061. [CrossRef]
29. Högberg, P.; Nordgren, A.; Ågren, G.I. Carbon allocation between tree root growth and root respiration in boreal pine forest. *Oecologia* **2002**, *132*, 579–581. [CrossRef]
30. Raich, J.W.; Tufekciogul, A. Vegetation and soil respiration: Correlations and controls. *Biogeochemistry* **2000**, *48*, 71–90. [CrossRef]
31. Kuzyakov, Y.; Gavrichkova, O. REVIEW: Time lag between photosynthesis and carbon dioxide efflux from soil: A review of mechanisms and controls. *Glob. Chang. Biol.* **2010**, *16*, 3386–3406. [CrossRef]
32. Atkin, O.K.; Edwards, E.J.; Loveys, B.R. Response of root respiration to changes in temperature and its relevance to global warming. *New Phytol.* **2000**, *147*, 141–154. [CrossRef]
33. Högberg, P.; Nordgren, A.; Buchmann, N.; Taylor, A.F.S.; Ekblad, A.; Högberg, M.N.; Nyberg, G.; Ottosson-Löfvenius, M.; Read, D.J. Large-scale forest girdling shows that current photosynthesis drives soil respiration. *Nature* **2001**, *411*, 789–792. [CrossRef] [PubMed]
34. Moyano, F.E.; Kutsch, W.L.; Schulze, E.-D. Response of mycorrhizal, rhizosphere and soil basal respiration to temperature and photosynthesis in a barley field. *Soil Biol. Biochem.* **2007**, *39*, 843–853. [CrossRef]
35. De Deyn, G.B.; Cornelissen, J.H.C.; Bardgett, R.D. Plant functional traits and soil carbon sequestration in contrasting biomes. *Ecol. Lett.* **2008**, *11*, 516–531. [CrossRef]
36. Kumari, T.; Singh, R.; Verma, P.; Raghubanshi, A.S. Monsoon-phase regulates the decoupling of auto- and heterotrophic respiration by mediating soil nutrient availability and root biomass in tropical grassland. *CATENA* **2021**, *209*, 105808. [CrossRef]

37. Shen, Y.; Gilbert, G.S.; Li, W.; Fang, M.; Lu, H.; Yu, S. Linking Aboveground Traits to Root Traits and Local Environment: Implications of the Plant Economics Spectrum. *Front. Plant Sci.* **2019**, *10*, 1412. [CrossRef]

38. Mao, W.; Felton, A.J.; Ma, Y.; Zhang, T.; Sun, Z.; Zhao, X.; Smith, M.D. Relationships between aboveground and belowground trait responses of a dominant plant species to alterations in watertable depth. *Land Degrad. Dev.* **2018**, *29*, 4015–4024. [CrossRef]

39. Raich, J.W. Temporal patterns of soil respiration in tropical forest plantations in lowland Costa Rica. In Proceedings of the 96th ESA Annual Meeting, Austin, TX, USA, 7–12 August 2011.

40. Jiang, Y.; Huang, X.; Zhang, X.; Zhang, X.; Zhang, Y.; Zheng, C.; Deng, A.; Zhang, J.; Wu, L.; Hu, S.; et al. Optimizing rice plant photosynthate allocation reduces N_2O emissions from paddy fields. *Sci. Rep.* **2016**, *6*, 29333. [CrossRef]

41. Martin, J.G.; Bolstad, P.V. Variation of soil respiration at three spatial scales: Components within measurements, intra-site variation and patterns on the landscape. *Soil Biol. Biochem.* **2009**, *41*, 530–543. [CrossRef]

42. Chen, Q.; Wang, Q.; Han, X.; Wan, S.; Li, L. Temporal and spatial variability and controls of soil respiration in a temperate steppe in northern China. *Glob. Biogeochem. Cycles* **2010**, *24*, GB2010. [CrossRef]

43. Buczko, U.; Bachmann, S.; Gropp, M.; Jurasinski, G.; Glatzel, S. Spatial variability at different scales and sampling requirements for in situ soil CO_2 efflux measurements on an arable soil. *CATENA* **2015**, *131*, 46–55. [CrossRef]

44. Wu, J.; Jones, B.; Li, H.; Loucks, O.L. *Scaling and Uncertainty Analysis in Ecology: Methods and Applications*; Springer: Dordrecht, The Netherlands, 2006.

45. Jenerette, D.; Wu, J. On the Definitions of Scale. *Bull. Ecol. Soc. Am.* **2000**, *81*, 104–105. [CrossRef]

46. Siefert, A.; Ravenscroft, C.; Althoff, D.; Alvarez-Yepiz, J.C.; Carter, B.E.; Glennon, K.L.; Heberling, M.; Jo, I.S.; Pontes, A.; Sauer, A.; et al. Scale dependence of vegetation-environment relationships: A meta-analysis of multivariate data. *J. Veg. Sci.* **2012**, *23*, 942–951. [CrossRef]

47. Tamme, R.; Hiiesalu, I.; Laanisto, L.; Szava-Kovats, R.; Pärtel, M. Environmental heterogeneity, species diversity and co-existence at different spatial scales. *J. Veg. Sci.* **2010**, *21*, 796–801. [CrossRef]

48. Kosugi, Y.; Mitani, T.; Itoh, M.; Noguchi, S.; Tani, M.; Matsuo, N.; Takanashi, S.; Ohkubo, S.; Nik, A.R. Spatial and temporal variation in soil respiration in a Southeast Asian tropical rainforest. *Agric. For. Meteorol.* **2007**, *147*, 35–47. [CrossRef]

49. Zhang, M.; Li, G.; Liu, B.; Liu, J.; Wang, L.; Wang, D. Effects of herbivore assemblage on the spatial heterogeneity of soil nitrogen in eastern Eurasian steppe. *J. Appl. Ecol.* **2020**, *57*, 1551–1560. [CrossRef]

50. Gong, G.L.X. Species diversity and dominant species' niches of eremophyte communities of the Tugai forest in the Ebinur basin of Xinjiang, China. *Biodivers. Sci.* **2017**, *25*, 29–35. [CrossRef]

51. Zerbe, S.; Halik, Ü.; Küchler, J. Urban greening in the oases of continental arid Southern Xinjiang (NW China)—An interdisciplinary approach. *Erde* **2005**, *136*, 245–266.

52. Thevs, N.; Zerbe, S.; Schnittler, M.; Abdusalih, N.; Succow, M. Structure, reproduction and flood-induced dynamics of riparian Tugai forests at the Tarim River in Xinjiang, NW China. *Forestry* **2008**, *81*, 45–57. [CrossRef]

53. Zhang, X.; Yang, X.; Lv, G. Diversity patterns and response mechanisms of desert plants to the soil environment along soil water and salinity gradients. *Acta Ecol. Sin.* **2016**, *36*, 3206–3215. [CrossRef]

54. Leemans, R.; Eickhout, B. Another reason for concern: Regional and global impacts on ecosystems for different levels of climate change. *Glob. Environ. Chang.* **2004**, *14*, 219–228. [CrossRef]

55. Gong, Y.; Ling, H.; Lv, G.; Chen, Y.; Guo, Z.; Cao, J. Disentangling the influence of aridity and salinity on community functional and phylogenetic diversity in local dryland vegetation. *Sci. Total Environ.* **2018**, *653*, 409–422. [CrossRef] [PubMed]

56. Zhang, X.-N.; Yang, X.-D.; Li, Y.; He, X.-M.; Lv, G.-H.; Yang, J.-J. Influence of edaphic factors on plant distribution and diversity in the arid area of Xinjiang, Northwest China. *Arid Land Res. Manag.* **2017**, *32*, 38–56. [CrossRef]

57. Wang, H.; Cai, Y.; Yang, Q.; Gong, Y.; Lv, G. Factors that alter the relative importance of abiotic and biotic drivers on the fertile island in a desert-oasis ecotone. *Sci. Total Environ.* **2019**, *697*, 134096. [CrossRef]

58. Shi, B.; Xu, W.; Zhu, Y.; Wang, C.; Loik, M.E.; Sun, W. Heterogeneity of grassland soil respiration: Antagonistic effects of grazing and nitrogen addition. *Agric. For. Meteorol.* **2019**, *268*, 215–223. [CrossRef]

59. Barba, J.; Yuste, J.C.; Martínez-Vilalta, J.; Lloret, F. Drought-induced tree species replacement is reflected in the spatial variability of soil respiration in a mixed Mediterranean forest. *For. Ecol. Manag.* **2013**, *306*, 79–87. [CrossRef]

60. Liu, C.; Song, X.; Wang, L.; Wang, D.; Zhou, X.; Liu, J.; Zhao, X.; Li, J.; Lin, H. Effects of grazing on soil nitrogen spatial heterogeneity depend on herbivore assemblage and pre-grazing plant diversity. *J. Appl. Ecol.* **2016**, *53*, 242–250. [CrossRef]

61. Yan, J.-X.; Sun, Q.; Li, J.-J.; Li, H.-J. Effect of the Sampling Scale and Number on the Heterogeneity of Soil Respiration in a Mixed Broadleaf-conifer Forest. *Huan Jing Ke Xue = Huanjing Kexue* **2019**, *40*, 383–391.

62. Jiang, Y.; Zhang, B.; Wang, W.; Li, B.; Wu, Z.; Chu, C. Topography and plant community structure contribute to spatial heterogeneity of soil respiration in a subtropical forest. *Sci. Total Environ.* **2020**, *733*, 139287. [CrossRef]

63. Han, M.; Shi, B.; Jin, G. Spatial patterns of soil respiration in a spruce-fir valley forest, Northeast China. *J. Soils Sediments* **2018**, *19*, 10–22. [CrossRef]

64. Darenova, E.; Čater, M. Effect of spatial scale and harvest on heterogeneity of forest floor CO_2 efflux in a sessile oak forest. *CATENA* **2020**, *188*, 104455. [CrossRef]

65. Russell, C.A.; Voroney, R.P. Carbon dioxide efflux from the floor of a boreal aspen forest. I. Relationship to environmental variables and estimates of C respired. *Can. J. Soil Sci.* **1998**, *78*, 301–310. [CrossRef]

66. Rayment, M.; Jarvis, P. Temporal and spatial variation of soil CO_2 efflux in a Canadian boreal forest. *Soil Biol. Biochem.* **2000**, *32*, 35–45. [CrossRef]
67. Adachi, M.; Bekku, Y.S.; Konuma, A.; Kadir, W.R.; Okuda, T.; Koizumi, H. Required sample size for estimating soil respiration rates in large areas of two tropical forests and of two types of plantation in Malaysia. *For. Ecol. Manag.* **2005**, *210*, 455–459. [CrossRef]
68. Shi, B.; Jin, G. Variability of soil respiration at different spatial scales in temperate forests. *Biol. Fertil. Soils* **2016**, *52*, 561–571. [CrossRef]
69. Subke, J.-A.; Inglima, I.; Cotrufo, M.F. Trends and methodological impacts in soil CO_2 efflux partitioning: A metaanalytical review. *Glob. Chang. Biol.* **2006**, *12*, 921–943. [CrossRef]
70. Tang, X.; Pei, X.; Lei, N.; Luo, X.; Liu, L.; Shi, L.; Chen, G.; Liang, J. Global patterns of soil autotrophic respiration and its relation to climate, soil and vegetation characteristics. *Geoderma* **2020**, *369*, 114339. [CrossRef]
71. Cassart, B.; Basia, A.A.; Jonard, M.; Ponette, Q. Functional traits drive the difference in soil respiration between Gilbertiodendron dewevrei monodominant forests patches and Scorodophloeus zenkeri mixed forests patches in the Central Congo basin. *Plant Soil* **2021**, *460*, 313–331. [CrossRef]
72. Garnier, E.; Cortez, J.; Billès, G.; Navas, M.-L.; Roumet, C.; Debussche, M.; Laurent, G.; Blanchard, A.; Aubry, D.; Bellmann, A.; et al. Plant functional markers capture ecosystem properties during secondary succession. *Ecology* **2004**, *85*, 2630–2637. [CrossRef]
73. Pei, Z.; Eichenberg, D.; Bruelheide, H.; Kröber, W.; Kühn, P.; Li, Y.; von Oheimb, G.; Purschke, O.; Scholten, T.; Buscot, F.; et al. Soil and tree species traits both shape soil microbial communities during early growth of Chinese subtropical forests. *Soil Biol. Biochem.* **2016**, *96*, 180–190. [CrossRef]
74. Fortunel, C.; Garnier, E.; Joffre, R.; Kazakou, E.; Quested, H.; Grigulis, K.; Lavorel, S.; Ansquer, P.; Castro, H.; Cruz, P.; et al. Leaf traits capture the effects of land use changes and climate on litter decomposability of grasslands across Europe. *Ecology* **2009**, *90*, 598–611. [CrossRef] [PubMed]
75. Conti, G.; Díaz, S. Plant functional diversity and carbon storage—An empirical test in semi-arid forest ecosystems. *J. Ecol.* **2012**, *101*, 18–28. [CrossRef]
76. Buzzard, V.; Michaletz, S.T.; Deng, Y.; He, Z.; Ning, D.; Shen, L.; Tu, Q.; Van Nostrand, J.D.; Voordeckers, J.W.; Wang, J.; et al. Continental scale structuring of forest and soil diversity via functional traits. *Nat. Ecol. Evol.* **2019**, *3*, 1298–1308. [CrossRef]
77. Fang, C.; Moncrieff, J.B.; Gholz, H.L.; Clark, K.L. Soil CO_2 efflux and its spatial variation in a Florida slash pine plantation. *Plant Soil* **1998**, *205*, 135–146. [CrossRef]
78. Frank, A. Carbon dioxide fluxes over a grazed prairie and seeded pasture in the Northern Great Plains. *Environ. Pollut.* **2001**, *116*, 397–403. [CrossRef]
79. Xu, X.; Shi, Z.; Li, D.; Zhou, X.; Sherry, R.A.; Luo, Y. Plant community structure regulates responses of prairie soil respiration to decadal experimental warming. *Glob. Chang. Biol.* **2015**, *21*, 3846–3853. [CrossRef]
80. Wofsy, S.C.; Harriss, R.C.; Kaplan, W.A. Carbon dioxide in the atmosphere over the Amazon Basin. *J. Geophys. Res. Earth Surf.* **1988**, *93*, 1377–1387. [CrossRef]
81. Yan, L.; Chen, S.; Huang, J.; Lin, G. Water regulated effects of photosynthetic substrate supply on soil respiration in a semiarid steppe. *Glob. Chang. Biol.* **2010**, *17*, 1990–2001. [CrossRef]
82. Liu, W.; Zhang, Z.; Wan, S. Predominant role of water in regulating soil and microbial respiration and their responses to climate change in a semiarid grassland. *Glob. Chang. Biol.* **2009**, *15*, 184–195. [CrossRef]
83. Hartman, K.; Tringe, S.G. Interactions between plants and soil shaping the root microbiome under abiotic stress. *Biochem. J.* **2019**, *476*, 2705–2724. [CrossRef] [PubMed]
84. Schwen, A.; Jeitler, E.; Böttcher, J. Spatial and temporal variability of soil gas diffusivity, its scaling and relevance for soil respiration under different tillage. *Geoderma* **2015**, *259*, 323–336. [CrossRef]
85. Pangle, R.; Seiler, J. Influence of seedling roots, environmental factors and soil characteristics on soil CO_2 efflux rates in a 2-year-old loblolly pine (*Pinus taeda* L.) plantation in the Virginia Piedmont. *Environ. Pollut.* **2001**, *116*, S85–S96. [CrossRef]
86. Yang, X.-D.; Ali, A.; Xu, Y.-L.; Jiang, L.-M.; Lv, G.-H. Soil moisture and salinity as main drivers of soil respiration across natural xeromorphic vegetation and agricultural lands in an arid desert region. *CATENA* **2019**, *177*, 126–133. [CrossRef]
87. Raich, J.W.; Potter, C.S. Global patterns of carbon dioxide emissions from soils. *Glob. Biogeochem. Cycles* **1995**, *9*, 23–36. [CrossRef]

Article

The Edaphic and Vegetational Properties Controlling Soil Aggregate Stability Vary with Plant Communities in an Arid Desert Region of Northwest China

Lamei Jiang [1], Dong Hu [2] and Guanghui Lv [1,*]

[1] College of Resources and Environmental Science, Xinjiang University, Urumqi 830017, China; jianglam0108@126.com
[2] College of Life Science, Northwest University, Xian 710069, China; dongh_xju@yeah.net
* Correspondence: guanghui_xju@sina.com; Tel.: +86-0991-2111427

Abstract: The stability of soil aggregates is the basis for supporting ecosystem functions and related services provided by the soil. In order to explore the mechanism of the influence of soil and vegetation properties on the stability of soil aggregates in desert communities, the particle size distribution and aggregate in different communities were compared, and the contribution of soil physical and chemical properties (soil salinity, soil water content, soil pH, soil organic carbon, soil total phosphorus, soil total nitrogen, etc.) and vegetation properties (species richness, phylogenetic richness, plant height and coverage, etc.) to the stability of soil aggregates was determined by using a structural equation model. The results show the following: Soil water content, organic carbon, and salt in river bank plant communities have significant direct positive effects on the mean weight diameter of soil, with path coefficients of 0.50, 0.11, and 0.24, respectively ($p < 0.01$). Water also indirectly affects soil stability by affecting plant height, soil salt, and soil organic carbon; species richness and vegetation coverage have significant direct positive effects on the soil stability index, with path coefficients of 0.13 and 0.11, respectively ($p < 0.01$). In the desert marginal plant community, the plant coverage and species richness have significant positive effects on soil stability, with path coefficients of 0.43 ($p < 0.001$) and 0.35 ($p < 0.001$), respectively. Phylogenetic richness has a significant direct negative effect on soil stability ($p < 0.05$), with an effect value of -0.27. Phylogenetic richness indirectly affects soil stability by adjusting the coverage, with an indirect effect value of 0.23. Moisture, ammonium nitrogen, and nitrate nitrogen have significant direct positive effects on soil stability, with effect values of 0.12, 0.09, and 0.15, respectively. Our research shows that the process of soil stabilization is mainly controlled by soil factors and vegetation characteristics, but its importance varies with different community types.

Keywords: soil aggregates; soil structure; soil properties; plant characteristics; arid desert region

Citation: Jiang, L.; Hu, D.; Lv, G. The Edaphic and Vegetational Properties Controlling Soil Aggregate Stability Vary with Plant Communities in an Arid Desert Region of Northwest China. *Forests* **2022**, *13*, 368. https://doi.org/10.3390/f13030368

Academic Editor: Richard Drew Bowden

Received: 25 January 2022
Accepted: 19 February 2022
Published: 22 February 2022

Publisher's Note: MDPI stays neutral with regard to jurisdictional claims in published maps and institutional affiliations.

1. Introduction

Soil is the main component of the terrestrial ecosystem, and soil degradation leads to a decline in the ecosystem services provided by the soil. Soil degradation weakens the soil structure and reduces soil productivity [1], thus aggravating soil desertification [2,3]. The soil structure is considered to control many processes in soil, regulating water retention and infiltration, gas exchange, the dynamics of soil organic matter and nutrients, root infiltration, and sensitivity to erosion. The soil structure also constitutes the habitat of countless species and the diversity of soil organisms, thus promoting their diversity and regulating their activities [4]. The soil structure is actively shaped by these organisms, in that they change the distribution of soil moisture and air [5–7]. Many processes have been proved to be related to the soil structure. One of the most important indicators of soil degradation is aggregate stability [8,9]. The stability of soil aggregates is closely related to the soil structure and can affect some soil physical and biogeochemical processes, such as the ability of soil to resist erosion, soil atmospheric exchange, water permeability, and nutrient availability [10–12].

Soil agglomeration is a complex process that involves the dynamic interaction between soil organic matter and mineral components and assembles micro-aggregates into macro-aggregates [13]. It is well known that soil stability is influenced by abiotic factors (e.g., soil organic matter, soil water content) and biological factors (e.g., vegetation type, plant roots). Previous studies have shown that soil properties closely related to the stability of soil aggregates include soil texture, organic matter content, soil cations, pH, and microbial activity [13–15]. SOM acts as the main binder of soil particles, while the mineralogy, cation ratio, and binder are closely related to the stability of micro-aggregates [16,17]. In this dynamic aggregation process, clay particles form an organic–mineral combination as a binder during the interaction with soil organic matter, thus reducing the wettability of aggregates and affecting the mechanical strength of soil aggregates [15,18]. However, studies have found that biological factors also play an important role. Plant diversity can buffer the influence of physical disturbances (such as raindrops) on soil aggregates, which is due to the increase in aboveground biomass [19,20], and promote root activities to enhance soil aggregation [21]. However, there are complex interactions between biotic and abiotic factors that will affect soil stability. For example, plant communities with a large number of species have a positive impact on the stability of soil aggregates by affecting soil carbon dynamics, soil microbial activity, and plant growth [20]. Fine roots can increase soil agglomeration by increasing the soil organic matter content and the soil drying–wetting alternation effect [22]. The combination of root biomass and soil water potential affects the production of microbial polysaccharides, and then plays an important role in controlling the formation of water-stable aggregates [23]. Additionally, different vegetation types have different effects on the soil structure [24,25]. On the one hand, different vegetation types affect the chemical composition of soil organic carbon, while SOC plays a key role in soil agglomeration [26–28]. On the other hand, the differences in plant coverage and composition in different communities also affect the stability of soil aggregates [29]. For example, soil aggregate stability under different rain conditions showed the order: forest land > forest–grass land > grass land [25]. Dou et al. found that the aggregate stability of natural shrub was significantly higher than that of natural grass [30]. Although many studies have been carried out regarding the influence of abiotic factors (soil organic carbon, soil water content) and biological factors (plant coverage, plant richness, and root biomass) on soil stability, their relative contribution and mechanism of interaction remain controversial and need to be discussed.

Arid and semi-arid areas are an important part of the global land area, accounting for approximately 20–25 percent of the total global land area, with scarce precipitation and a lack of water resources, and the areas are most sensitive to global climate change [31]. In arid areas, the change in the soil structure has a particularly significant impact on the change in the plant community, which can also improve the soil stability. Yang et al. [32] compared the differences in soil aggregate under four typical halophyte communities (*Karelinia caspia*, *Bassia dasyphylla*, *Haloxylon ammodendron*, and *Tamarix ramosissim*) in an arid area, and found that the percentage of soil aggregate in the >0.25 mm fraction is significantly higher under the *H. ammodendron* community. Abdi et al. [33] found that soil fixation and erosion are controlled by *Haloxylon persicum* roots in arid lands. However, these studies only focused on the percentage of aggregate under different communities and the contribution of roots to soil fixation. There are relatively few studies on the stability of soil aggregates in arid areas, and there are few reports on the relationship between soil aggregate stability and plant community characteristics, structure, and function. Studying the relationship between soil stability and plant distribution and composition and analyzing the mechanism of influence to improve community resistance have become problems to be solved. Based on this, we selected one plant community on the river bank and the desert margin in the Ebinur Lake Wetland National Nature Reserve to compare the differences in soil physical and chemical properties and vegetation properties under different community types, as well as the changes in the stability of aggregates with different community types. In this study, we tried to clarify the direct and indirect roles of abiotic and biological factors in determining the stability of soil aggregates under different vegetation types in arid areas (the plant communities of the river bank and desert margin). In order to verify this hypothesis, this study aimed to: (1) quantify the contribution of abiotic factors (soil physicochemical properties) and biological factors (vegetation properties) to the stability of soil aggregates; and (2) describe the complex interaction between these factors.

2. Materials and Methods

2.1. Study Area, Sample Layout, and Plant Sample Collection

The study area is located in the Ebinur Lake Wetland National Nature Reserve in the northwest of Jinghe County, Xinjiang and the southwest of Junggar Basin (Figure 1). It is the lowest depression and the center of water and salt collection in the southwest margin of Junggar Basin. The basin has a typical temperate continental arid climate [34], with an annual evaporation of over 1600 mm, an annual rainfall of approximately 100 mm, annual sunshine of approximately 2800 h, an extreme maximum temperature of 44 °C, and an extreme minimum temperature of -33 °C [35]. There are various types of sandy vegetation, mesophytic vegetation, and aquatic vegetation in the area. The main dominant plants in the sample plot were *Populus euphratica, Haloxylon ammodendron, Halimodendron halodendron, Alhagi sparsifolia, Reaumuria soongarica, Nitraria roborowskii, Apocynum venetum, Seriphidium terrae-albae, Phragmites australis, Halocnemum strobilaceum, Salsola collina*, and *Suaeda glauca*.

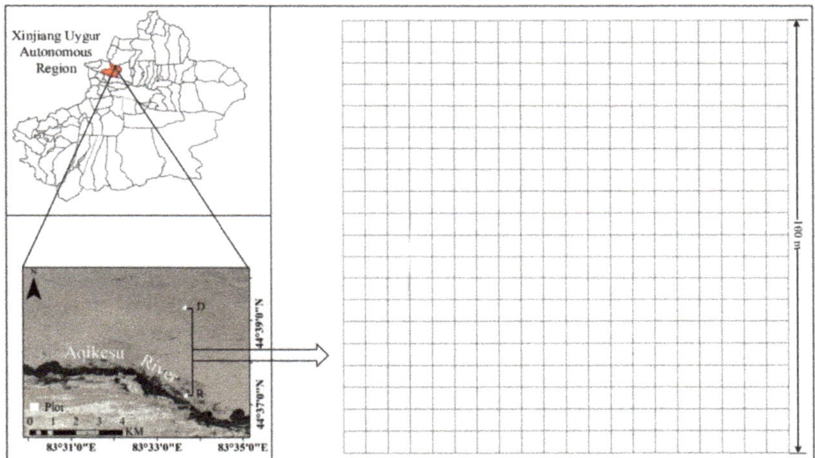

Figure 1. The study area and location of the plots. Note: R and D represent the plant communities of the river bank and desert margin, respectively.

Two large sample plots of 1 hm^2 (100 m × 100 m) (Figure 1) were set up on the river bank and in the desert margin to the north of Aqiksu River near the East Bridge Management and Protection Station in the reserve. Based on a scale of 5 m × 5 m, we studied the community, and each sample plot had 400 quadrats (totaling 800 quadrats). After the establishment of the quadrats, we investigated plant community characteristics in each quadrat, including species composition, species abundance, plant crown, and plant height.

2.2. Collection and Measurement of Soil Samples

Within each sample plot of 5 m × 5 m, the diagonal method was used to select the center point, and 0–20 cm of topsoil was taken in two soil samples. Each soil sample was collected in an aluminum box (the quality of the aluminum box was determined in advance).

The aluminum box was numbered after the sample was collected, and the sample's fresh weight was measured. Then, the sample was taken back to the laboratory and dried in an oven in order to calculate the soil water content (SWC). In addition, undisturbed soil samples were taken and brought back to the laboratory. The soil samples were peeled into small clods with a diameter of about 1 cm, and the visible pebbles and animal and plant residues were picked out. Each sample was then air-dried indoors and, after being mixed evenly, one part (kept the same) was classified as aggregate, and the other part was ground with a 100-mesh nylon sieve and naturally air-dried for later soil index determination.

The collected soil samples were measured in the laboratory, and SWC was determined by the drying method at 105 °C for 48 h to a constant quality [36]. Soil salinity content (SA) was determined by the weight method, soil pH value was determined by potentiometry, soil organic carbon (SOC) was determined by the potassium dichromate dilution heat method, and AN was determined by indophenol blue colorimetry [37]. NN was determined by dual-wavelength ultraviolet spectrophotometry [38]. Total nitrogen (TN) in the soil was determined by the Kjeldahl digestion method [39]. The soil's total phosphorus (TP) and available phosphorus (AP) were determined by the Mo-Sb anti-spectrophotography method [40]. The particle size distribution and stability of soil aggregates were measured by the dry screening method and the wet screening method, respectively [41].

2.3. Calculations for Soil Aggregates and Plant Diversity

2.3.1. Calculations for Soil Aggregates

The following formulas were used to calculate the mean weight diameter (MWD), the percentage of soil aggregates with a particle size larger than 0.25 mm (WSR), and the percentage of aggregate destruction (PAD).

The formula for the mass percentage of aggregate content at all levels is as follows [42,43]:

The mass percentage of aggregates in each grade = the mass of aggregates in this grade/the total mass of soil samples × 100%.

The average mass diameter was calculated using the formula:

$$\mathrm{MWD} = \frac{\sum_{i=1}^{n} X_i W_i}{\sum_{i=1}^{n} W_i}$$

where n is the number of particle size groupings; X_i is the average diameter of this particle size component; and W_i is the mass fraction of this size aggregate.

The percentage of soil aggregates with a grain size greater than 0.25 mm was calculated using the formula:

$$\mathrm{WSR} = \frac{M_{r>0.25}}{M_t}$$

where WSR is the percentage of soil aggregates with a particle size greater than 0.25 mm, Mr > 0.25 is the cumulative mass of aggregates with a particle size greater than 0.25 mm, and Mt is the sum of the masses of aggregates with different particle sizes.

We used the following formula to calculate the aggregate failure rate:

$$\mathrm{PAD} = (W_d - W_w)/W_d$$

where PAD is the aggregate failure rate, W_d is the sum of the mass percentages of aggregates with particle sizes larger than 0.25 mm in the dry screening method, and W_w is the sum of the mass percentages of aggregates with sizes larger than 0.25 mm in the wet screening method.

2.3.2. Calculation of Plant Diversity

The vegan package was used to calculate the weighted average of the plant community species diversity index and plant height. The picante package was used to calculate the phylogenetic richness index. The above-mentioned index calculations were all performed in R version 4.1.1.

2.4. Data Analysis

In R version 4.1.1, the t-test was used to analyze the differences in the proportion and stability of soil aggregates in different vegetation communities, and the cor.test function was used to analyze the correlation between vegetation and soil characteristics and soil stability. In SPSS Amos version 24.0, the structural equation model was used to analyze the relative effects of soil factors (soil water content, salt content, organic carbon, pH, and total nitrogen) and plant characteristics (species richness, phylogenetic richness, plant height, and plant coverage) on the stability of soil aggregates in different plant communities.

3. Results

3.1. Differences in the Proportion and Stability of Soil Aggregates in Different Community Types

A t-test showed that the percentages of soil water-stable large aggregate (>2 mm) and middle aggregate (2–0.25 mm) fractions in the river bank were larger than those in the desert margin, and the percentage of micro-aggregate fractions (<0.25 mm) on the river bank was lower than that on the desert margin (Figure 2). The MWD of the soil and the WSR in the desert margin plant community were significantly smaller than those in the river bank plant community. However, the river bank plant community had a significantly lower aggregate destruction rate than the desert margin plant community (Figure 3).

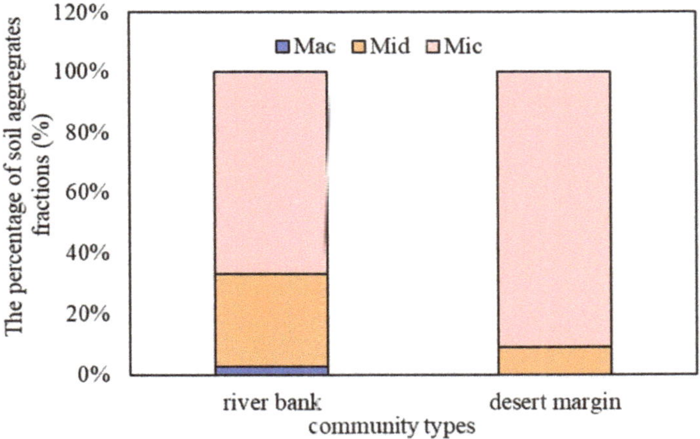

Figure 2. The difference in the soil water-stable aggregate proportion of different communities. Note: Mac, Mid, and Mic represent soil macro-aggregates, intermediate aggregates, and micro-aggregates, respectively.

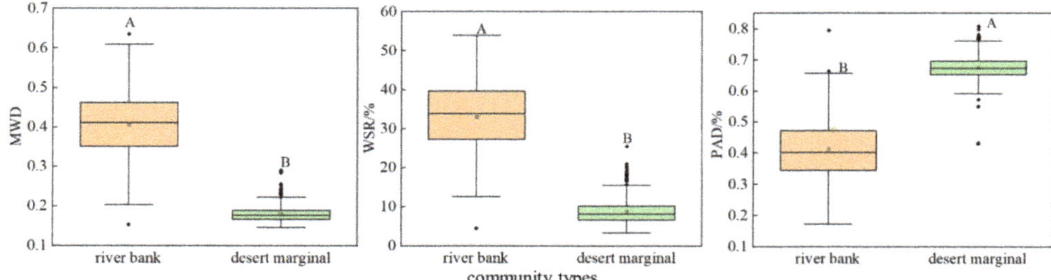

Figure 3. The difference in the soil aggregate stability of different community types. Note: MWD, mean weight diameter; WSR, the percentage of soil aggregates with a particle size larger than 0.25 mm; PAD, the percentage of aggregate destruction. The different letters (A and B) indicate statistically significant differences between the two communities ($p < 0.05$).

3.2. Influencing Factors of Soil Aggregates

Soil physicochemical and vegetation properties (species richness, phylogenetic richness, plant height, etc.) varied in the different communities (Table 1). There were 14 species in the river bank community, with a height range of 4–1350 cm, and 10 species in the desert margin community, with a height range of 3.5–420 cm. The Maglef species richness index of the river bank plant community was significantly larger than that of the desert margin plant community, while the Simpson and Shannon–Weiner indices showed no significant differences. The plant phylogenetic richness was significantly larger in the river bank plant community than in the desert margin plant community. The soil pH, water content, salt content, organic carbon, total nitrogen, total phosphorus, available phosphorus, nitrate nitrogen, and ammonium nitrogen of the river bank plant community were significantly higher than those of the desert margin plant community.

Table 1. The vegetation and soil characteristics.

	River Bank	Desert Margin
Vegetation		
Plant height/cm	4–1350	3.5–420
Species number	14	10
Simpson	1.52 ± 0.03 [a]	1.54 ± 0.03 [a]
Maglef	0.59 ± 0.02 [a]	0.53 ± 0.02 [b]
Shannon–Weiner	0.52 ± 0.02 [a]	0.50 ± 0.02 [a]
Phylogenetic richness	384.24 ± 0.09 [a]	187.82 ± 0.11 [b]
Soil		
pH	8.07 ± 0.02 [a]	7.38 ± 0.02 [b]
SWC %	13.12 ± 0.19 [a]	1.04 ± 0.02 [b]
SA (g/kg)	5.58 ± 0.12 [a]	1 ± 0.02 [b]
SOC (g/kg)	9.57 ± 0.28 [a]	1.35 ± 0.03 [b]
TN (g/kg)	1.31 ± 0.01 [a]	0.59 ± 0.01 [b]
AP (g/kg)	38.19 ± 0.75 [a]	7.96 ± 0.18 [b]
TP (mg/kg)	2.05 ± 0.04 [a]	0.63 ± 0.01 [b]
AN (mg/kg)	2.47 ± 0.06 [a]	1.39 ± 0.03 [b]
NN (mg/kg)	12.51 ± 0.36 [a]	2.76 ± 0.06 [b]

Note: The different letters indicate statistically significant differences between the two plant communities ($p < 0.05$). SWC, SA, SOC, TN, AP, TP, AN, and NN represent soil water content, soil salinity content, soil organic carbon, soil total nitrogen, soil available phosphorus, soil total phosphorus, soil ammonium nitrogen, and soil nitrate nitrogen, respectively.

Correlation analysis showed that the factors that determine the stability of soil aggregates were significantly different among the different vegetation types (Figure 4). Soil aggregates in the river bank community were jointly influenced by soil physical and chemical properties (SWC, salinity, organic matter, total nitrogen, nitrate nitrogen, and soil pH), plant coverage, and plant diversity (plant species richness, plant height, and phylogenetic richness). The intermediate aggregates were not significantly affected by ammonium nitrogen, water, total phosphorus, and soil pH, while soil water, salt, organic matter, total nitrogen, nitrate nitrogen, plant coverage, species richness, plant height, and phylogenetic richness were negatively correlated with micro-aggregate content and aggregate destruction rate and positively correlated with the mass fraction of soil aggregates with a particle size larger than 0.25 mm. The mean weight diameter of soil had no significant correlation with ammonium nitrogen, available phosphorus, total phosphorus, or soil pH. Soil macro-aggregates in desert margin plant communities were significantly positively correlated with soil nitrate nitrogen and vegetation coverage, while soil moisture, total nitrogen, ammonium nitrogen, nitrate nitrogen, plant coverage, species richness, plant height, and phylogenetic richness were significantly positively correlated with the mass fraction of intermediate aggregates and soil aggregates with a particle size >0.25 mm, and significantly negatively correlated with the content of micro-aggregates. The destruction rate of soil aggregates was negatively correlated with species richness, phylogenetic richness, and soil available phosphorus.

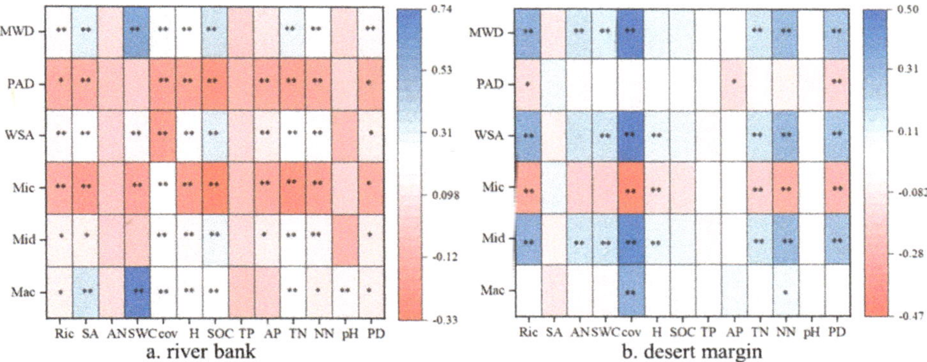

Figure 4. The correlation between vegetation and soil characteristics and soil stability in different communities. Note: MWD, PAD, and WSR represent mean weight diameter, the percentage of soil aggregates with a particle size larger than 0.25 mm, and the percentage of aggregate destruction, respectively. Mac, Mid, and Mic represent soil macro-aggregates, intermediate aggregates, and micro-aggregates, respectively. Ric, SA, AN, SWC, cov, H, SOC, TP, AP, TN, NN, pH, and PD represent species richness, soil salinity, ammonium nitrogen, soil water content, plant coverage, plant height, organic carbon, total phosphorus, available phosphorus, total nitrogen, nitrate nitrogen, soil acidity and alkalinity, and plant phylogenetic richness, respectively. * and ** represent $p < 0.05$ and $p < 0.01$, respectively. The blue grid represents a positive correlation and the red grid represents a negative correlation.

3.3. Effects of Soil Factors and Plant Characteristics on the Stability of Soil Aggregates

The results of the structural equation model show that under different vegetation types, the effects and mechanisms of soil factors and plant characteristics on the stability of soil aggregates were different (Figures 5 and 6). Soil water content, organic carbon, and salt in the river bank community had significant direct positive effects on the mean weight diameter of soil, with path coefficients of 0.50, 0.11, and 0.24, respectively ($p < 0.01$). Water also indirectly affected soil stability by affecting plant height, soil salt, and soil organic carbon. Species richness and vegetation coverage had significant direct positive effects on MWD, with path coefficients of 0.13 and 0.11, respectively ($p < 0.01$). However, the effect of vegetation height on soil stability was not significant. The effect of soil factors on the soil stability of the river bank plant community was significantly greater than that of plant characteristics. In the desert margin plant community, compared with the river bank plant community, the effect of soil factors on soil stability was weaker, and the effect of vegetation characteristics was stronger. Plant coverage and species richness had significant positive effects on soil stability, with path coefficients of 0.43 ($p < 0.001$) and 0.35 ($p < 0.001$), respectively. Phylogenetic richness had a significant direct negative effect on soil stability ($p < 0.05$), with an effect value of -0.27. Phylogenetic richness indirectly affected soil stability by adjusting the plant coverage, with an indirect effect value of 0.23. Moisture, ammonium nitrogen, and nitrate nitrogen all had significant direct positive effects on soil stability, with effect values of 0.12, 0.09, and 0.15, respectively.

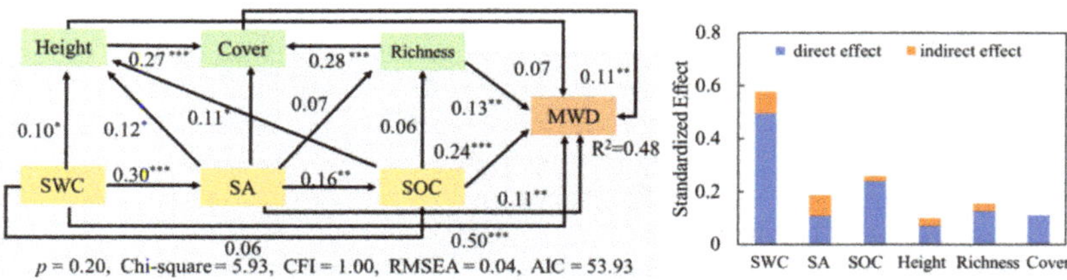

Figure 5. Effects of plant characteristics and soil factors on soil aggregate stability in the river bank. Note: SWC, SA, SOC, Height, Richness, and cover represent soil water content, soil salinity, soil organic carbon, plant height, species richness, and plant coverage, respectively. Black solid lines indicate a positive effect and red dashed lines indicate a negative effect. Values on lines denote the standardized effect size and significance (* $p < 0.05$; ** $p < 0.01$; *** $p < 0.001$).

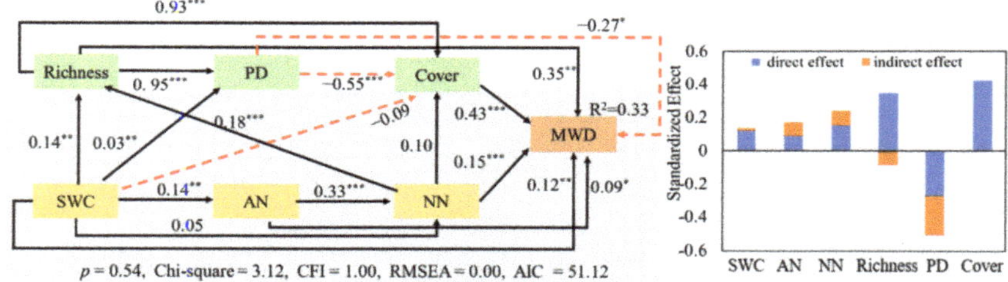

Figure 6. Effects of plant characteristics and soil factors on soil aggregate stability in the desert margin. Note: SWC, AN, NN, Richness, PD, and cover represent soil water content, soil ammonium nitrogen, soil nitrate nitrogen, species richness, phylogenetic richness, and plant coverage, respectively. Black solid lines indicate a positive effect and red dashed lines indicate a negative effect. Values on lines denote the standardized effect size and significance (* $p < 0.05$; ** $p < 0.01$; *** $p < 0.001$).

4. Discussion

4.1. Differences in the Proportion and Stability of Soil Aggregates in Different Community Types

Soil structure stability is an important foundation for the maintenance of soil functions and the growth of animals and plants, and the water stability of soil aggregates is an important index that reflects the stability of the soil structure [26]. Macro-aggregates are the basis for maintaining the stability of the soil structure, and their content can reflect the quality of the soil structure to a certain extent. MWD has been widely used to measure the stability of aggregates, and it is acknowledged that a larger MWD value indicates higher aggregate stability [28]. In this study, the percentage of water-stable aggregates with a particle size >0.25 mm in the river bank was found to be significantly higher than that in the desert margin, and the aggregate destruction rate was lower than that in the desert margin. Although the MWD values in both locations were small, the soil of the river bank showed higher water stability. The reason for this result may be that the soil of the river bank had high soil water content and a large amount of organic matter. Higher soil water content can increase the content of large aggregates that improve the soil stability [10], so the soil stability in the river bank is higher. The reason for this result may be the high degree of vegetation coverage on the river bank, which may promote soil agglomeration by increasing the input of organic matter above and below the ground, reducing the erosion due to wind and rain, and protecting the stability of soil aggregates [44]. It can be seen that the different vegetation types will significantly change the aggregate content, but the particle size and stability of soil aggregates are affected by many factors, such as soil physical and chemical properties, plant characteristics, and climate factors, and show different distribution characteristics. Therefore, further research is needed on the influence of different soil conditions on aggregate properties.

The correlation between SOC content and MWD, WSA, and PAD was significant in the plant community of the river bank, indicating that soil organic carbon content affects the stability of soil aggregates in the river bank (Figure 4). This finding is consistent with the research results of Zhao et al. (2018) and Wang (2019) on soil aggregates and their stability in different plant communities [45,46]. The correlation between SOC content and WSR showed that the soil organic carbon content had a significant influence on the formation of soil aggregates in the river bank, which promoted the formation of large aggregates as SOC is the main cementing material of large aggregates. This is because, in general, large aggregates (>2 mm) can accumulate more carbon than micro-aggregates (<0.25 mm) [1]. Micro-aggregates form large aggregates through organic matter bonding, and an increase in the soil organic carbon content creates favorable conditions for smaller particles in the soil to bond into larger aggregates [13]. The content of intermediate aggregates and MWD were significantly correlated with pH, SWC, NN, TN, and AP, indicating that soil water content, nutrient content, and acid–base level had an important influence on the formation of large aggregates and the stability of the soil structure in the river bank. The relationship observed between soil water-stable aggregates and soil total nitrogen content was consistent with the research results of An et al. This is because the addition of N aided the chemical reaction inside the aggregates to form a stable soil structure, and the smaller aggregates re-cemented to form larger aggregates [47]. The pH mainly affected the decomposition of soil organic matter by affecting the type, quantity, and activity of soil microorganisms [48], and changed the content of soil aggregates. In alkaline soil, with an increase in the pH value, the mineralization intensity gradually weakened, while the nitrification intensity gradually strengthened, which led to a decrease in the amount of ammonium nitrogen generated by mineralization in the soil, and, at the same time, more ammonium nitrogen was converted into nitrate nitrogen. Finally, under alkaline conditions, the amount of ammonium nitrogen in the soil decreased [49], which led to ammonium nitrogen having an insignificant effect on aggregates in the alkaline soil, while the effect of nitrate nitrogen was more significant. Soil available phosphorus in the river bank had a significant effect on the stability of soil aggregates, while soil phosphorus in the desert margin had no significant effect on the soil stability. This may be because a high pH value can enhance the fixation of soil phosphorus, and more phosphorus can promote the aggregation of soil particles [50]. Plant species richness had a significant positive impact on the stability of soil aggregates; for example, a *Robinia pseudoacacia* community with an abundance of species was found to be beneficial for the stability of soil aggregates [51].

4.2. Influence of Soil Factors on the Stability of Soil Aggregates

Water is an important factor affecting soil aggregates. Water can better promote the formation of soil aggregates, while less or excess soil water will destroy the formation of aggregates. At the same time, the formation of aggregates is also beneficial to the maintenance of soil moisture [52]. The results of the two structural equation models show that water had a significant impact on soil aggregates (Figures 5 and 6), but the mechanisms and sizes of action were different, which may have been caused by different water and nutrient levels and microbial environments under different vegetation types. Water is an important factor affecting the soil structure and function in arid areas. A change in soil moisture will also cause a series of changes in plants and microorganisms [53]. Plants and microorganisms are important factors affecting soil aggregates. The effects of various factors on soil aggregates are superimposed, and it is often difficult to distinguish the effects of water alone. Soil water content had a significant direct positive effect on the MWD of soil in the two communities, with path coefficients of 0.50 and 0.12, respectively. In the river bank, soil water also indirectly affected soil stability by affecting soil organic carbon, because water addition can increase the content of soil organic carbon. SOC can enhance the agglomeration of aggregates and promote the formation of an aggregate structure [54,55], which led to the increase in large soil aggregates. Soil moisture in the desert margin affected the soil stability by adjusting the nitrogen content. This may be because soil moisture is also an important factor affecting soil nitrogen mineralization. Soil moisture can regulate the population of nitrifying bacteria. Moreover, in a certain range of soil moisture values, an increase in soil moisture is beneficial to the growth of nitrifying bacteria, but not conducive to the growth of denitrifying bacteria, which makes the amount of ammonium nitrogen that is retained less than the amount of nitrification that occurs, and then increases the content of nitrate nitrogen [56] and promotes the stability of soil aggregates.

In the river bank, salt had significant direct and indirect effects on soil aggregates. The direct mechanism may be that the river bank had relatively high soil salt, and its soil contains more calcium ions and magnesium ions. Ca^{2+} has a strong ion bridge function, which is beneficial to the combination of soil clay and organic matter [57]. The soil salinity had an indirect effect on the soil stability. This may be because the river bank had higher soil salinity, which slowed down the decomposition rate, improved the accumulation and storage of soil organic carbon [58], and promoted the adsorption of more soil particles [58] as the accumulation of organic carbon content promotes the adsorption of soil particles. These two aspects jointly promote the formation and stability of aggregates, so the content of soil aggregates in a riparian forest is significantly positively correlated with salt. Yu et al. (2016) also pointed out that the content of water-stable aggregates in salinized soil increased significantly [59]. Soil salinity is also one of the important factors that control the changes in plant diversity in arid areas, and plant diversity is closely related to the accumulation of soil organic carbon. Therefore, soil salinity can indirectly affect soil aggregates by controlling the decomposition rate of soil organic carbon, vegetation distribution, and plant diversity.

4.3. Effects of Plant Characteristics on the Stability of Soil Aggregates

Soil aggregates are the basic units of the soil structure and an important factor in the maintenance of soil fertility. A stable soil aggregate is beneficial to the balance between soil nutrient retention and release, while the stability of a soil aggregate is positively influenced by the characteristics of plant communities. Plant characteristics are particularly important for soil aggregation in ecosystems that have been seriously disturbed by the external environment [51,60]. Some studies that have reported on how plant characteristics influence soil aggregation emphasized the important roles of root traits, shoot biomass, and niche complementarity [61–63]. In this study, we found that species richness has a positive effect on soil stability. This finding is in line with the insurance hypothesis. Higher biodiversity ensures that the ecosystem is protected from environmental fluctuations and maintains its functions [64]. However, in this study, plant diversity, including species richness and phylogenetic richness, was found to have different impacts on soil stability (Figures 5 and 6). In the correlation analysis, the species richness and phylogenetic richness in the plant communities of the river bank and desert margin had significant impacts on the soil stability; however, in the comprehensive analysis of soil factors and plant characteristics, phylogenetic richness did not enter into the structural equation model. However, phylogenetic richness had a negative effect on soil stability in the desert margin plant community, which may be due to the fact that when soil factors and plant characteristics were integrated, the influence of genealogy on plants was genetically different from the level of conservation of plant characteristics [65], which means that when plants and soil are combined, genealogy may not play a significant role, or it may play an opposing role. Phylogenetic richness in the desert margin plant community had a negative effect on soil stability, which shows that the closer the phylogenetic distance between species is, the more stable the soil structure can be. This means that plants belonging to the same family and genus can promote soil stability to a greater degree than a diversity of plant families and genera. For example, Leguminosae plants have higher rhizosphere microbial biomass than non-Leguminosae plants, thus improving soil stability [66], and the glommycin-related proteins secreted by arbuscular mycorrhizal fungi (AMF) have a 3–10 times stronger ability to adhere soil particles than other sugar substances [67]. Inoculation with rhizobia and mycorrhizal fungi can significantly increase the proportion of macro-aggregates [66]. Studies have also shown that the mechanism by which plants increase the stability of soil aggregates may be their roots [68]. Roots play a key role in soil aggregation because they transform loose soil particles into stable aggregates through root exudates. The fine roots of plants affect soil aggregation through the entanglement of fine root hyphae. The growth of fine roots and hyphae can stimulate microbial activity and promote the formation of large soil aggregates [69].

Most studies regard plant coverage as the key factor in soil stability [70]. A dense plant canopy will increase the surface roughness, act as a windbreak net and sediment trap, intercept raindrops, and reduce the evaporation of soil moisture [71]. In this study, we found that the coverage of plant communities can promote the stability of soil, and the intensity of the effect varies with different communities. This may be due to the increase in the surface vegetation coverage, which can reduce the damage to the soil epidermis and the loss of surface soil and nutrients, increase the organic matter input into the soil by litter, underground roots, and root exudates, and significantly increase the soil organic carbon and total nitrogen content [72–74], which in turn leads to an increase in the number of large soil aggregates. However, our findings indicate that, besides plant coverage, biodiversity may play a key role in the stability of soil aggregates. However, it is difficult to distinguish causality among these correlations. It is hard to say whether soil with high aggregate stability supports more species, or vice versa.

Wind-blown sand deposition is a common phenomenon in arid and semi-arid ecosystems. A large number of field observations and indoor simulation experiments have proved that plants are a necessary condition for wind-blown sand deposition. It has been shown that tufted plants can effectively influence the near-surface airflow and cause sedimentable particles carried in the wind-blown sand flow to be deposited among the plants [75,76]. The main component of the sedimentable particles carried by the wind-blown sand flow is soil clay particles [77]. Plants will lose water during transpiration and other processes, and some of the water may be dispersed among the soil clay particles at the base of plants, where it will act as an adhesive to bind the clay particles together and gradually form large-particle aggregates.

We preliminarily explored the mechanism of the influence of soil and vegetation properties on the stability of soil aggregates in desert plant communities. However, we did not include some of the important factors in this research. Vegetation roots contribute to soil fixation and reinforcement in arid areas, thus improving the soil's resistance to erosion [33]. Studies have shown that microbial diversity can promote soil stability [66,67]. Soil microbial diversity and plant roots also jointly influence soil aggregates. Our future studies on soil aggregates in arid regions will incorporate these two factors, helping us to better understand the plant–soil–microbe relationship.

5. Conclusions

The proportion and stability of soil aggregates in different plant communities were found to be obviously different. Soil physicochemical properties and vegetation properties varied in the different plant communities. Soil water content in the river bank and desert margin had a significant and direct positive effect on the MWD of the soil. The soil moisture also indirectly affected the soil stability by affecting the plant height, soil salinity, and organic carbon in the river bank. Salt and organic carbon had significant direct effects on soil aggregates, and soil salinity was found to indirectly affect soil aggregates by controlling the decomposition rate of soil organic carbon, vegetation distribution, and plant diversity. Phylogenetic richness had a negative effect on soil stability in the desert margin plant community, and plant community coverage had a positive effect on soil stability. The effect's intensity varied with the different communities, and the coverage of the plant communities had a greater effect on the soil stability in the desert margin plant community. In short, both soil properties and plant diversity had a significant impact on the soil stability of desert margin plant communities, but the mechanism of action changed with the type of plant community.

Author Contributions: Conceptualization, L.J. and G.L.; methodology, L.J.; software, L.J. and D.H.; writing—original draft preparation, L.J.; writing—review and editing, L.J.; supervision, G.L. funding acquisition, G.L. All authors have read and agreed to the published version of the manuscript.

Funding: This research was supported by the National Natural Science Foundation of China (42171026) and the Xinjiang Uygur Autonomous Region Graduate Research and Innovation Project (XJ2020G011).

Institutional Review Board Statement: Not applicable.

Informed Consent Statement: Not applicable.

Data Availability Statement: Not available.

Conflicts of Interest: The authors declare that they have no conflicts of interest.

References

1. Yılmaz, E.; Çanakcı, M.; Topakcı, M.; Sönmez, S. Effect of vineyard pruning residue application on soil aggregate formation, aggregate stability and carbon content in different aggregate sizes. *Catena* **2019**, *183*, 104219. [CrossRef]
2. Fialho, R.C.; Zinn, Y.L. Changes in soil organic carbon under eucalyptus plantations in Brazil: A comparative analysis. *Land Degrad. Dev.* **2014**, *25*, 428–437. [CrossRef]
3. Parras-Alcántara, L.; Díaz-Jaimes, L.; Lozano-García, B. Management effects on soil organic carbon stock in Mediterranean open rangelands-treeless grasslands. *Land Degrad. Dev.* **2015**, *26*, 22–34. [CrossRef]
4. Elliott, E.T.; Coleman, D.C. Let the soil work for us. *Ecol. Bull.* **1988**, *39*, 23–32.
5. Feeney, D.S.; Crawford, J.W.; Daniell, T.; Hallett, P.D.; Nunan, N.; Ritz, K.; Rivers, M.; Young, I.M. Three-dimensional microorganization of the soil-root-microbe system. *Microbic. Ecol.* **2006**, *52*, 151–158. [CrossRef] [PubMed]
6. Young, I.M.; Crawford, J.W.; Rappoldt, C. New methods and models for characterizing structural heterogeneity of soil. *Soil Tillage Res.* **2001**, *61*, 33–45. [CrossRef]
7. Bottinelli, N.; Jouquet, P.; Capowiez, Y.; Podwojewski, P.; Grimaldi, M.; Peng, X. Why is the influence of soil macrofauna on soil structure only considered by soil ecologists? *Soil Tillage Res.* **2015**, *146*, 118–124. [CrossRef]
8. An, S.; Mentler, A.; Mayer, H.; Blum, W.E.H. Soil aggregation, aggregate stability, organic carbon and nitrogen in different soil aggregate fractions under forest and shrub vegetation on the Loess Plateau, China. *Catena* **2010**, *81*, 226–233. [CrossRef]
9. Deviren, S.S.; Cornelis, W.M.; Erpul, G.; Gabriels, D. Comparison of different aggregate stability approaches for loamy sand soils. *Appl. Soil Ecol.* **2012**, *54*, 1–6. [CrossRef]
10. Blankinship, J.C.; Fonte, S.J.; Six, J.; Schimela, J.D. Plant versus microbial controls on soil aggregate stability in a seasonally dry ecosystem. *Geoderma* **2016**, *272*, 39–50. [CrossRef]
11. Bedel, L.; Legout, A.; Poszwa, A.; Heijden, G.V.D.; Court, M.; Goutal-Pousse, N.; Montarges-Pelletier, E.; Ranger, J. Soil aggregation may be a relevant indicator of nutrient cation availability. *Ann. For. Sci.* **2018**, *75*, 103. [CrossRef]
12. Egan, G.; Crawley, M.J.; Fornara, D.A. Effects of long-term grassland management on the carbon and nitrogen pools of different soil aggregate fractions. *Sci. Total Environ.* **2018**, *613*, 810–819. [CrossRef]
13. Six, J. A history of research on the link between (micro) aggregates, soil biota, and soil organic matter dynamics. *Soil Tillage Res.* **2004**, *79*, 7–31. [CrossRef]
14. Dimoyiannis, D. Wet aggregate stability as affected by excess carbonate and other soil properties. *Land Degrad. Dev.* **2012**, *23*, 450–455. [CrossRef]
15. Regelink, I.C.; Stoof, C.R.; Rousseva, S.; Weng, L.P.; Lairef, G.J.; Kramg, P.; Nikolaos, P.; Nikolaidish, N.P.; Kerchevaf, M.; Banwartgi, S.; et al. Linkages between aggregate formation, porosity and soil chemical properties. *Geoderma* **2015**, *247*, 24–37. [CrossRef]
16. Duchicela, J.; Sullivan, T.; Bontti, E.; Bever, J.D. Soil aggregate stability increase is strongly related to fungal community succession along an abandoned agricultural field chronosequence in the Bolivian Altiplano. *J. Appl. Ecol.* **2013**, *50*, 1266–1273. [CrossRef]

17. Paul, B.K.; Vanlauwe, B.; Ayuke, F.; Gassner, A.; Hoogmoed, M.; Hurisso, T.T.; Koala, S.; Lelei, D.; Ndabamenye, T.; Six, J.; et al. Medium-term impact of tillage and residue management on soil aggregate stability, soil carbon and crop productivity. *Agric. Ecosyst. Environ.* **2013**, *164*, 14–22. [CrossRef]

18. Onweremadu, E.U.; Onyia, V.N.; Anikwe, M. Carbon and nitrogen distribution in water-stable aggregates under two tillage techniques in Fluvisols of Owerri area, southeastern Nigeria. *Soil Tillage Res.* **2007**, *97*, 195–206. [CrossRef]

19. Reich, P.B.; Tilman, D.; Isbell, F.; Mueller, K.; Hobbie, S.E.; Flynn, D.F.B.; Eisenhauer, N. Impacts of biodiversity loss escalate through time as redundancy fades. *Science* **2012**, *336*, 589–592. [CrossRef]

20. Peres, G.; Cluzeau, D.; Menasseri, S.; Soussana, J.F.; Bessler, H.; Engels, C.; Habekost, M.; Gleixner, G.; Weigelt, A.; Weisser, W.W.; et al. Mechanisms linking plant community properties to soil aggregate stability in an experimental grassland plant diversity gradient. *Plant Soil* **2013**, *373*, 285–299. [CrossRef]

21. Eisenhauer, N.; Lanoue, A.; Strecker, T.; Scheu, S.; Steinauer, K.; Thakur, M.P.; Mommer, L. Root biomass and exudates link plant diversity with soil bacterial and fungal biomass. *Sci. Rep.* **2017**, *7*, 44641. [CrossRef]

22. Stokes, A.; Atger, C.; Bengough, A.G.; Fourcaud, T.; Sidle, R.C. Desirable plant root traits for protecting natural and engineered slopes against landslides. *Plant Soil* **2009**, *324*, 1–30. [CrossRef]

23. Sher, Y.; Baker, N.R.; Herman, D.; Fossum, C.; Hale, L.; Zhang, X.X.; Nuccio, E.; Saha, M.; Zhou, J.H.; Pett-Ridge, J. Microbial extracellular polysaccharide production and aggregate stability controlled by switchgrass (*Panicum virgatum*) root biomass and soil water potential. *Soil Biol. Biochem.* **2020**, *143*, 107742. [CrossRef]

24. Zhao, D.; Xu, M.; Liu, G.; Ma, L.; Zhang, S.; Xiao, T.; Peng, G. Effect of vegetation type on microstructure of soil aggregates on the Loess Plateau, China. *Agric. Ecosyst. Environ.* **2017**, *242*, 1–8. [CrossRef]

25. Zeng, Q.; Darboux, F.; Man, C.; Zhu, Z.; An, S. Soil aggregate stability under different rain conditions for three vegetation types on the Loess Plateau, China. *Catena* **2018**, *167*, 276–283. [CrossRef]

26. Bronick, C.J.; Lal, R. Soil Structure and Management: A Review. *Geoderma.* **2005**, *124*, 3–22. [CrossRef]

27. Liang, C.; Schimel, J.P.; Jastrow, J.D. The importance of anabolism in microbial control over soil carbon storage. *Nat. Microbiol.* **2017**, *2*, 17105. [CrossRef] [PubMed]

28. Sarker, T.C.; Incerti, G.; Spaccini, R.O.; Piccolo, A.; Mazzoleni, S.; Bonanomi, G. Linking organic matter chemistry with soil aggregate stability: Insight from 13 C NMR spectroscopy. *Soil Biol. Biochem.* **2018**, *117*, 175–184. [CrossRef]

29. Demenois, J.; Carriconde, F.; Rey, F.; Stokes, A. Tropical plant communities modify soil aggregate stability along a successional vegetation gradient on a Ferralsol. *Ecol. Eng.* **2017**, *109*, 161–168. [CrossRef]

30. Dou, Y.; Yang, Y.; An, S.; Zhu, Z. Effects of different vegetation restoration measures on soil aggregate stability and erodibility on the Loess Plateau, China. *Catena* **2019**, *185*, 104294. [CrossRef]

31. Whitford, W.G. *Ecology Desert Systems*; Academic Press: San Diego, CA, USA, 2002.

32. Yang, H.; Wang, J.; Zhang, F. Soil aggregation and aggregate-associated carbon under four typical halophyte communities in an arid area. *Environ. Sci. Pollut. Res.* **2016**, *23*, 1–10. [CrossRef]

33. Abdi, E.; Saleh, H.R.; Majnonian, B.; Deljouei, A. Soil fixation and erosion control by *Haloxylon persicum* roots in arid lands, Iran. *J. Arid. Land* **2019**, *11*, 86–96. [CrossRef]

34. Gong, Y.; Lv, G.H.; Guo, Z.J.; Chen, Y. Influence of aridity and salinity on plant nutrient s scales up from species to community level in a desert ecosystem. *Sci. Rep.* **2017**, *7*, 6811. [CrossRef]

35. Yang, X.D.; Zhang, X.N.; Lv, G.H.; Arshad, A. Linking *Populus euphratica* hydraulic redistribution to diversity assembly in the arid desert zone of Xin jiang, China. *PLoS ONE* **2014**, *9*, e109071.

36. Zhang, Z.S.; Dong, X.J.; Xu, B.X.; Dong, X.J.; Zhang, Z.S.; Gao, Y.H; Hu, Y.G.; Huang, L. Soil respiration sensitivities to water and temperature in a revegetated desert. *J. Geophys. Res. Biogeosci.* **2015**, *120*, 773–787. [CrossRef]

37. Bao, S.D. *Soil and Agricultural Chemistry Analysis*, 3rd ed.; China Agriculture Press: Beijing, China, 2000.

38. Norman, R.; Edberg, J.; Stucki, J. Determination of Nitrate in Soil Extracts by Dual-wavelength Ultraviolet Spectrophotometry. *Soil Sci. Soc. Am. J.* **1985**, *49*, 1182–1185. [CrossRef]

39. Dalai, R.C.; Sahrawat, K.L.; Myers, R.J.K. Inclusion of nitrate and nitrite in the Kjeldahl nitrogen determination of soils and plant materials using sodium thiosulphate. *Commun. Soil Sci. Plant Anal.* **1984**, *15*, 1453–1461. [CrossRef]

40. Li, Q.; Song, X.; Chang, S.X.; Peng, C.G.; Xiao, W.F.; Zhang, G.B.; Xiang, W.H.; Li, Y.; Wang, W.F. Nitrogen depositions increase soil respiration and decrease temperature sensitivity in a Moso bamboo forest. *Agric. For. Meteorol.* **2019**, *268*, 48–54. [CrossRef]

41. Zhao, Q.G. *Nanjing Institute of Soil, Chinese Academy of Sciences Soil Physical and Chemical Analysis*; Shanghai Science and Technology Press: Shanghai, China, 1983; pp. 62–126.

42. Zhao, J.; Chen, S.; Hu, R.; Li, Y. Aggregate stability and size distribution of red soils under different land uses integrally regulated by soil organic matter, and iron and aluminum oxides. *Soil Tillage Res.* **2017**, *167*, 73–79. [CrossRef]

43. Mengke, Z.; Siqian, Y.; Shenghao, A.; Xiaoyan, A.; Xue, J.; Jiao, C.; Ruirui, L.; Yingwei, A. Artificial soil nutrient, aggregate stability and soil quality index of restored cut slopes along altitude gradient in southwest China. *Chemosphere* **2020**, *246*, 125687. [CrossRef]

44. Mulumba, L.N.; Lal, R. Mulching effects on selected soil physical properties. *Soil Tillage Res.* **2008**, *98*, 106–111. [CrossRef]

45. Zhao, Y.P.; Meng, M.J.; Zhang, J.C.; Ma, J.Y.; Liu, S.L. Study on the composition and stability of soil aggregates of the main forest stands in Fengyang Mountain, Zhejiang Province. *J. Nanjing For. Univ.* **2018**, *42*, 84–90.

46. Wang, X.Y.; Zhou, C.; Feng, W.H.; Zhang, Y.; Cheng, J.L.; Jiang, X.H. Changes of soil aggregates and organic carbon in Chinese Fir Plantation with different forest ages. *J. Soil Water Conserv.* **2019**, *33*, 126–131.

47. Chen, H.; Ma, W.M.; Zhou, Q.P.; Yang, Y.; Liu, C.W.; Liu, J.Q.; Du, Z.M. Shrub encroachement effects on the stability of soil aggregates and the differeneiation of Fe and Al oxides in Qinghai-Tibet alpine grassland. *Acta Pratacult. Sin.* **2020**, *29*, 73–84.

48. Dai, W.H.; Huang, Y.; Wu, L.; Yu, J. Relationship between soil organic matter content and pH in topsoil of zone soils in China. *Acta Pedol. Sin.* **2009**, *46*, 851–860.

49. Wang, X.X.; Dong, S.K.; Gao, Q.Z.; Zhang, Y.; Hu, G.Z.; Luo, W.R. The rate of soil nitrogen transformation decreased by the degradation of alpine grassland in Qinghai Tibet Plateau. *Acta Pratacult. Sin.* **2018**, *27*, 1–9.

50. Liao, J.J.; Huang, B.; Sun, W.X.; Zou, Z.; Su, J.P.; Ding, F.; Huang, Y. Spatio-Temporal variation of soil available phosphorus and its influencing factors—Acasestudy of rugao county, Jiangsu province. *Acta Pedol. Sin.* **2007**, *4*, 620–628.

51. Pohl, M.; Stroude, R.; Buttler, A.; Rixen, C. Functional traits and root morphology of alpine plants. *Ann. Bot.* **2011**, *108*, 537–545. [CrossRef]

52. Leffelaar, P.A. Water movement, oxygen supply and biological processes on the aggregate scale. *Geoderma* **1993**, *57*, 143–165. [CrossRef]

53. He, X.L.; Gao, L.; Zhao, L. Effects of AM fungi on the growth and drought resistance of Seriphidium minchünense under water stress. *Acta Ecol. Sin.* **2011**, *31*, 1029–1037.

54. Pulleman, M.M.; Marinissen, J. Physical protection of mineralizable C in aggregates from long-term pasture and arable soil. *Geoderma* **2004**, *120*, 273–282. [CrossRef]

55. Liu, W.T.; Wang, T.L.; Zhang, S.; Ding, L.J.; Lv, S.J.; Wei, Z.J. Effects of grazing on edificators and soil aggregate characteristics in Stipa breviflora desert steppe. *Ecol. Environ. Sci.* **2017**, *26*, 978–984.

56. Chen, F.S.; Yu, K.; Gan, L.; Liu, Y.; Hu, X.F.; Ge, G. Effects of temperature moisture and forests succession on nitrogen mineralization in hillside red soils in mid-subtropical region China. *Chin. J. Appl. Ecol.* **2009**, *20*, 1529–1535.

57. Totsche, K.U.; Amelung, W.; Gerzabek, M.H.; Guggenberger, G.; Klumpp, E.; Knief, C.; Lehndorff, E.; Mikutta, R.; Peth, S.; Prechtel, A. Microaggregates in soils. *J. Plant Nutr. Soil Sci.* **2018**, *181*, 104–136. [CrossRef]

58. Chambers, L.G.; Osborne, T.Z.; Reddy, K.R. Effect of salinity-altering pulsing events on soil organic carbon loss along an intertidal wetland gradient: A laboratory experiment. *Biogeochemistry* **2013**, *115*, 363–383. [CrossRef]

59. Yu, H.Y.; Li, T.X.; Zhou, J.M. Secondary Salinization of Greenhouse Soil and Its Effects on Soil Properties. *Soils* **2005**, *6*, 581–586.

60. Erktan, A.L.; Cécillon, L.; Graf, F.; Roumet, C.; Legout, C.; Rey, F. Increase in soil aggregate stability along a Mediterranean successional gradient in severely eroded gully bed ecosystems: Combined effects of soil, root traits and plant community characteristics. *Plant Soil* **2016**, *398*, 121–137. [CrossRef]

61. Bardgett, R.D.; Mommer, L.; Vries, F. Going underground: Root traits as drivers of ecosystem processes. *Trends Ecol. Evol.* **2014**, *29*, 692–699. [CrossRef]

62. Gould, I.J.; Quinton, J.N.; Weigelt, A.; Deyn, G.B.D.; Bardgett, R.D.; Seabloom, E. Plant diversity and root traits benefit physical properties key to soil function in grasslands. *Ecol. Lett.* **2016**, *19*, 1140–1149. [CrossRef]

63. Liu, R.; Zhou, X.; Wang, J.; Shao, J.; Fu, Y.; Liang, C.; Yan, E.; Chen, X.; Wang, X.; Bai, S.H. Differential magnitude of rhizosphere effects on soil aggregation at three stages of subtropical secondary forest successions. *Plant Soil* **2019**, *436*, 365–380. [CrossRef]

64. Yachi, S.; Loreau, M. Biodiversity and ecosystem productivity in a fluctuating environment: The insurance hypothesis. *Proc. Natl. Acad. Sci. USA* **1999**, *96*, 1463–1468. [CrossRef]

65. Winter, M.; Devictor, V.; Schweiger, O. Phylogenetic diversity and nature conservation: Where are we? *Trends Ecol. Evol.* **2013**, *28*, 199–204. [CrossRef]

66. Haynes, R.J.; Beare, M.H. Influence of six crop species on aggregate stability and some labile organic matter fractions. *Soil Biol. Biochem.* **1997**, *29*, 1647–1653. [CrossRef]

67. Wright, S.F.; Upadhyaya, A. Extraction of an abundant and unusual protein from soil and comparison with hyphal protein of Arbuscular Mycorrhizal Fungi. *Soil Sci.* **1996**, *161*, 575–586. [CrossRef]

68. Pohl, M.; Alig, D.; Krner, C.; Rixen, C. Higher plant diversity enhances soil stability in disturbed alpine ecosystems. *Plant Soil* **2009**, *324*, 91–102. [CrossRef]

69. Liu, J.Y.; Zhou, Z.C.; Su, X.M. Review of the mechanism of root systems on the formation of soil aggregate. *J. Soil Water Conserv.* **2020**, *34*, 267–273, 298.

70. Bird, S.B.; Herrick, J.E.; Wander, M.M.; Murray, L. Multi-scale variability in soil aggregate stability: Implications for understanding and predicting semi-arid grassland degradation. *Geoderma* **2007**, *140*, 106–118. [CrossRef]

71. Zuazo, V.H.D.; Pleguezuelo, C.R.R. Soil-erosion and run off prevention by plant covers. A review. *Agron. Sustain. Dev.* **2008**, *28*, 65–86. [CrossRef]

72. Agata, K.; Renée, M.B.; Rudy, V.D.; Kotowski, W. Species trait shifts in vegetation and soil seed bank during fen degradation. *Plant Ecol.* **2010**, *206*, 59–82.

73. Lu, Q.; Ma, H.; Yu, H.; Wang, L.; Shen, Y.; Xu, D.M.; Xie, Y.Z. Effects of rotational grazing methods on soil aggregates and organic carbon characteristics in desert steppe. *Chin. J. Appl. Ecol.* **2019**, *30*, 3028–3038.

74. Zhang, S.J.; He, X.B.; Bao, Y.H.; Tang, Q. Change characteristics of soil aggregates at different water levels in the water-level fluctuation zone of the three Gorges reservoir. *Res. Soil Water Conserv.* **2021**, *28*, 25–30.

75. Du, H.D.; Jiao, J.Y.; Jia, Y.F.; Wang, N.; Wang, D.L. Phytogenic mounds of four typical shoot architecture species at different slope gradients on the Loess Plateau of China. *Geomorphology* **2013**, *193*, 57–64. [CrossRef]

Forests **2022**, *13*, 368

76. Pastrán, G.; Carretero, E.M. Phytogenic Mounds (Nebkhas): Effect of Tricomaria usillo on Sand Entrapment in Central-West of Argentina. *J. Geogr. Inf. Syst.* **2016**, *8*, 429–437.
77. Xiao, H.L.; Zhang, J.X.; Li, J.H. Dustfall Particle size and sedimentation rate at the Souther edge of Tengger Desert. *J. Desert Res.* **1997**, *17*, 127–132.

forests

Article

Effects of Salinity and Oil Contamination on the Soil Seed Banks of Three Dominant Vegetation Communities in the Coastal Wetland of the Yellow River Delta

Zhaoyang Fu [1], Xiuli Ge [1,*], Yongchao Gao [2,*], Jian Liu [3], Yuhong Ma [4], Xiaodong Yang [5] and Fanbo Meng [2]

1   Department of Environmental Science and Technology, Shandong Academy of Science, Qilu University of Technology, Jinan 250353, China; zhaoyangfu1122@gmail.com
2   Shandong Provincial Key Laboratory of Applied Microbiology, Ecology Institute, Shandong Academy of Sciences, Qilu University of Technology, 28789 East Jingshi Road, Jinan 250103, China; mengfanbo1010@126.com
3   Environment Research Institute, Shandong University, Qingdao 266237, China; ecology@sdu.edu.cn
4   Shandong Institute of Geological Survey, Jinan 250014, China; ddymayuhong@shandong.cn
5   Department of Geography and Spatial Information Technology, Ningbo University, Ningbo 315211, China; yangxiaodong@nbu.edu.cn
*   Correspondence: gexiuli@qlu.edu.cn (X.G.); gaoyc@sdas.org (Y.G.); Tel.: +86-137-0541-7155 (X.G.); +86-137-0640-5409 (Y.G.)

Abstract: In view of the important role of vegetation in the integrity of structures and functions of coastal wetland ecosystems, the restoration of degraded coastal wetland vegetation has attracted increased attention. In this paper, the newborn coastal wetland in the Yellow River Delta (YRD) of China was selected to research the effect of salinity and oil exploitation on the germination of soil seed banks of three dominant vegetation communities. The germination experiment with three concentration gradients of NaCl and three concentration gradients of diesel treatments showed that there were 14 species present in the soil seed bank of the multi-species community: three species in the *Phragmites australis* community, and five species in the *P. australis—Suaeda glauca* community. The species in the seed bank of the three communities were much richer than the above-ground vegetation in this study. Soil salinity had a significant inhibitory effect on the seedling numbers of germinated species, the seedling density, and the species diversity of the soil seed banks, while the inhibitory effect of diesel was indistinctive under the designed concentrations. There existed significant interactions between the vegetation community type and soil salinity on the number of germinated species, the seedling density, and the Margalef index. Soil salinity is considered an important factor for wetland vegetation restoration in the YRD, but its effect had species-specific differences. Soil seed banks of the present three communities could be used to promote the restoration of degraded wetlands within certain soil salinity and oil concentration ranges.

Keywords: wetlands; Yellow River Delta; soil seed bank; restoration; oil contamination; salinity

Citation: Fu, Z.; Ge, X.; Gao, Y.; Liu, J.; Ma, Y.; Yang, X.; Meng, F. Effects of Salinity and Oil Contamination on the Soil Seed Banks of Three Dominant Vegetation Communities in the Coastal Wetland of the Yellow River Delta. *Forests* **2022**, *13*, 615. https://doi.org/10.3390/f13040615

Academic Editor: Pablo Ferrandis Gotor

Received: 16 February 2022
Accepted: 12 April 2022
Published: 14 April 2022

Publisher's Note: MDPI stays neutral with regard to jurisdictional claims in published maps and institutional affiliations.

1. Introduction

Coastal wetlands are an important ecological barrier between the sea and the land [1]. However, wetland areas around the world have degraded by approximately 87% since 1700 [2]. The total area of China's wetlands larger than 1 km^2 decreased from 3.85×10^5 km^2 in the First National Inventory of Wetland Resources (1999–2001) to 3.51×10^5 km^2 in the Second National Inventory of Wetland Resources (2009–2011), and the natural wetland accounted for 99.4% of the total degraded wetland area [3]. Pollution, excessive exploitation of biological resources, reclamation, bio-invasion, and infrastructure construction are considered the main drivers of wetland degradation [4]. Many wetland-dependent species, including 21% of bird species, 37% of mammal species, and 20% of freshwater fish species, are either extinct or globally threatened [5]. The loss of area and biodiversity has had a large adverse impact on the integrity of structures and functions of wetland ecosystems [6].

A soil seed bank is the sum of all surviving seeds present on the soil surface and in the soil. Seeds can spread in soils over time to maintain species populations in the face of harsh environments [7]. The bank has the potential to restore the vegetation of damaged wetlands and regain some species that have disappeared from the surface vegetation [8–10]. The soil seed bank has been regarded as an important source of regenerative material in wetland restoration [11–13]. The assessment of the restoration potential and the impact of environmental conditions on the seed bank have great importance for the vegetation restoration of degraded wetlands. Although most studies are focused on seed germination potential [14], growth inhibition effects of contaminants [15], or the relationships between the soil seed bank and above-ground vegetation [16], little information is available regarding the effects of intense human disturbances on wetland soil seed banks in coastal regions.

The coastal wetland in the Yellow River Delta (YRD) is an important site for birds to migrate, overwinter and breed in Northeast Asia and the Pacific Rim. The natural wetland area of the YRD region decreased by 991.1 km^2 from 1986 to 2008 due to economic development and human activities [17]. The YRD region is also the third-largest oil field in China, the average total petroleum hydrocarbon (TPH) concentration is more than 2100 mg kg^{-1} in the soil surrounding the oil wells due to long term oil exploitation [18,19]. The light fraction of the crude oil is toxic to the growth of vegetation and the germination of seeds, but the toxic effects vary greatly and are species-specific [15,20]. However, the effects of crude oil contamination on seed banks in coastal wetlands are not well studied. Moreover, due to the large evaporation and precipitation ratio, the soil salinity in the YRD region is very serious, with a mean soil salt content of approximately 10.45 g kg^{-1} [21]. High salinity has a serious inhibitory effect on the germination of seeds and the growth of vegetation [22]. However, little is known about the seeds' germination under the dual environmental stresses of oil contamination and salinity. In the present study, we measured the above-ground vegetation characteristics of three dominant wetland plant communities in the YRD region; analyzed the soil's physical and chemical parameters; and evaluated the germination ability of the sampled soil seed banks under different salinity and diesel concentration conditions. We hypothesize that: (1) the inhibitory effects of diesel and salinity on the germination of different seeds vary; and (2), there exists an interaction between diesel and salinity on the germination of soil seed banks. The aim of this study was to evaluate the potential of soil seed banks for vegetation restoration in the degraded wetlands of the YRD.

2. Materials and Methods

2.1. Site Description

The present study site is located in the Yellow River Delta Nature Reserve (YRDNR), Dongying City, Shandong Province, China. It is located in the warm temperate zone and belongs to the temperate semi-humid continental monsoon climate zone. The annual average temperature of the YRD is 11.7–12.6 °C; the frost-free period is 211 days; the annual average precipitation is 530–630 mm, 70% of which is distributed in summer. The annual evaporation is about 1962 mm. The vegetation coverage is about 53.7%, and the main plants, such as *Phragmites australis* and *Suaeda glauca* are saline-tolerant coastal wetland species [23].

2.2. Sampling Method

Three dominant vegetation communities named: the multi-species community, *P. australis* community, and *P. australis—S. glauca* community, were selected based on the above-ground vegetation in the wetlands of the YRD (Table 1). For each community type, four replicates with an area of 50 m^2 were selected as the sampling plots (Table 1). In each sampling plot, 5 sampling sites with a size of 1 m × 1 m were set randomly. The above-ground vegetation communities were investigated firstly to record the species of plants in each site. Then, 6 soil cores (Φ = 10 cm) of the surface layer soil (0–10 cm) were sampled randomly at each sample site. All 30 soil cores were mixed as one seed bank soil sample. Therefore, there were 12 seed bank soil samples altogether. At the same time, in each sampling plot, soil with a depth of 0–10 cm was sampled at the four corners and at the center and mixed together to determine the soil's physical and chemical parameters of the sampling plot. Therefore, there were 12 soil samples collected for the following physical and chemical parameters test altogether.

Table 1. Geographic information of the sampling plots in the Yellow River Delta.

Community Type	Code of Sampling Plots	Longitude and Latitude of The Sampling Plots
Multi-Species Community	1	37°44′16.46″ N, 119°9′38.05″ E
	2	37°44′13.44″ N, 119°9′46.43″ E
	3	37°44′4.18″ N, 119°10′6.48″ E
	4	37°43′50.69″ N, 119°10′53.69″ E
Phragmites Australis Community	1	37°45′47.33″ N, 119°3′16.76″ E
	2	37°45′49.92″ N, 119°4′35.92″ E
	3	37°45′37.36″ N, 119°5′17.1″ E
	4	37°45′35.67″ N, 119°5′22.05″ E
Phragmites Australis–Suaeda Glauca Community	1	37°44′52.99″ N, 119°7′38.76″ E
	2	37°44′58.62″ N, 119°7′58.62″ E
	3	37°45′48.58″ N, 119°3′1.51″ E
	4	37°45′50.84″ N, 119°4′6.43″ E

2.3. Analysis of Soil Physical and Chemical Parameters

The soil samples were air-dried and passed through a 2-mm sieve prior to analysis of the soil's physical and chemical parameters. The pH and electrical conductivity (EC) were determined in a 1:5 sample/water mixture by using a SevenExcellence^{TM} S-470K multifunction pH meter (Mettler Toledo, Schwerzenbach, Switzerland) at 25 °C after shaking for 30 min. Soil total carbon (TC), total organic matter (TOM), total nitrogen (TN), available nitrogen (AN), and available phosphorus (AP), were analyzed at the laboratory of Suez NWS Limited, China. The analysis methods include TC: ISO 10694-1995; TOM: NY/T 1121.6-2006; TN: APHA 4500 N_{org}/NO_3; AN: NH_3-N: APHA 4500 NH_3-H; NO_3-N: APHA 4500 NO_3-I; NO_2-N: APHA 4500 NO_2-I; AP: HJ 704-2014.

2.4. Soil Seed Germination Experiment

Firstly, the broken plant roots, leaf litter, and stones in the soil seed bank samples were picked out. Then, the soil seed bank samples were sieved (0.178 mm) and washed using distilled water to remove the soil. Finally, the treated seed bank samples were sealed in separate bags and stored in a refrigerator at 4 °C for 1 week for vernalizing. To investigate the effects of salinity and oil contamination on seed germination, an experiment with three concentration gradients of NaCl (0%, 1%, and 2%) and three concentration gradients of diesel (0%, 1%, and 2%) were designed, there were nine treatments altogether, and each treatment had four replicates. The seed bank germination experiment was carried out in plastic germination boxes the size of 12 cm × 12 cm × 5 cm. According to the design, different amounts of diesel and NaCl were weighed out separately according to the weight of each soil seed bank sample. The sterilized commercial nutrient soil was dissolved with the weighed NaCl and the same amount of distilled water. The weighed NaCl and diesel were then added to the nutrient soil and mixed thoroughly. Each seed bank sample was mixed thoroughly and divided equally into four portions and spread on the surface of the treated commercial nutrient soil with a thickness of ≤1 cm, respectively. The treated commercial nutrient soil was spread in the box to a height of 2–3 cm. The germination experiment was carried out at room temperature in a sunny greenhouse, and the soil in the germination box was kept moist with regular replenishment of water. The counting of germinated seedlings and the species identification were carried out regularly in the following 2 months.

2.5. Data Analysis

2.5.1. Diversity and Similarity Indicators Calculation of the Soil Seed Bank and Vegetation Community

The biodiversity indicators, such as the Margalef index, Simpson index, and the Shannon–Wiener index, of the soil seed bank, were characterized using the following Equations (1)–(3).

$$\text{Margalef index} = \frac{S-1}{\ln N} \tag{1}$$

$$\text{Simpson index} = 1 - \sum P_i^2 \tag{2}$$

$$\text{Shannon–Wiener index} = -\sum (P_i \times \ln P_i) \tag{3}$$

$$P_i = \frac{N_i}{N} \tag{4}$$

where S is the total number of species in the vegetation community, N is the total number of all species, and N_i is the number of individuals in species i.

The Sørensen index was used to characterize the species similarity between the above-ground vegetation community and the below-ground seed bank Formula (5).

$$\text{Sorensen Index} = \frac{2j}{a+b} \tag{5}$$

where a is the number of species present in the above-ground vegetation community, b is the number of species present in the seed bank, and j is the number of species present in both the vegetation community and its seed bank.

2.5.2. Statistical Analysis of Data

One-way ANOVA was used to analyze the physical and chemical properties of the sampled soil, respectively (SPSS Version 18.0, SPSS Inc., Chicago, IL, USA). The difference in the AP, TC, TN, AN, TOM, EC, pH, and C/N ratio was performed using Duncan's multiple comparison test at the significance level of $p = 0.05$. Three-way ANOVAs followed by Duncan's multiple range test ($p = 0.05$) were performed to identify the differences among the germination characteristics of the soil seed banks, vegetation community types, and the environmental stress conditions based on the normal distribution of the variables (SPSS Version 18.0, SPSS Inc., Chicago, IL, USA). The germination characteristics of the soil seed banks included the number of germinated species, number of seedlings, the Margalef index, Simpson index, and the Shannon–Wiener index. The species-level analysis was limited to the plant species (seedling numbers > 20).

3. Results

3.1. Soil Physical and Chemical Parameters of the Three Dominant Vegetation Communities in the YRD

The main environmental parameters of the soil in the three dominant vegetation communities are presented in Table 2. The TC concentration of the soil in the multi-species community was significantly higher than the *P. australis–S. glauca* community. Although the TN concentration of the soil in the *P. australis–S. glauca* community was significantly higher than the multi-species community and the *P. australis* community. In addition, the C/N ratio of the soil in the multi-species community was significantly higher than the *P. australis–S. glauca* community. The EC in the soil of the multi-species community was significantly lower than the *P. australis* community and the *P. australis–S. glauca* community.

There were no significant differences between the TOM, AN, AP, and pH, among the three vegetation communities.

Table 2. Soil physical and chemical parameters of the three dominant vegetation communities in the YRD.

	Multi-Species Community	*Phragmites Australis* Community	*Phragmites Australis–Suaeda glauca* Community
AP (mg/kg)	2.30 ± 0.41 a	1.75 ± 0.21 a	1.80 ± 0.11 a
TC (g/kg)	13.03 ± 0.81 b	11.53 ± 0.29 ab	10.90 ± 0.20 a
TN (mg/kg)	743.75 ± 90.32 a	775.75 ± 76.45 a	1008.75 ± 27.48 b
AN (mg/kg)	24.00 ± 0.82 a	25.25 ± 1.97 a	24.50 ± 0.29 a
TOM (g/kg)	8.20 ± 0.74 a	9.30 ± 0.38 a	8.38 ± 0.23 a
EC (µS/cm)	557.18 ± 260.70 a	4303.28 ± 842.28 b	3658.63 ± 649.75 b
pH	7.99 ± 0.04 a	7.91 ± 0.04 a	7.90 ± 0.03 a
C/N ratio	18.17 ± 1.95 b	15.31 ± 1.26 ab	10.83 ± 0.40 a

Notes: The variables: mean $\pm$ S. E.; AP: available phosphorus; TC: total carbon; TN: total nitrogen; AN: available nitrogen; TOM: total organic matter; EC: electrical conductivity; C/N ratio: ratio of total carbon and total nitrogen; different letters (a, b) within the row indicate a significant difference at $p = 0.05$ (Duncan test, $N = 4$).

3.2. The Main Plant Species Presented in the Three Dominant Soil Seed Banks and the Above-ground Vegetation Communities

There were 14 species present in the multi-species community, 3 species present in the *P. australis* community, and 5 species present in the *P. australis–S. glauca* community (Table 3). *P. australis*, *Artemisia mongolica*, and *Setaria viridis* were the three dominant species in the multi-species community. *P. australis* was the monodominant species in the *P. australis* community. *P. australis* and *S. glauca* were the two dominant species in the *P. australis–S. glauca* community. The species composition had significant differences among the three dominant vegetation communities in the soil seed germination experiment (Table 4). In addition, the seedling density in the multi-species community, the *P. australis* community, and the *P. australis–S. glauca* community were $19.60 \pm 5.90 \times 10^3$ seedlings m^{-2}, $3.01 \pm 0.55 \times 10^3$ seedlings m^{-2}, and $3.85 \pm 0.60 \times 10^3$ seedlings m^{-2}, separately, under 0 salinity and 0 diesel concentration germination condition (Table 4).

The survey of the above-ground community species showed that there were more plant species in the soil seed banks than in the above-ground communities. The Sørensen similarity index of the above-ground species and the soil seed bank species was significantly higher in the *P. australis–S. glauca* community (0.75 ± 0.17) than the multi-species community (0.55 ± 0.08) and the *P. australis* community (0.54 ± 0.09) (Table 4).

3.3. Effects of Salinity and Diesel Contamination on the Germination of Seeds

The result of the seedling density of all species under different salinity and diesel treatment has been shown in Table 3. The inhibitory effects of salinity and diesel on the germination of the seeds increased with the increase in the concentration of salinity and diesel (Figure 1). Furthermore, the number of germinated species, seedling density, and species diversity (Margalef index, Simpson index, Shannon–Wiener index) had a marked difference among the three dominant vegetation communities (Table 5). Soil salinity had a significant inhibitory effect on the above-mentioned germination characteristics, while the inhibitory effect of diesel was indistinctive (Table 5). The interactions between the vegetation community type and soil salinity significantly affected the number of germinated species, seedling numbers, and the Margalef index (Table 5).

Table 3. Seedling number in the seed banks of the three dominant vegetation communities in the YRD.

Community Type	Species	Seedling Density (Seedlings/m^2)								
		0 (S), 0 (D)	1% (S), 0 (D)	2% (S), 0 (D)	0 (S), 1% (D)	1% (S), 1% (D)	2% (S), 1% (D)	0 (S), 2% (D)	1% (S), 2% (D)	2% (S), 2%(D)
Multi-Species Community	*Phragmites australis*	4404 ± 1269	1165 ± 263	459 ± 87	5579 ± 1689	2895 ± 1654	630 ± 369	2369 ±655	1662 ± 251	678 ± 186
	Suaeda glauca	258 ± 65	153 ± 35	115 ± 41	267 ± 91	134 ± 59	105 ± 42	344 ± 102	210 ± 123	163 ± 29
	Sonchus arvensis	363 ± 236	48 ± 36		115 ± 66	29 ± 18		124 ± 61	48 ± 29	
	Typha orientalis	38 ± 27			248 ± 171	19 ± 19		229 ± 122	10 ± 10	
	Atriplex patens	105 ± 80	96 ± 63		267 ± 131	29 ± 18		86 ± 74	29 ± 18	10 ± 10
	Artemisia mongolica	10976 ± 4116	3353 ± 1483	640 ± 301	7499 ± 3728	2417 ± 1006	564 ± 180	7002 ± 3459	2350 ± 1237	736 ± 421
	Setaria viridis	2723 ± 1452	659 ± 290	220 ± 153	2025 ± 1142	927 ± 630	191 ± 191	2302 ± 911	1395 ± 489	115 ± 115
	Glycine soja	19 ± 19			38 ± 16			10 ± 10	10 ± 10	
	Tripolium vulgare		10 ± 10							
	Artemisia fauriei	172 ± 89			57 ± 37			67 ± 67		
	Suaeda salsa	287 ± 250	325 ± 226	220 ± 207	143 ± 83	392 ± 366	229 ± 182	172 ± 135	287 ± 225	210 ± 150
	Chloris virgata	86 ± 86	48 ± 48		57 ± 57	19 ± 19				
	Capsella bursa -pastoris		10 ± 10							
	Cirsium arvense var. integrifolium	115 ± 102			86 ± 74					
	Ixeris polycephala	19 ± 19						19 ± 19	19 ± 11	
	Conyza canadensis	29 ± 18	10 ± 10					29 ± 29		
	Ranunculus sceleratus				10 ± 10					
	Artemisia capillaris		10 ± 10					29 ± 29		
Phragmites Australis Community	*Phragmites australis*	2961 ± 571	1404 ± 362	95 ± 57	2293 ± 356	1079 ± 223	76 ± 27	2178 ± 421	1013 ± 123	191 ± 68
	Sonchus arvensis	105 ± 24	19 ± 19		57 ± 25	29 ± 18	10 ± 10	57 ± 25	29 ± 18	19 ± 19
	Typha orientalis				19 ± 11			29 ± 18		
Phragmites Australis– Suaeda Glauca Community	*Phragmites australis*	1824 ± 696	831 ± 366	152 ± 128	1652 ± 409	888 ± 441	162 ± 83	1757 ± 583	898 ± 456	487 ± 231
	Suaeda glauca	1948 ± 801	2407 ± 740	1538 ± 508	2502 ± 709	1719 ± 510	1136 ± 404	2502 ± 986	1242 ± 525	850 ± 366
	Sonchus arvensis	57 ± 37	10 ± 10		67 ± 45	29 ± 29		48 ± 48		
	Typha orientalis	19 ± 19			10 ± 10			27 ± 27		
	Suaeda salsa			38 ± 38	10 ± 10					

Notes: The variables: mean ± S. E.; S: soil salinity; D: diesel concentration; $N = 4$.

Table 4. Seedling density in the seed banks of the three dominant vegetation communities in the YRD.

Community Type	Seedling Density (Seedlings/m^2)	Sørensen Index
Multi-Species Community	$19.60 \times 10^3 \pm 15.90 \times 10^3$ b	0.55 ± 0.06 a
P. Australis Community	$3.01 \times 10^3 \pm 0.55 \times 10^3$ a	0.54 ± 0.09 a
P. Australis–S. Glauca Community	$3.85 \times 10^3 \pm 0.60 \times 10^3$ a	0.75 ± 0.17 b

Notes: The variables: mean $\pm$ S. E.; different letters (a, b) within the column indicate a significant difference at $p = 0.05$ (Duncan test, $N = 4$).

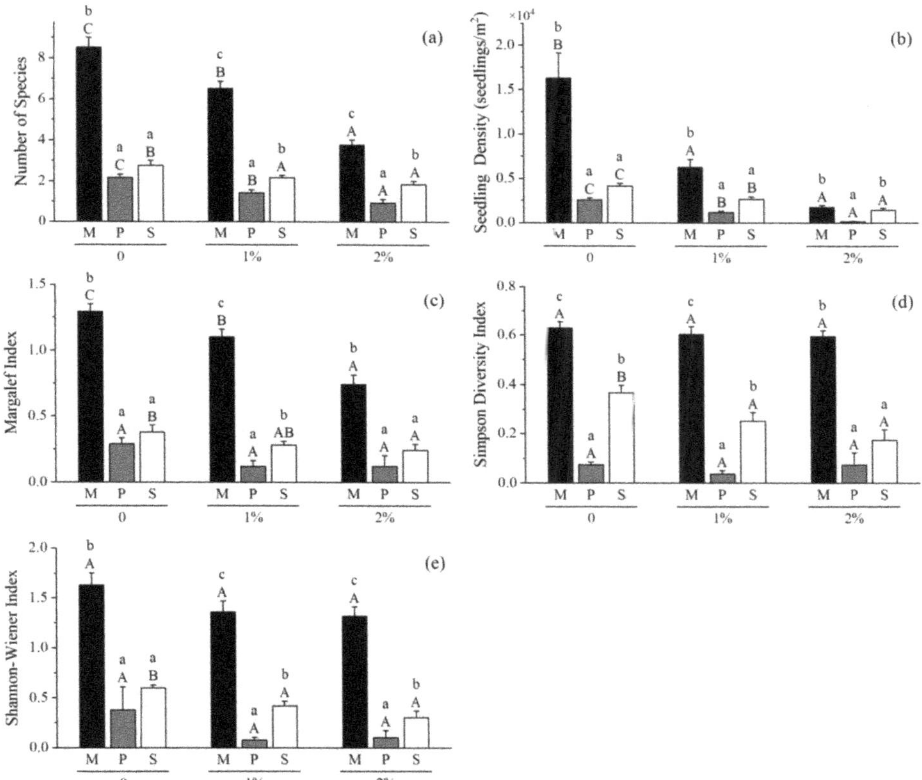

Figure 1. Effects of salinity on the seed germination of the three dominant communities in the YRD. (**a–e**) are the different parameters to describe the seed germination characteristics under different salinity. The columns are the mean values of the parameters, and the error bars are the standard error of the values, N = 4, $p = 0.05$. M: Multi-species community; P: *Phragmites australis* community; S: *Phragmites australis–Suaeda glauca* community. The 0%, 1% and 2% refers to the different soil salinities; a, b and c are used to describe the significance of the differences among the three communities under the same soil salinity; A, B and C are used to describe the differences among the three soil salinities under the same community type.

The effect of salinity on seed germination was further analyzed at the community level (Figure 1). It can be concluded from the analysis that the number of germinated species, seedling density, and the species diversity of the seed bank in the multi-species community are the highest in the three dominant vegetation communities under the same soil salinity. In addition, the adaptive ability of the *P. australis–S. glauca* community to soil salinity is better than that of the *P. australis* community (Figure 1). With the increase in soil salinity, the number of germinated species, seedling numbers, and species diversity of the three vegetation communities all show a decreasing trend; the number of germinated species and the seedling numbers have significant differences among the three dominant vegetation communities (Figure 1a,b). At the same time, the effects of salinity and diesel on seedling numbers of specific species were also analyzed using a three-way ANOVA (Table 6). The result proved that the seedling numbers of *P. australis, S. glauca, S. arvensis, T. orientalis, A. patens, A. mongolica, S. viridis, A. fauriei, S. salsa* and *C. virgata* had significant difference among the three vegetation communities (Table 5). The seedling numbers of *P. australis, S. glauca, S. arvensis, T. orientalis, A. patens, A. mongolica, S. viridis,* and *A. fauriei* were significantly higher than *S. salsa, C. virgata, Cirsium arvense,* and *Ixeris polycephala* (Table 3). There existed a significant interaction effect between the vegetation community type and the soil salinity on the seedling numbers of *T. orientalis, A. mongolica, S. viridis, A. fauriei,* and *Cirsium arvense* (Table 6).

Table 5. Three-way ANOVA analysis of vegetation community types, salinity, and diesel concentration on the germination and diversity of soil seed banks in the YRD.

	Number of Germinated Species		Seedling Number		Margalef Index		Simpson Index		Shannon–Wiener Index	
	F	P	F	P	F	P	F	P	F	P
C	282.060	0.000	36.253	0.000	159.960	0.000	206.764	0.000	110.522	0.000
S	57.070	0.000	33.251	0.000	20.161	0.000	5.499	0.006	5.207	0.007
D	0.169	0.845	0.627	0.537	0.338	0.714	1.586	0.211	0.348	0.707
C × S	17.517	0.000	11.959	0.000	4.507	0.002	1.892	0.119	0.137	0.968
C × D	0.938	0.446	0.293	0.882	1.662	0.166	0.440	0.779	0.300	0.877
S × D	0.548	0.701	0.389	0.816	0.342	0.849	0.347	0.845	1.949	0.109
C × S × D	0.488	0.862	0.516	0.841	0.428	0.901	0.303	0.963	0.742	0.655

Notes: C: vegetation community types; S: soil salinity; D: diesel concentration; $N = 4$.

Table 6. Three-way ANOVA of the effects of community type, salinity, and diesel concentration on the seedling numbers of the dominant species in the YRD.

Species Name	C	S	D	C × S	C × D	S × D	C × S × D
Phragmites australis	0.000	0.000	0.262	0.067	0.114	0.326	0.609
Suaeda glauca	0.000	0.031	0.750	0.055	0.769	0.650	0.802
Sonchus arvensis	0.038	0.000	0.346	0.070	0.421	0.292	0.671
Typha orientalis	0.003	0.000	0.256	0.001	0.380	0.339	0.566
Atriplex patens	0.022	0.003	0.730	0.168	0.851	0.837	0.633
Artemisia mongolica	0.000	0.000	0.579	0.000	0.712	0.886	0.972
Setaria viridis	0.000	0.001	0.920	0.000	0.988	0.915	0.984
Artemisia fauriei	0.001	0.001	0.353	0.000	0.401	0.384	0.426
Suaeda salsa	0.000	0.701	0.919	0.804	0.996	0.992	0.999
Chloris virgata	0.025	0.257	0.304	0.263	0.330	0.808	0.925
Cirsium arvense	0.059	0.054	0.453	0.025	0.545	0.529	0.626
Ixeris polycephala	0.014	0.222	0.222	0.214	0.214	0.551	0.654

Notes: Only the species' with seedling numbers greater than 20 were included. C: Community type; S: Salinity; D: Diesel concentration, $N = 4$, $p = 0.05$.

4. Discussion

4.1. The Species Diversity of Above-Ground and Under-Ground Vegetation in the YRD

The wetlands in the YRD region are typical saline wetlands, and the dominant vegetation species are mainly salt-tolerant species such as *S. salsa, S. glauca,* and *P. australis* [24]. *S. salsa* and *S. glauca* are the two common *Suaeda* species that coexist in the YRD. *S. salsa* is mainly distributed near the muddy tidal flat and high tide line of the coast, whereas *S. glauca* mainly exists in the inland close to the sea [25]. *P. australis* is another most important and widespread wetland plant species [26]. There is about 2600 ha of *P. australis* wetlands that form monodominant communities or mix with other plant species such as *Triarrhena lutarioriparia, Typha orientalis, Sonchus arvensis*, and *S. salsa* [27]. The species of vegetation are usually low in the low-nutrient and high-salinity coastal wetlands. In the present study, there were only 14 species found from the investigation of the soil seed banks, and the above-ground species were fewer than the species of the seed banks (Table 3). *P. australis* was always the dominant species in the multi-species community, *P. australis* community, and the *P. australis–S. glauca* community. A previous study on the above-ground vegetation and soil seed banks of the new Yellow River course (NYR) and the abandoned Yellow River course (OYR) in the YRD wetlands had similar results. There were only 17 plant species across both sites, 9 species were in OYR, and 16 species were in the NYR [28].

4.2. Effects of Soil Salinity and Diesel Contamination on the Germination of Seeds

Soil salinity is an important environmental factor affecting the physiological conditions of plants. High salinity can inhibit the germination of seeds and the elongation of leaves [29], reduce the photosynthetic rate [30], and decrease the root uptake capacity [31], thereby affecting the composition of vegetation communities [32,33]. In the present study, soil salinity significantly inhibits the number of species, seedling numbers, and the species biodiversity of the soil seed banks (Table 5, Figure 1). However, the reactions of specific species to salinity are different; *P. australis, S. glauca, S. arvensis, T. orientalis, A. patens, A. mongolica, S. viridis*, and *A. fauriei* are more tolerant of soil salinity, especially *P. australis, S. glauca, A. mongolica* and *S. viridis* (Tables 3 and 6). *P. australis* is one of the dominant species in the YRD, and its habitats span from freshwater swamps to salt marshes with a salinity ranging from 0 to 20% [22]. However, a soil salinity of 10% significantly decreases the growth of *P. australis* [32]. Moreover, the density and species richness of the seed banks are also significantly lower under saline-alkaline stresses [34]. *S. glauca* is a succulent halophyte that is highly resistant to salt and alkali stresses [25,35]. The seedlings of *S. glauca* do not show significant symptoms of injury after imposing 5.85–17.55% of NaCl stress, and the low concentration of NaCl (5.85%) even promoted its growth [36]. Increasing shoot–root ratio, maintaining leaf succulence and relying on proline and metal ions as osmotic regulators are regarded as the main regulation strategies of *Suaeda* species to adapt to high salinity environment conditions.

Diesel oil is a complex mixture produced by the distillation of crude oil. It consists of hydrocarbons with carbon numbers in the range of C_9–C_{38} [37]. Diesel is more toxic than crude oil due to the easy uptake and bioavailability of lower molecular weight over higher molecular weight compounds [38]. The research of Adam and Duncan (2002) shows that diesel oil postpones the germination time, but it has no significant inhibitory effect on the germination rate of seeds [20]. The hard seed coat is the main barrier to resisting the damage of oil penetration; therefore, the embryo of the seed cannot easily be injured or killed. The inhibitory effect of diesel fuel on seed germination is due to its water repellent property, the film of diesel fuel around the seeds may prevent or reduce both water and oxygen from entering the seeds; therefore, the lag phase preceding germination is increased many times [20,39]. Furthermore, the germination rates vary greatly with plant species when affected by hydrocarbons. Zhu et al. evaluated the seed germination rates of 65 grass species in North Dakota affected by crude oil [15]. The results showed that the germination rate of all species reduced from 4.3 to 100% under the inhibition of crude oil, 28 were tolerant species, 29 were moderately tolerant species, 6 were moderately sensitive species, and 2 were sensitive species. The pot experiment carried out by Wei et al., shows that crude oil has a negative effect on the germination rate, plant height, and biomass of six indigenous plant species in the semi-arid loess area, and 2% is the crucial level that has a significant inhibitory effect on the growth of plants [39]. The above studies are inconsistent with the results of our present study on soil seed banks.

4.3. Potential of Soil Seed Banks for the Restoration of the Degraded Coastal Wetland

The wetlands in China account for 5.58% of the whole national territorial area, which plays an important role in maintaining the ecosystem function [4]. Although the wetlands have been severely disturbed due to multiple influencing factors, such as urbanization, oil exploitation, agriculture activities, etc., soil seed banks are a germplasm reservoir that can directly affect the species composition and plant community structure. It is considered an important method in wetland restoration [26]. Guan et al. (2019), studied the above-ground vegetation and soil seed banks of the NYR course and the abandoned OYR course in the YRD wetlands; there were 17 species with a seed density of $2.06 \pm 1.25 \times 10^3$ seeds m^{-2} [28]. In the present study, we found similar results; there were 14 species found in the three common vegetation communities altogether, the seed density ranged from $3.01 \pm 0.55 \times 10^3$ to $19.60 \pm 5.90 \times 10^3$ seeds m^{-2}. The multi-species vegetation community had the highest seed bank storage, species richness, and diversity. Moreover, the species number of the seed banks was closely related to its above-ground vegetation. The species number of the multi-species vegetation community was also the highest in the three vegetation communities (Tables 3 and 5).

The ultimate goal of wetland restoration is to create a self-supporting ecosystem that can resist perturbation without further assistance [17]. In the present study, soil salinity was identified as an important factor that influences the number of germinated species, seedling density, and the species diversity of the vegetation communities in the YRD (Tables 5 and 6). The multi-species vegetation community was mainly distributed in the low salinity soil, while the *P. australis* community and the *P. australis–S. glauca* communities were more suitable to living in the high salinity soil. Diesel has no significant inhibitory effect on seed germination under 2% concentration (Tables 5 and 6). Therefore, soil salinity should be one of the most important factors to consider prior to wetland restoration using topsoil transplantation. The *P. australis* community and *P. australis–S. glauca* communities can be used in degraded wetlands with 1–2% soil salinity and 0–2% oil contamination, while multi-species communities are more suitable for the restoration of wetlands with 0–1% salinity and 0–2% oil contamination.

5. Conclusions

The soil seed banks and above-ground vegetation of three dominant vegetation communities (multi-species community, *P. australis* community, and *P. australis–S. glauca* community) in the coastal wetlands of the YRD were investigated. The number of above-ground species was fewer than the species of seed banks, and *P. australis* was the most dominant species among the 14 species in the soil seed banks. The tolerance of different plant species to salinity and diesel is species-specific, and the inhibitory effect of salinity is more significant than diesel. The seed coat can resist the penetration of diesel but cannot resist the stress of salt, and the inhibition starts in the initial stage of germination. Therefore, soil salinity is considered a more important factor than diesel for wetland restoration in the YRD. To facilitate improved germination and growth, it is recommended that soil seed banks are transplanted into the soil conditions for which they are better adapted. The soil seed banks of the *P. australis* and *P. australis–S. glauca* communities are suitable to be used in degraded wetlands with a soil salinity of between 1% and 2%, while multi-species communities are more suitable for use in a soil salinity of between 0% and 1%.

Author Contributions: Conceptualization, X.G. and Y.G.; methodology, X.G. and Y.G.; software, Z.F.; validation, X.G., Y.G. and Z.F.; formal analysis, X.G. and Z.F.; investigation, Z.F.; resources, X.G. and Y.G.; data curation, Z.F.; writing—original draft preparation, Z.F.; writing—review and editing, X.G., Y.G., J.L., Y.M., X.Y. and F.M. All authors have read and agreed to the published version of the manuscript.

Funding: This research was funded by the Joint Funds of the National Natural Science Foundation of China (No. U1806217), Shandong Provincial Key Research and Development Program (Major

Scientific and Technological Innovation Project) (No. 2021CXGC011201) and the Shandong Provincial Natural Science Foundation (ZR2019BD012, ZR2021MD126, ZR2014CM32).

Institutional Review Board Statement: Not applicable.

Informed Consent Statement: Not applicable.

Data Availability Statement: The data presented in this study are available on request from the corresponding author.

Conflicts of Interest: The authors declare no conflict of interest.

References

1. Engelhardt, K.A.M.; Ritchie, M.E. The effect of aquatic plant species richness on wetland ecosystem processes. *Ecology* **2002**, *83*, 2911–2924. [CrossRef]
2. Davidson, N.C. How much wetland has the world lost? Long-term and recent trends in global wetland area. *Mar. Freshw. Res.* **2014**, *65*, 934–941. [CrossRef]
3. Bian, H.; Li, W.; Li, Y.; Ren, B.; Niu, Y.; Zeng, Z. Driving forces of changes in China's wetland area from the first (1999–2001) to second (2009–2011) National Inventory of Wetland Resources. *Glob. Ecol Conserv.* **2019**, *21*, e00867. [CrossRef]
4. Meng, W.; He, M.; Hu, B.; Mo, X.; Li, H.; Liu, B.; Wang, Z. Status of wetlands in China: A review of extent, degradation, issues and recommendations for improvement. *Ocean Coast. Manag.* **2017**, *146*, 50–59. [CrossRef]
5. Millennium Ecosystem Assessment (MEA). *Ecosystems and Human Wellbeing: Wetlands and Water Synthesis*; World Resources Institute: Washington, DC, USA, 2005.
6. Gedan, K.B.; Kirwan, M.L.; Wolanski, E.; Barbier, E.B.; Silliman, B. The present and future role of coastal wetland vegetation in protecting shorelines: Answering recent challenges to the paradigm. *Clim. Chang.* **2011**, *106*, 7–29. [CrossRef]
7. Ooi, M.K. Seed bank persistence and climate change. *Seed Sci. Res.* **2012**, *22*, S53–S60. [CrossRef]
8. Tellier, A. Persistent seed banking as eco-evolutionary determinant of plant nucleotide diversity: Novel population genetics insights. *New Phytol.* **2018**, *221*, 725–730. [CrossRef]
9. Nishihiro, J.; Nishihiro, M.A.; Washitani, I. Restoration of wetland vegetation using soil seed banks: Lessons from a project in Lake Kasumigaura, Japan. *Landsc. Ecol. Eng.* **2006**, *2*, 171–176. [CrossRef]
10. Saatkamp, A.; Poschlod, P.; Venable, L. *The Functional Role of Soil seed Banks in Natural Communities*, 3rd ed.; Gallagher, R., Ed.; CABI: Wallingford, UK, 2014; pp. 263–295.
11. Wang, X.; Zhang, D.; Qi, Q.; Tong, S.; An, Y.; Lu, X.; Liu, Y. The restoration feasibility of degraded Carex Tussock in soda-salinization area in arid region. *Ecol. Indic.* **2018**, *98*, 131–136. [CrossRef]
12. Morimoto, J.; Shibata, M.; Shida, Y.; Nakamura, F. Wetland restoration by natural succession in abandoned pastures with a degraded soil seed bank. *Restor. Ecol.* **2017**, *25*, 1005–1014. [CrossRef]
13. Kiss, R.; Deák, B.; Török, P.; Tóthmérész, B.; Valkó, O. Grassland seed bank and community resilience in a changing climate. *Restor. Ecol.* **2018**, *26*, S141–S150. [CrossRef]
14. Bai, J.; Huang, L.; Gao, Z.; Lu, Q.; Wang, J.; Zhao, Q. Soil seed banks and their germination responses to cadmium and salinity stresses in coastal wetlands affected by reclamation and urbanization based on indoor and outdoor experiments. *J. Hazard. Mater.* **2014**, *280*, 295–303. [CrossRef] [PubMed]
15. Zhu, H.; Gao, Y.; Li, D. Germination of grass species in soil affected by crude oil contamination. *Int. J. Phytoremediation* **2018**, *20*, 567–573. [CrossRef] [PubMed]
16. Touzard, B.; Amiaud, B.; Langlois, E.; Lemauviel, S.; Clément, B. The relationships between soil seed bank, aboveground vegetation and disturbances in an eutrophic alluvial wetland of Western France. *Flora—Morphol. Distrib. Funct. Ecol. Plants* **2002**, *197*, 175–185. [CrossRef]
17. Wang, M.; Qi, S.; Zhang, X. Wetland loss and degradation in the Yellow River Delta, Shandong Province of China. *Environ. Earth Sci.* **2012**, *67*, 185–188. [CrossRef]
18. Wang, Z.-Y.; Gao, D.-M.; Li, F.-M.; Zhao, J.; Xin, Y.-Z.; Simkins, S.; Xing, B.-S. Petroleum Hydrocarbon Degradation Potential of Soil Bacteria Native to the Yellow River Delta. *Pedosphere* **2008**, *18*, 707–716. [CrossRef]
19. Wu, B.; Guo, S.; Wang, J. Spatial ecological risk assessment for contaminated soil in oiled fields. *J. Hazard. Mater.* **2021**, *403*, 123984. [CrossRef]
20. Adam, G.; Duncan, H. Influence of diesel fuel on seed germination. *Environ. Pollut.* **2002**, *120*, 363–370. [CrossRef]
21. Yang, J.; Yao, R. Spatial variability of soil water and salt characteristics in the Yellow River Delta. *Sci. Geogr. Sin.* **2007**, *27*, 348–353.
22. Sun, X.-S.; Chen, Y.-H.; Zhuo, N.; Cui, Y.; Luo, F.-L.; Zhang, M.-X. Effects of salinity and concomitant species on growth of Phragmites australis populations at different levels of genetic diversity. *Sci. Total Environ.* **2021**, *780*, 146516. [CrossRef]
23. Liang, K.; Fan, Y.; Meki, K.; Meng, D.; Yan, Q.; Zheng, H.; Li, F.; Luo, X. The seasonal dynamics of nitrogen and rhizosphere effects in the typical saline-alkali vegetation communities of the Yellow River Estuary wetland. *Environ. Chem.* **2019**, *38*, 2327–2335.
24. Xia, J.; Zhang, S.; Guo, J.; Rong, Q.; Zhang, G. Critical effects of gas exchange parameters in Tamarix chinensis Lour on soil water and its relevant environmental factors on a shell ridge island in China's Yellow River Delta. *Ecol. Eng.* **2014**, *76*, 36–46. [CrossRef]

25. Zhang, Q.-H.; Sairebieli, K.; Zhao, M.-M.; Sun, X.-H.; Wang, W.; Yu, X.-N.; Du, N.; Guo, W.-H. Nutrients Have a Different Impact on the Salt Tolerance of Two Coexisting Suaeda Species in the Yellow River Delta. *Wetlands* **2020**, *40*, 2811–2823. [CrossRef]
26. Wang, X.; Yu, J.; Zhou, D.; Dong, H.; Li, Y.; Lin, Q.; Guan, B.; Wang, Y. Vegetative Ecological Characteristics of Restored Reed (Phragmites australis) Wetlands in the Yellow River Delta, China. *Environ. Manag.* **2011**, *49*, 325–333. [CrossRef]
27. Guan, B.; Yu, J.; Hou, A.; Han, G.; Wang, G.; Qu, F.; Xia, J.; Wang, X. The ecological adaptability of Phragmites australis to interactive effects of water level and salt stress in the Yellow River Delta. *Aquat. Ecol.* **2016**, *51*, 107–116. [CrossRef]
28. Guan, B.; Chen, M.; Elsey-Quirk, T.; Yang, S.; Shang, W.; Li, Y.; Tian, X.; Han, G. Soil seed bank and vegetation differences following channel diversion in the Yellow River Delta. *Sci. Total Environ.* **2019**, *693*, 133600. [CrossRef]
29. Taleisnik, E.; Rodríguez, A.A.; Bustos, D.; Erdei, L.; Ortega, L.; Senn, M.E. Leaf expansion in grasses under salt stress. *J. Plant Physiol.* **2009**, *166*, 1123–1140. [CrossRef]
30. Flexas, J.; Bota, J.; Galmés, J.; Medrano, H.; Ribas-Carbó, M. Keeping a positive carbon balance under adverse conditions: Responses of photosynthesis and respiration to water stress. *Physiol. Plant.* **2006**, *127*, 343–352. [CrossRef]
31. Attia, H.; Karray, N.; Rabhi, M.; Lachaâl, M. Salt-imposed restrictions on the uptake of macroelements by roots of Arabidopsis thaliana. *Acta Physiol. Plant.* **2008**, *30*, 723–727. [CrossRef]
32. James, K.; Hart, B. Effect of salinity on four freshwater macrophytes. *Mar. Freshw. Res.* **1993**, *44*, 769–777. [CrossRef]
33. Hasegawa, P.M.; Bressan, R.A.; Zhu, J.-K.; Bohnert, H.J. Plant cellular and molecular responses to high salinity. *Annu. Rev. Plant Physiol. Plant Mol. Biol.* **2000**, *51*, 463–499. [CrossRef] [PubMed]
34. Zhao, Y.; Wang, G.; Zhao, M.; Wang, M.; Jiang, M. Direct and indirect effects of soil salinization on soil seed banks in salinizing wetlands in the Songnen Plain, China. *Sci. Total Environ.* **2021**, *819*, 152035. [CrossRef] [PubMed]
35. Duan, H.; Ma, Y.; Liu, R.; Li, Q.; Yang, Y.; Song, J. Effect of combined waterlogging and salinity stresses on euhalophyte *Suaeda glauca*. *Plant Physiol. Biochem.* **2018**, *127*, 231–237. [CrossRef] [PubMed]
36. Jin, H.; Dong, D.; Yang, Q.; Zhu, D. Salt-Responsive Transcriptome Profiling of *Suaeda glauca* via RNA Sequencing. *PLoS ONE* **2016**, *11*, e0150504. [CrossRef] [PubMed]
37. Gao, Y.; Du, J.; Bahar, M.; Wang, H.; Subashchandrabose, S.; Duan, L.; Yang, X.; Megharaj, M.; Zhao, Q.; Zhang, W.; et al. Metagenomics analysis identifies nitrogen metabolic pathway in bioremediation of diesel contaminated soil. *Chemosphere* **2021**, *271*, 129566. [CrossRef]
38. Dorn, P.B.; Salanitro, J.P. Temporal ecological assessment of oil contaminated soils before and after bioremediation. *Chemosphere* **2000**, *40*, 419–426. [CrossRef]
39. Wei, Y.; Wang, Y.; Duan, M.; Han, J.; Li, G. Growth tolerance and remediation potential of six plants in oil-polluted soil. *J. Soils Sediments* **2019**, *19*, 3773–3785. [CrossRef]

MDPI

St. Alban-Anlage 66

4052 Basel

Switzerland

Tel. +41 61 683 77 34

Fax +41 61 302 89 18

www.mdpi.com

Forests Editorial Office

E-mail: forests@mdpi.com

www.mdpi.com/journal/forests